Studies in Computational Intelligence

Volume 642

Series editor

Janusz Kacprzyk, Polish Academy of Sciences, Warsaw, Poland
e-mail: kacprzyk@ibspan.waw.pl

About this Series

The series "Studies in Computational Intelligence" (SCI) publishes new developments and advances in the various areas of computational intelligence—quickly and with a high quality. The intent is to cover the theory, applications, and design methods of computational intelligence, as embedded in the fields of engineering, computer science, physics and life sciences, as well as the methodologies behind them. The series contains monographs, lecture notes and edited volumes in computational intelligence spanning the areas of neural networks, connectionist systems, genetic algorithms, evolutionary computation, artificial intelligence, cellular automata, self-organizing systems, soft computing, fuzzy systems, and hybrid intelligent systems. Of particular value to both the contributors and the readership are the short publication timeframe and the worldwide distribution, which enable both wide and rapid dissemination of research output.

More information about this series at http://www.springer.com/series/7092

Dariusz Król · Lech Madeyski
Ngoc Thanh Nguyen
Editors

Recent Developments in Intelligent Information and Database Systems

 Springer

Editors
Dariusz Król
Faculty of Computer Science and
  Management
Wrocław University of Technology
Wrocław
Poland

Lech Madeyski
Faculty of Computer Science and
  Management
Wrocław University of Technology
Wrocław
Poland

Ngoc Thanh Nguyen
Division of Knowledge and System
  Engineering for ICT, Faculty of
  Information Technology
Ton Duc Thang University
Ho Chi Minh City
Vietnam

and

Faculty of Computer Science
  and Management
Wrocław University of Technology
Wrocław
Poland

ISSN 1860-949X   ISSN 1860-9503 (electronic)
Studies in Computational Intelligence
ISBN 978-3-319-81004-1   ISBN 978-3-319-31277-4 (eBook)
DOI 10.1007/978-3-319-31277-4

Printed on acid-free paper

This Springer imprint is published by Springer Nature
The registered company is Springer International Publishing AG Switzerland

Preface

More than three decades of research in intelligent information and database systems has principally led from theory to new practical developments. Research issues on big data to gain and share knowledge is probably the most significant example of this. The other example concerns system integrity issues. The industry needs to address these issues through applications focusing on long-term integrity and operational reliability. The complexity of modern information and database systems requires a deep understanding of old issues and the adoption of new insights and experiences to the development of such systems.

Fortunately, recent years have seen remarkable progress on both intelligent information and database systems. These systems invariably involve complex, data-intensive and resource-consuming tasks, which often carry heavy losses in terms of cost in the event of failure. They have regained attention because in many cases large-scale and long-lived applications have to be updated and several new methods have to be developed. Some of the most important of these updates and developments include increased attention to the integration of artificial intelligence, multimedia, social media and database technologies towards the next generation computer systems and services.

The main objective of this book is to contribute to the development of the intelligent information and database systems with the essentials of current knowledge, experience and know-how. Over the last decade new roles for these systems have been discovered, particularly their role as social data and service integrator. Understanding and combining data from different sources has now become a standard practice. Also, the use of statistical methods in a large corpus of data is more efficient and therefore more frequent. But there are expected yet more and different roles to be discovered in the near future.

The scope represented here, relevant to the study of intelligent information and database systems, encompasses a wide spectrum of research topics discussed both from the theoretical and the practical points of view. There are fundamental issues such as algorithmics, artificial and computational intelligence, nature-inspired paradigms, ontologies, collective knowledge, natural language processing, image processing and temporal databases. On the other hand, there are also a number of interdisciplinary topics outside or close to end-user applications. These concern, for example, heterogeneous and distributed databases, social networks, recommendations systems, web services, etc.

This brand-new volume in the well-established "Studies in Computational Intelligence" series provides a valuable compendium of current and potential problems in the field. It contains a selection of 40 chapters based on the original research accepted for the presentation as posters during *the Asian Conference on Intelligent Information and Database Systems (ACIIDS 2016)* held on 14–16 March 2016 in Da Nang, Vietnam. This is the eighth, in the order, conference jointly organized by Wroclaw University of Technology and its partners.

The selected papers to some extent reflect the achievements of scientific teams from 17 countries in five continents, and report on the progress in theory and application of three main areas: Intelligent information systems, intelligent database systems and tools and applications. More precisely, the book is divided into parts related to six primary topics: Part I. Computational Intelligence in Data Mining and Machine Learning, Part II. Ontologies, Social Networks and Recommendation Systems, Part III. Web Services, Cloud Computing, Security and Intelligent Internet Systems, Part IV. Knowledge Management and Language Processing, Part V. Image, Video, Motion Analysis and Recognition and finally, Part VI. Advanced Computing Applications and Technologies. Each part deals with somewhat different aspects. Let us now consider it in more detail. In Part I, we start with the genetic and memetic algorithms, nature-inspired heuristics, cloud computing, clustering-based classifications, deep learning, artificial neural networks and wi-fi networks. Part II deals with behaviour ontologies, P2P social networks, customer management strategies and recommendations systems. Part III contains six papers about web services, secure communications and integrity of private information. The next part covers collective knowledge management, Internet of Things, natural language processing, sentiment analysis and event detection. In Part V, we cover image processing, human motion capturing, humanoid robots, gesture recognition and emotion detection. The volume closes with Part VI. It presents various practical applications of assistive devices for elderly people, biomedical data integration, epidemiological cancer studies, temporal educational databases, location-based service applications, smart electronic wallet, CT image processing, supply chain management and logistics, and learning management systems.

The book will be an excellent resource for researchers, who are working in the merging of artificial intelligence, multimedia, networks and big data technologies, as well as for students who are interested in computer science, computer engineering, management science, ontological engineering and other related fields.

There are several people whose help has been invaluable in the preparation of this volume. First of all, on behalf of the Steering Committee, the Program Committee and the Organizing Committee we would like to thank all participants, among others, computer scientists, mathematicians, engineers, logicians and other researchers who found it worthwhile to travel to Da Nang from around the world, and who prepared first-rate contributions to these publications. Warm thanks are also due to the referees who reviewed the chapters with remarkable expertise and engagement. We express a special gratitude to Prof. Janusz Kacprzyk, the editor of this series, and Dr. Thomas Ditzinger from Springer for their interest and support of our project.

Our last but not least observation is that the rise of new generation systems and services is leading to profound changes in overall world functioning. In particular, the Internet is undergoing a transition from the global network to the permanent interconnection between human beings and everyday devices equipped with ubiquitous intelligence. It opens tremendous opportunities for a large number of novel applications that promise to improve the quality of our lives.

We sincerely hope that this volume will be a valuable reference work in your study and research, and you enjoy reading it.

March 2016   Dariusz Król
                                                                                               Lech Madeyski
                                                                                        Ngoc Thanh Nguyen

Contents

Part IV Knowledge Management and Language Processing

Part V Image, Video, Motion Analysis and Recognition

Part I
Computational Intelligence in Data Mining and Machine Learning

Cryptanalysis of SDES Using Genetic and Memetic Algorithms

Kamil Dworak, Jakub Nalepa, Urszula Boryczka
and Michal Kawulok

Abstract In this paper, we exploit evolutionary algorithms for cryptanalysis and we focus on a chosen-plaintext attack model, in which the attacker is able to access both the ciphertext and the plaintext. The aim of this attack is to determine the decryption key for the Simplified Data Encryption Standard, so that other encrypted texts can be easily deciphered. We propose to extract the key using genetic and memetic algorithms (the latter being a hybrid of the evolutionary techniques and some refinement procedures). An extensive experimental study, coupled with the sensitivity analysis on method components and statistical tests, show the convergence capabilities of our approaches and prove they are very competitive compared with other state-of-the-art algorithms.

Keywords Memetic algorithm · Genetic algorithm · Cryptanalysis · SDES

1 Introduction

Data security plays a pivotal role in all computer systems nowadays [1]. It concerns not only secure data storage, but—most importantly—secure data exchange. There exist various cryptography algorithms which help maintain a desired security level

K. Dworak · U. Boryczka
University of Silesia, Sosnowiec, Poland
e-mail: kamil.dworak@us.edu.pl; kdworak@future-processing.com

U. Boryczka
e-mail: urszula.boryczka@us.edu.pl

J. Nalepa (✉) · M. Kawulok
Silesian University of Technology, Gliwice, Poland
e-mail: jakub.nalepa@polsl.pl; jnalepa@future-processing.com

M. Kawulok
e-mail: michal.kawulok@polsl.pl; mkawulok@future-processing.com

K. Dworak · J. Nalepa · M. Kawulok
Future Processing, Gliwice, Poland

© Springer International Publishing Switzerland 2016
D. Król et al. (eds.), *Recent Developments in Intelligent Information
and Database Systems*, Studies in Computational Intelligence 642,
DOI 10.1007/978-3-319-31277-4_1

3

of data being transferred between users. In the key-based techniques, the plaintext is ciphered (transformed into the ciphertext) using an encryption key—this ciphertext may be read only by intended users which possess the decryption key [2]. The algorithms which utilize keys for securing the information include, among others, the Simplified Data Encryption Standard (SDES), Data Encryption Standard (DES), and Advanced Encryption Standard (AES).

The aim of many emerging cryptanalysis techniques is to decipher the encoded text and to extract the corresponding decryption key as well, so that other (unseen) pieces of text can be later deciphered. These techniques are very useful to assess the quality of the security systems and to find their potential drawbacks and backdoors. In a chosen-plaintext attack, the attacker is able to access both the plaintext and the encrypted information to retrieve the decryption key.

1.1 Related Work

Since the size of the key is large in modern encryption algorithms (and the number of possible keys to be reviewed during the attack rapidly grows), the approximate heuristic methods for determining the key are attracting the research attention. Such meta-heuristics encompass population-based techniques, including genetic [3–6] and memetic algorithms [7], simulated annealing, tabu and guided searches, particle swarm optimization approaches [8], hybrid systems [9], and other [10, 11]. The current advances on the analysis of various ciphers have been summarized in many thorough surveys and reviews [12].

Evolutionary algorithms (EAs) are built upon the principles of natural evolution [13]. In genetic algorithms (GAs), a population of individuals (which encode the solutions for a problem being tackled) evolves in time to improve their quality. The evolution involves applying genetic operators including selection, crossover, and mutation. Memetic algoritms (also referred to as hybrid GAs) enhance standard EAs with some additional refinement procedures, including intensive local searches, hill climbing procedures, archiving the knowledge attained during the search [14], guided searches, and many others. These memetic operators aim at boosting the convergence capabilities of these algorithms, and at guiding the search efficiently towards most promising parts of the solution space. Genetic and memetic techniques (both sequential and parallel) proved to be very efficient and were applied in many fields of science and engineering to tackle difficult optimization and pattern recognition tasks [14–19].

1.2 Contribution

In this paper, we introduce genetic and memetic algorithms to perform a chosen-plaintext attack on the data encrypted using the SDES. The memetic algorithm enhances a standard GA with additional refinement procedures which aim at improving the individuals using both local optimization and the historical data acquired during the evolution (i.e., the information concerning the best-fitted individuals found up to date). Therefore, these techniques help boost the algorithm convergence capabilities. Although the evolutionary algorithms have been applied to break the SDES [7], they have not been intensively studied to verify how different algorithm components affect the search. Here, we perform an extensive experimental study to investigate these issues. The analysis is complemented with the two-tailed Wilcoxon tests for verifying the statistical significance of the results. Also, we compare the proposed algorithm with other state-of-the-art techniques, including binary particle swarm optimization [8], simple random walk and the naive brute force approach.

1.3 Paper Structure

The remaining of this paper is organized as follows. The problem is formulated in Sect. 2. Section 3 discusses the proposed genetic and memetic algorithms for cryptanalysis. In Sect. 4, we discuss our extensive experimental study. Section 5 concludes the paper and highlights the directions of our future work.

2 Problem Formulation

In this paper, we consider a chosen-plaintext attack, which is one of the most popular and well-known cryptanalytical attacks [20]. Is it assumed that the crypt-analyst (the attacker) retrieved a fragment of the ciphertext, encrypted by a given encryption algorithm, along with the corresponding piece of the plaintext. With this information, the attacker aims at finding a valid decryption key. Section 2.1 discusses the background behind the Simplified Data Encryption Standard which was used in our attack model.

2.1 Simplified Data Encryption Standard

The Simplified Data Encryption Standard has been developed by Schaefer in late 90s [21]. In general, it has been created for educational and experimental purposes, and it is not recommended to use it as a secure encryption algorithm in the production environment. SDES is a symmetric block cipher, which takes an 8-bit block of the plaintext (denoted as P), written as a binary string, and a 10-bit binary key as an input. Then, it generates an 8-bit block of the ciphertext (C). This algorithm requires two additional subkeys marked as SK_1 and SK_2, for two algorithm rounds.

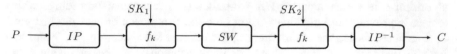

Fig. 1 SDES encryption algorithm

The encryption process is presented in Fig. 1. First, the input data is passed to the initial permutation procedure, which is defined as $IP = [2\,6\,3\,1\,4\,8\,5\,7]$. The first bit of a message will be relocated to the second position, the second one to the sixth position and so forth. In the next step, the transformed string and the subkey SK_1 are passed to the *complex function* f_k (see Sect. 2.3). The result is relocated by a special swapping permutation *SW*. This swap function reorders the first four bits of a string to the last four positions. Later, the f_k function is performed again, but with the second subkey SK_2. Finally, the data go through the inverse permutation, defined as $IP^{-1} = [4\,1\,3\,5\,7\,2\,8\,6]$ (the meanings of these values are as for *IP*). The encryption process is repeated for every 8-bit data block, until the entire ciphertext is generated.

2.2 Generation of the Subkeys

The process of generating the subkeys is presented in Fig. 2. First, the encryption key K is passed to the permutation $P10 = [3\,5\,2\,7\,4\,10\,1\,9\,8\,6]$. The generated string is divided into two 5-bits parts. For every substring, a single cyclic shift (shifting to the left) is performed (*Shift1*). To retrieve the first subkey, two 5-bit parts are concatenated into one binary string. The result is passed to the reduction permutation: $P8 = [6\,3\,7\,4\,8\,5\,10\,9]$. To produce the second subkey, two 5-bit parts are shifted again, but by two positions to the left (*Shift2*). After this step, the reordered parts are concatenated. The generated string is finally reordered by the permutation *P8*.

2.3 Complex Function f_k

The f_k function is the "complex component" of SDES. It is executed by the *Feistal network* [21]. The subkey K_i is divided into two 4-bits parts. This function (note that it is non-linear) is defined as follows:

Fig. 2 Generating the subkeys for SDES

Fig. 3 S-Boxes: S_1 and S_2

$$S_1 = \begin{bmatrix} 01 & 00 & 11 & 10 \\ 11 & 10 & 01 & 00 \\ 00 & 10 & 01 & 11 \\ 11 & 01 & 11 & 10 \end{bmatrix}, \quad S_2 = \begin{bmatrix} 00 & 01 & 10 & 11 \\ 10 & 00 & 01 & 11 \\ 11 & 00 & 01 & 00 \\ 10 & 01 & 00 & 11 \end{bmatrix}$$

$$f_{k_i}(L_i, R_i) = (L_{i-1} \oplus f(R_{i-1}, SK_i), R_{i-1}), \tag{1}$$

where L denotes the 4 leftmost bits of data block, R corresponds to 4 rightmost bits, and i is the round identifier. Initially, the right part of the data is extended by the permutation $E = [4\,1\,2\,3\,2\,3\,4\,1]$ to the 8-bit string (it is represented as the f function). This expanded string is XORed with the subkey SK_i. The first 4 bits of the result are passed to the S-Box (S_1), whereas the rest of bits to the S-Box (S_2). The first and the last bit of each substring represents the number of a row, and the second and third bit represents the number of a column. The S-Boxes are visualized in Fig. 3. Finally, the retrieved substrings are concatenated and subjected to the permutation $P4 = [2\,4\,3\,1]$. The result of the f function is XORed with the L part of the data block, being the input of the f_k function.

$$S_1 = \begin{bmatrix} 01 & 00 & 11 & 10 \\ 11 & 10 & 01 & 00 \\ 00 & 10 & 01 & 11 \\ 11 & 01 & 11 & 10 \end{bmatrix}, \quad S_2 = \begin{bmatrix} 00 & 01 & 10 & 11 \\ 10 & 00 & 01 & 11 \\ 11 & 00 & 01 & 00 \\ 10 & 01 & 00 & 11 \end{bmatrix}$$

3 Evolutionary Algorithms for Cryptanalysis Processes

In this paper, we propose genetic and memetic algorithms (GA and MA, respectively) for the cryptanalysis process. As already mentioned, we exploit the chosen-plaintext attack, and aim at finding the 10-bit decryption key. Thus, each individual (a *chromosome*) in the population (in both GA and MA) represents the decryption key. The initial population, which consists of N randomly generated SDES decryption keys, is evolved to boost the quality of the solutions, using standard genetic operators (enhanced by the memetic operators in the MA). It is worth noting that the total number of possible individuals is $2^{10} = 1024$, thus the solution space encompasses a relatively small number of possible keys (it can be traversed using the brute force technique in a reasonable time). However, it is much easier to observe and investigate the behavior and search capabilities of the GA and MA in this case (the proposed algorithms may be later utilized to retrieve

significantly larger decryption keys which could not be found using exact
approaches due to their execution times).

Each chromosome in a population is assessed using the fitness function:

$$F_f = H(P, D) = \sum_{i=1}^{n} P_i \oplus D_i, \tag{2}$$

where H denotes the Hamming distance, P_i is the single character of the plaintext P,
D_i is a single character of the decrypted text D, and n is the message (data) length.
Therefore, the fitness function counts the number of differences between the piece
of the captured plaintext and the decrypted text (generated during the decryption
process) for each key from the population. The lower value of the fitness function
corresponds to the higher-quality individual (thus, we aim at minimizing the values
of F_f during the evolution).

3.1 Genetic Algorithm Attack

Once the initial population is generated—denoted as $P(0)$—it undergoes the evolution process. This evolution employs standard genetic operators: *selection*,
crossover, and *mutation* applied to the current population $P(t)$. In this work, we
utilize the tournament selection to retrieve two parents to be crossed over. In each
tournament, two individuals are randomly drawn from the population, and the better
one (i.e., with a smaller value of the fitness function) becomes a parent. Then, the
parents are crossed over using a single point crossover operator. Here, a randomly
chosen bit (at the position 1–10) is the crossover point. The bits after this point are
swapped between the parents, and two children solutions are generated. Finally,
each child individual is mutated. This operator involves swapping the bit values (0
is changed to 1, and vice versa) with the probability $p_m = 0.1$ for each bit. The next
generation ($P(t + 1)$) is composed of N child individuals. In this scenario, we do not
incorporate any mechanisms for avoiding the premature convergence of the search
in order to investigate the GA capabilities. Also, the elitism approach is not applied
in this basic GA variant.

An exemplary GA attack is visualized in Fig. 4. The red dots represent the
"active" individuals which take part in each algorithm step. Here, the individuals
K_{55} and K_{13} won two tournaments and became the parents. Then, the fifth bit was
selected as the crossover point to generate two children, which were finally mutated
and transferred to the next generation.

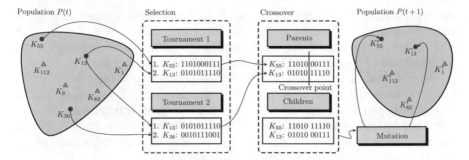

Fig. 4 Steps of the genetic algorithm attack

3.2 Memetic Algorithm Attack

The memetic algorithm attack incorporates the same selection, crossover and mutation operators as the GA discussed in Sect. 3.1. In the MA, we introduce two additional memetic steps (presented in Fig. 5) which help boost the convergence capabilities of the algorithm. These operations include the *education* process (which can be considered as the hill climbing procedure), and creating and exploiting a pool of the best individuals. These solutions were annotated as the best ones in certain generations during the evolution. Thus, we utilize the historical information to guide the search efficiently.

The education procedure is executed after creating children for each pair of parents. In this step, we replace each bit of the best child (selected from all N children) with the probability $p_e = 0.5$, and verify the value of the fitness function. If this value is decreased (i.e., the change of the bit value resulted in a higher-quality individual—green arrows in Fig. 5), this modified solution replaces

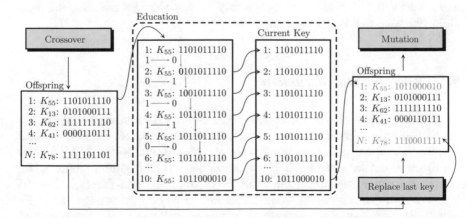

Fig. 5 Steps of the memetic algorithm attack

the child, and the next bit is investigated. If the fitness value becomes larger, then the individual is backtracked to the previous state (red arrows in Fig. 5). This process can be considered as a hill-climbing operation, since we explore the neighborhood of the best child by the local refinements of its bits.

Once N children are generated, we sort them according to their fitness values (thus, the worst individual is the last one in the sorted list). Then, we replace the worst-fitted child with a random individual from the pool. If this solution already exists in the population, then we select another one for inclusion in $P(t + 1)$. Finally, the mutation is performed for each child in order to diversify the search, and to escape local minima during the optimization process.

4   Experimental Results

All algorithms were implemented using the C++ programming language, and were executed on a computer equipped with an Intel i7 processor clocked at 2.1 Ghz. Texts (of a 500-bytes size) came from standard English novels and were encrypted using the SDES algorithm with randomly generated encryption keys. The maximum number of generations of all attacks was fixed to 30. Our algorithms were compared with the binary particle swarm optimization (BPSO) [8], simple random walk (SRW), and the brute force (BF) approaches. For BPSO algorithm the cognitive coefficent c_1 and the social factor c_2 were set to 2. The V_{MAX} particle speed was adjusted to 4. We investigated various population sizes N, where $N = \{10, 20, 30, 50\}$ (for all of the mentioned heuristic algorithms) to verify how this value influences the search (note that we ran the SRW approach N times for a fair comparison). Finally, we examined three variants of our MA: (a) the full MA (both education and replacing the worst child are switched on), (b) the variant in which we employ only the education (replacing the worst child is switched off)—MA(E), and (c) the variant in which we employ only the procedure for replacing the worst child (the education is switched off)—MA(R).

4.1   Analysis and Discussion

In this section, we compare the results obtained using the investigated algorithms which perform the chosen-plaintext attack. Table 1 summarizes the fitness values (the best, worst and average across 30 runs of each technique)—for SRW we show the fitness calculated using Eq. 2. It is worth noting that all algorithms retrieved the correct encryption key during the experiments (thus, the best fitness is $F_f = 0$). However, the evolutionary techniques appeared to be best on average and in the worst case. Therefore, they converge to the higher-quality solutions compared with PSO and SRW. Increasing the population size helps improve the solutions in most cases. It indicates that exploring larger regions of the solution space (which is not

Table 1 The fitness of the best individuals retrieved using the investigated algorithms for various N

$N \rightarrow$	10			20			30			50		
Alg.	Best	Worst	Avg.	Best	Worst	Avg.	Best	Worst	Avg.	Best	Worst	Avg.
SRW	**0**	458	244.3	**0**	407	177.1	**0**	380	70.5	**0**	394	65.3
PSO	**0**	465	266	**0**	438	167.9	**0**	410	129.2	**0**	410	95.9
GA	**0**	462	231.5	**0**	428	136.0	**0**	380	92.9	**0**	**353**	70.1
MA(E)	**0**	446	**154.4**	**0**	384	125.4	**0**	**378**	88.6	**0**	372	**12.4**
MA(R)	**0**	451	176.5	**0**	462	129.3	**0**	393	**68.8**	**0**	395	36.4
MA	**0**	**445**	162.6	**0**	**353**	**44.9**	**0**	397	84.3	**0**	395	31.9

The best fitness (the lowest across all techniques) is boldfaced

extremely large considering the possible keys applied in SDES) is very beneficial. Since it also increases the processing time of a single generation, the decision on the population size should be undertaken very carefully. Alternatively, N can be adapted on the fly [19].

Although it is the PSO which converged quickly (see Table 2—the convergence time, after which the best solution could not be further improved), the solutions retrieved using this algorithm were not of the highest quality. It, in turn, suggests that the swarm optimization arrived too quickly to the local minima of the solution space. On the other hand, the evolutionary techniques kept improving the best solutions for a longer time (thus the convergence speed is lower). The convergence time was enlarged with the increase of N. Since the solution space is relatively small (1024 possible keys), it is possible to run the brute force algorithm which traverses through all keys in a reasonable time. In the worst case (in which $K = 11...1$), the time of finding a valid key using the BF was $\tau = 180.72$ s. Albeit it is feasible for SDES, for more challenging encryption techniques applying the BF for the cryptanalysis task is impossible due to a rapid growth of the solution space size (and its execution time).

Finally, we performed the two-tailed Wilcoxon tests (performed for the best-quality keys found for each run, and for each population size) in order to verify the statistical significance of the results. In Table 3, we gather the p-values—those which are lower than 0.05 (thus indicating the statistically important differences between the best results retrieved using the corresponding techniques) are boldfaced. It is worth noting that our MA improved the best solutions significantly when other state-of-the-art algorithms are considered. The differences between various MA variants are not necessarily statistically important. Therefore, it would be beneficial to utilize the variant which has the best convergence capabilities. However, the impact of the proposed memetic operations is still to be investigated for more challenging cryptanalysis problems (in which the solution space encompasses a notably larger number of possible encryption keys).

Table 2 The convergence time (τ_c) of the investigated algorithms for various N

$N \rightarrow$	10			20			30			50		
Alg.	Best	Worst	Avg.	Best	Worst	Avg.	Best	Worst	Avg.	Best	Worst	Avg.
SRW	**0.18**	26.55	51.25	**0.51**	39.01	99.85	2.55	57.09	146.99	**1.90**	106.3	241.75
PSO	1.23	**15.36**	**39.10**	0.78	**38.81**	**93.77**	**0.38**	**36.15**	**98.50**	3.48	**74.33**	**203.04**
GA	3.52	26.40	53.38	3.47	42.09	104.38	5.25	95.13	156.55	17.04	122.30	256.01
MA(E)	1.73	52.67	121.99	3.54	106.94	230.53	5.18	124.29	331.34	8.51	232.85	508.44
MA(R)	5.02	52.82	103.43	3.29	81.35	148.19	14.98	129.93	286.49	8.43	201.41	497.67
MA	14.01	63.48	127.69	11.16	112.65	230.18	5.29	139.24	323.38	8.60	157.57	369.46

The best τ_c (the lowest across all techniques) is boldfaced

Table 3 The level of statistical significance obtained using the two-tailed Wilcoxon test for each pair of the investigated algorithms

	PSO	GA	MA(E)	MA(R)	MA
SRW	0.1707	0.749	**0.0183**	0.0767	**0.0088**
PSO		0.1118	**0.0007**	**0.0056**	**0.0007**
GA		–	0.0574	0.1362	**0.0078**
MA(E)			–	0.7114	0.5029
MA(R)				–	0.2187

The *p*-values which are less than 0.05 (thus statistically significant) are boldfaced

5 Conclusions and Future Work

In this paper, we proposed the genetic and memetic algorithms for the cryptanalysis of texts ciphered using the SDES algorithm. In the MA, we enhanced a standard GA using two refinement procedures. An extensive experimental study (which also included the sensitivity analysis of the algorithm components) showed the convergence capabilities of the proposed techniques. Also, we compared them with other state-of-the-art methods. The analysis was complemented with the statistical tests to verify the significance of the results.

Our ongoing research encompasses investigating the EAs for the cryptanalysis of challenging approaches using benchmark datasets. Also, we aim at designing adaptive techniques to modify the crucial algorithm parameters during the execution [19]. We plan to design and implement a parallel EAs for finding valid encryption keys in order to further speed up the execution, and compare it with other EAs (using the same number of fitness evaluations).

Acknowledgments This research was performed using the infrastructure supported by the POIG.02.03.01-24-099/13 grant: "GeCONiI—Upper Silesian Center for Computational Science and Engineering".

References

1. Chase, M., Chow, S.S.: Improving privacy and security in multi-authority attribute-based encryption. In: Proceedings of the ACM Conference on Computer and Communications Security. CCS 09, pp. 121–130. ACM, New York, USA (2009)
2. Stinson, D.R.: Cryptography: Theory and Practice. CRC Press, Inc., Boca Raton (1995)
3. Spillman, R., Janssen, M., Nelson, B., Kepner, M.: Use of a genetic algorithm in the cryptanalysis of simple substitution ciphers. Cryptologia **17**(1), 31–44 (1993)
4. Song, J., Zhang, H., Meng, Q., Wang, Z.: Cryptanalysis of two-round DES using genetic algorithms. In: Kang, L., Liu, Y., Zeng, S. (eds.) Advances in Computation and Intelligence, vol. 4683, pp. 583–590. LNCS, Springer, Berlin (2007)
5. Boryczka, U., Dworak, K.: Genetic transformation techniques in cryptanalysis. In: Nguyen, N., Attachoo, B., Trawiński, B., Somboonviwat, K. (eds.) Intelligent Information and Database Systems, vol. 8398, pp. 147–156. LNCS, Springer, Berlin (2014)

6. Boryczka, U., Dworak, K.: Cryptanalysis of transposition cipher using evolutionary algorithms. In: Hwang, D., Jung, J., Nguyen, N.T. (eds.) Proceedings of ICCCI, vol. 8733, pp. 623–632. LNCS, Springer, Berlin (2014)
7. Garg, P.: Cryptanalysis of SDES via evolutionary computation techniques. CoRR abs/0906.5123 (2009)
8. Dworak, K., Boryczka, U.: Cryptanalysis of SDES using modified version of binary particle swarm optimization. In: Proceedings of ICCCI. LNCS, Springer (2015) (in press)
9. Youssef, A.M.: Cryptanalysis of a quadratic knapsack cryptosystem. Comp. Math. App. **61**(4), 1261–1265 (2011)
10. Russell, M., Clark, J., Stepney, S.: Using ants to attack a classical cipher. In: Cantu-Paz, E., et al. (eds.) Proceedings of GECCO, vol. 2723, pp. 146–147. Lecture Notes in Computer Science, Springer, Berlin (2003)
11. Garici, M., Drias, H.: Cryptanalysis of substitution ciphers using scatter search. In Mira, J., Alvarez, J. (eds.) Artificial Intelligence and Knowledge Engineering Applications: A Bioinspired Approach, vol. 3562, pp. 31–40. Lecture Notes in Computer Science, Springer, Berlin (2005)
12. Dewu, X., Wei, C.: A survey on cryptanalysis of block ciphers. In: 2010 International Conference on Computer Application and System Modeling (ICCASM), vol. 8, pp. 218–220 (2010)
13. Michalewicz, Z.: Genetic Algorithms + Data Structures = Evolution Programs, 3rd edn. Springer, London (1996)
14. Nalepa, J., Kawulok, M.: A memetic algorithm to select training data for support vector machines. In: Proceedings of GECCO 2014, pp. 573–580. ACM, NY, USA (2014)
15. Chen, X., Ong, Y.S., Lim, M.H., Tan, K.C.: A multi-facet survey on memetic computation. IEEE Trans. Evol. Comp. **15**(5), 591–607 (2011)
16. Kawulok, M., Nalepa, J.: Support vector machines training data selection using a genetic algorithm. In: Gimel'farb, G., Hancock, E., Imiya, A., Kuijper, A., Kudo, M., Omachi, S., Windeatt, T., Yamada, K. (eds.) S + SSPR 2012, vol. 7626, pp. 557–565, LNCS, Springer, Berlin (2012)
17. Nalepa, J., Kawulok, M.: Adaptive genetic algorithm to select training data for support vector machines. In: Esparcia-Alcazar, A.I., Mora, A.M. (eds.) Applications of Evolutionary Computation, vol. 8602, pp. 514–525. LNCS, Springer, Berlin (2014)–525
18. Cekala, T., Telec, Z., Trawinski, B.: Truck loading schedule optimization using genetic algorithm for yard management. In: Nguyen, N.T., Trawiński, B., Kosala, R. (eds.) Intelligent Information and Database Systems, vol. 9011, pp. 536–548. LNCS, Springer, Berlin (2015)
19. Nalepa, J., Blocho, M.: Adaptive memetic algorithm for minimizing distance in the vehicle routing problem with time windows. Soft Comput 1–19 (2015)
20. Knudsen, L., Mathiassen, J.: A chosen-plaintext linear attack on DES. In: Goos, G., Hartmanis, J., van Leeuwen, J., Schneier, B. (eds.) Fast Software Encryption, vol. 1978, pp. 262–272. LNCS, Springer, Berlin (2001)
21. Schaefer, E.F.: A simplified data encryption standard algorithm. Cryptologia **20**(1), 77–84 (1996)

Admission Control and Scheduling Algorithms Based on ACO and PSO Heuristic for Optimizing Cost in Cloud Computing

Ha Nguyen Hoang, Son Le Van, Han Nguyen Maue
and Cuong Phan Nhat Bien

Abstract Scheduling problem for user requests in cloud computing environment is NP-complete. This problem is usually solved by using heuristic methods in order to reduce to polynomial complexity. In this paper, heuristic ACO (Ant Colony Optimization) and PSO (Particle Swarm Optimization) are used to propose algorithms admission control, then building a scheduling based on the overlapping time between requests. The goal of this paper is (1) to minimize the total cost of the system, (2) satisfy QoS (Quality of Service) constraints for users, and (3) provide the greatest returned profit for SaaS providers. These algorithms are set up and run a complete test on CloudSim, the experimental results are compared with a sequential and EDF algorithms.

Keywords Admission control · Scheduling algorithms · Qos constraint · Resource allocation

1 Introduction

Cloud computing is the development of distributed computing, parallel computing and grid computing [1]. Resources in cloud computing including virtual machines, storage space, network bandwidth, etc., in which virtual machines have speed,

H.N. Hoang (✉) · H.N. Maue · C.P.N. Bien
Hue University of Sciences, Hue, Vietnam
e-mail: nhha76@gmail.com

H.N. Maue
e-mail: nmhan2009@gmail.com

C.P.N. Bien
e-mail: pnbckhmtpy2@gmail.com

S. Le Van
Da Nang University of Education, Da Nang, Vietnam
e-mail: levansupham2004@yahoo.com

© Springer International Publishing Switzerland 2016   15
D. Król et al. (eds.), *Recent Developments in Intelligent Information
and Database Systems*, Studies in Computational Intelligence 642,
DOI 10.1007/978-3-319-31277-4_2

bandwidth and costs vary. Users must pay for the entire time the rent for each virtual machine, although they may not use up this time.

The previous studies mainly focus on scheduling for user requests on multi-processor system with a fixed number of processors. For scalable computing, the virtual machines (VMs) are rented, and can be scaled up to any number. This creates some fundamental changes in the problem: (1) the requests can be done in parallel, its deadline and budget can be always guaranteed; (2) the number of VMs ready to serve is huge, so at every moment one can choose the number and type of VMs appropriately. (3) VMs can be rented with a fixed cost in a certain period of time. If the user does not use up the amount of time, other users can take advantage of it.

Generally, the admission control and scheduling request with parameters such as arrival time, deadline, budget, workload, and penalty rate, etc. is an NP-complete problem [2]. Therefore, to give an optimal solution one must often do exhaustive search while complexity is exponential, so this method cannot be applied. To overcome this disadvantage, we study the heuristic methods to produce a near optimal solution such as ACO method [1, 3, 4], PSO method [5–7]. After that, we take advantage of the remaining period of this request to implement for the next request with the purpose of greatest benefit for service providers.

In cloud computing environment, users rent through Internet and pay a fee for use. Therefore, the scheduling algorithms based on constraints are often used. In this case, the user's parameters such as time, users' service fees, providers' service fees, reliability, etc., are given priority when scheduling. Lee [8] made scheduling model for the requests on the cloud computing environment with the goal of bringing the highest profit for the service provider but looking in detail at the two participating elements of budget and deadline requirements. The studies [9, 10] focus on the scheduling requests for power savings on data center. The recent study by Ramkumar [11] of schedule in real-time requests used for priority queues mapped into resource requests but focused to solve scheduling tasks quickly satisfy most of the requests deadline regardless of cost and its budget. Irugurala [12] make scheduling algorithm with the objective to bring the highest return for SaaS providers but considering between the two types of costs: the cost of initializing virtual machine and the fee of virtual machine which are used to select resources. The studies in [1, 7], using the heuristic PSO, ACO to propose scheduling algorithms but only focus on the minimal makespan of the tasks.

In this paper, ACACO and ACPSO algorithms are proposed with the goal of are making the smallest cost for the system and combining with these algorithms for proposing Mprofit algorithm to bring big profits to SaaS providers.

The article includes: building system model (Sect. 2), building algorithm, introducing three ACACO, ACPSO and Mprofit algorithms then simulating, evaluating between the algorithms (Sect. 3) and conclusions (Sect. 4).

2 System Model

Systems in cloud computing environment consist of the following components [13]: Users, SaaS, PaaS and IaaS providers. Users send requests to the SaaS provider. PaaS providers use component admission control here to analyze the QoS parameters and to decide acceptance or rejection of the request based on the user's abilities, the availability and cost of virtual machines. If the request is accepted, the scheduling component is responsible for locating the resources for the user's requests.

Users send N service requests $\{t_1, t_2,..., t_N\}$ to SaaS providers [13], each request t_i (a_i, d_i, b_i, α_i, w_i, in_i, out_i) includes the following constraints: a_i: arrival time of request; d_i: deadline of request; b_i (budget): the maximum cost users will pay for the services; α_i (penalty rate): a ratio of compensation to the user if the SaaS provider does not provide timely; w_i (workload): how many MI (million instruction) are required to meet the request; Size of input and output file: in_i and out_i.

In cloud computing environment with Y IaaS providers $\{x_1, x_2,..., x_Y\}$, each IaaS provider provides M virtual machines $\{vm_1, vm_2,..., vm_m\}$ for SaaS providers and is responsible for coordinating the VMs which runs on the their physical resources, each virtual machine $vm_{jx}(p_{jx}, s_{jx}, Dtp_{jx}, Dts_{jx})$ of the provider x attributes includes: p_{jx}(price): pricing depends on per hour that SaaS providers must pay for IaaS providers using VMs; s_{jx}: processor speed of virtual machines (MIPS); Dtp_{jx}: the price SaaS providers must pay to transport data from resource provider to user's computer; Dts_{jx}: data transporting speed depends on network performance

In this session, we focus on PaaS provider model. All IaaS's resource providers are not related to one another, can be executed in parallel and are represented by R. We set schedule for N requests independently not to follow any particular order of priority (non-preemptive) on Y providers. The requirements are denoted $npmtn$. The aim is to find the minimum cost but still satisfying deadline and budget of the requests, it means that C_{min} must be found. So the model is $R \mid npmtn \mid C_{min}$

Call C_{ijx} the cost of executing the request i on the virtual machine j of the resource provider x. Mean while C_{ijx} costs include costs:

- The cost of executing request (CP_{ijx}):

$$CP_{ijx} = p_{jx} * \frac{w_i}{s_{jx}} \tag{1}$$

- The cost of data transmission (CTD_{ijx}):

$$CTD_{ijx} = Dtp_{jx} * \frac{in_i + out_i}{Dts_{jx}} \tag{2}$$

- Costs of the SaaS provider must be returned to the users if not meeting the deadline (CR_{ijx}), depending on the penalty rate (α_i) and exceeded time deadline β_{ijx}:

$$CR_{ijx} = \alpha_i * \beta_{ijx} \tag{3}$$

Let T_{ijx} is the time to process the request i on the virtual machine j of resource providers x. T_{ijx} is determined as follows:

$$T_{ijx} = \frac{w_i}{s_{jx}} + \frac{in_i + out_i}{Dts_{jx}} + \beta_{ijx} \tag{4}$$

Therein: $\frac{w_i}{s_{jx}}$: time to process the requests; $\frac{in_i + out_i}{Dts_{jx}}$: time to transfer data; β_{ijx}: exceeded time deadline.

The goal of the paper is to construct algorithms to find the virtual machine in the data center to minimize the cost, such as:

$$\underset{1 \leq i \leq N}{\text{Max}} \left(\sum_{x=1}^{Y} \sum_{j=1}^{M} (b_i - C_{ijx}) \right) \tag{5}$$

- For the profit of SaaS provider, the cost of request i must satisfy the requests of its budget that is:

$$C_{ijx} < b_i \tag{6}$$

- To satisfy the constraints of user, the execution time of request i must meet the deadline itself:

$$T_{ijx} \leq d_i + \beta_{ijx} \tag{7}$$

Thus, to achieve the proposed goals (5), it must satisfy two constraints (6) and (7).

3   Construction of Algorithm

3.1   ACACO Algorithm

The ACO heuristic and system model (session 2) are used to make a scheduling algorithm with the objective of making the total cost to the minimum but still meet the budget and deadline of the requests. To apply the ACO, one must determine the minimum function F, heuristic information η_i, pheromone update and probability P [3, 14]. We construct minima function F and heuristic information η_i to find the best IaaS provider as follows:

$$F = \text{Max}(C_{jx}), \quad j = 1 \ldots M, x = 1 \ldots Y \tag{8}$$

$$\eta_i = \frac{1}{C_{jx}}, \quad i = 1 \ldots N, j = 1 \ldots M, x = 1 \ldots Y \tag{9}$$

Every ant starts from the resource provider IaaS and requests resources randomly. At each iteration of the ants; find the minima function and pheromone update as follows:

$$\tau_{ijx} = \rho * \tau_{ijx} + \Delta\tau_{ijx} \tag{10}$$

Therein: $\Delta\tau_{ijx} = \frac{1-\rho}{F_k}$: with F_k is a minima function of the ant k; $\Delta\tau_{ijx}$: was added to the pheromone; ρ is the evaporation rate is determined in the range (0,1).

According to [4] first request is done and it selects providers randomly. The next request will be processing and it selects the next provider with the probability:

$$P_{ij} = \frac{\tau_{ij} * \eta_{ij}}{\sum_{j=1}^{M} \tau_{ijx} * \eta_{ijx}} \tag{11}$$

In this paper, we consider on multiple providers, each provider offers multiple virtual machines, so the probability to select the next request t_i on the virtual machine j of the resource provider x is defined as follows:

$$P_{ijx} = \frac{\tau_{ijx} * \eta_{ijx}}{\sum_{x=1}^{Y} \sum_{j=1}^{M} \tau_{ijx} * \eta_{ijx}} \tag{12}$$

Therein: η_{ijx}: heuristic information, τ_{ijx} pheromone rate left when moving

ACACO algorithm
Input: $\rho = 0.05$, $\tau_{ijx} = 0.01$, number of ant $k = 10$; T, X, VMx: The set of user requests, the set of IaaS providers and the set of VMs.
2**Output**: The scheduling list S contains all approved requests by SaaS provider, each request i is mapped to the virtual machine j of provider x.

Description algorithm:

FOR EACH t_i **in** T **DO**
  FOR EACH ant k **DO**
  FOR EACH x **in** X **DO**
    Calculating heuristic information for request t_i on virtual machines vm_{i_x} as the formula (8);
    Find the value of current pheromone τ_{ijx};
    Calculated cost and execution time as the formula (1), (2), (3), (4);
    Pheromone update as the formula (10);
    Calculate the probability for request t_i map into virtual machine vm_{i_x} as the formula (12);
   END FOR
  END FOR
  From the probability on virtual machines vm_{j_x}, find the virtual machine which has the highest probability, but the $cost \leq b_i$ and processing time $\leq d_i + \beta_{ijx}$ if found then $S=S+\{t_i \rightarrow vm_{j_x}\}$ else inform the users that the request has been reject;

END FOR

3.2 ACPSO Algorithm

Eberhart and Kennedy introduced PSO algorithm based on the experience of swarm in 1995 [5]. It simulates the social behavior of birds or fish searching for food. In every generation, each particle can change its position from time to time and find local optimal solution (*Pbest*) in D-dimension search space. Then *Pbest* is compared with global optimization solution (*Gbest*) of the swarm to update the value for *Gbest*. Based on *Gbest*, the best optimal solution is found. To apply the PSO algorithm, one must determine the position, velocity, *Pbest* and *Gbest* [5–7]. Each particle is based on current velocity and distance from *Pbest* to *Gbest* to change position and speed as follows [6]:

$$v_x^{j+1} = \omega * v_x^{j+1} + c_1 * r_1 * (Pbest_x - pos_x^{j+1}) + c_2 * r_2 * (Gbest - pos_x^{j+1}) \quad (13)$$

$$pos_x^{j+1} = pos_x^j + v_x^{j+1} \quad (14)$$

Therein: pos_x^j: position of particle x in dimension j; pos_x^{j+1}: position of particle x in dimension $j + 1$; ω: inertia weight (value: 0.1 ... 0.9); c_1, c_2: acceleration coefficient (value: 1...2); r_1, r_2: random number between 0 and 1; v_x^j: velocity of particle x in dimension j; $Pbest_x$: local best position of particle x; $Gbest$: global best position of the entire swarm.

We call P the population consisting of Y particles: $P = \{X_1, \ldots, X_Y\}$. Each particle X_k $(k = 1 \ldots Y)$ searches food in M_k dimension space and is defined: $X_k = \{pos_{ix}^1, \ldots, pos_{ix}^{M_k}\}$, where in pos_{ix}^j is the position of the particle x at the loop i in dimension j $(j = 1 \ldots M_k)$.

Velocity V_k of particle X_k is shown as follows: $V_k = \{v_{ix}^1, \ldots, v_{ix}^{M_k}\}$, wherein v_{ix}^j is the velocity of the particle x at the loop i in dimension j.

In the model at Sect. 2, we have N requests of users and Y providers, each particle X_k $(k = 1 \ldots Y)$ loop M_k times to find resources for the request t_i $(i = 1 \ldots N)$, each particle is equivalent with each providers. The value of pos_{ix}^j is virtual machine j of provider x, which is mapped to request t_i. This value is taken from 1 to M_k, and the value of v_{ix}^j is taken from $-M_k$ to M_k randomly, wherein M_k is the number of VMs in each provider x. To achieve the objective of SaaS providers as in formula (6) we construct fitness function for particle x to select virtual machine j for request i as follows:

$$f(pos_{ix}^j) = \frac{1}{C_{ijx}} \tag{15}$$

Therein, C_{ijx} is defined as in formula (1)–(3).

Based on the fitness function in formula (15), the local optimal position of particle x is as follows:

$$pb_{ix}^{j+1} = \begin{cases} p_{ix}^{j+1} & \text{if } f(pos_{ix}^{j+1}) \geq f(pos_{ix}^j) \text{ and } C_{ijx} \leq b_i \text{ and } T_{ijx} \leq d_i + \beta_{ijx} \\ pb_{ix}^j & \text{Other cases} \end{cases} \tag{16}$$

Therein, $C_{ijx} \leq b_i$ and $T_{ijx} \leq d_i + \beta_{ijx}$ as formula (6) and (7)

Local best position of particle x ($Pbest_x$) and global best position ($Gbest$) are calculated as:

$$Pbest_x = \max_{1 \leq j \leq =M_x} (pb_{ix}^j) \tag{17}$$

$$Gbest = \max_{1 \leq x \leq Y} (Pbest_x) \tag{18}$$

ACPSO algorithm

Input: T, P, VM_j: The set of user requests, the set of IaaS providers, the set of VMs.

Output: The scheduling list S contains all approved requests by SaaS provider, each request $t_i \in S$ is mapped to the virtual machine j of provider x.

Description algorithm:

Initialization: pos_{ix}^j is a random number from 1 to M_k; v_{ix}^j is a random number from

$-M_k$ to M_k; Pbest$_x$= pos_{1x}^1 ; $Gbest = \max_{1 \leq x \leq Y}(Pbest_x)$;

FOR EACH t_i **in** T **DO**
  FOR EACH X_i **in** P **DO**{
    Calculate fitness function as formula (15);
    Calculate $Pbest_i$ as formula (16) and (17);
    Calculate $Gbest_i$ as formula (18);
  END FOR
  Based on $Gbest_i$, find the virtual machine which has the *cost* $<b_i$ and processing
  time $\leq d_i + \beta_{ijx}$, if found then $S=S+\{ t_i \rightarrow vm_{jx} \}$ else inform the users that the
  request has been reject;
  Update the position and velocity of the particles as formula (13),(14);
END FOR

3.3 Mprofit Algorithm

Each virtual machine of IaaS providers is hired for hours and SaaS vendors must pay a fixed fee for the rental hours, if they do not use all their one-hour of hiring time, they also have to pay for a whole hour. This promotes a demand for effective positioning of costs for requests. Each vendor x can accept multiple requests, the advantage of validity period of the lease within 1 h of request is taken in the same vendor to provide the highest return for SaaS providers. The period of validity within an hired hour is called as the advance time of both requests and defines the set T_i including every request of the same provider with request t_i and put the advantage on request t_i. All these requests can share the same virtual machine.

$$T_i = \{t_l | d_l \geq d_i \text{ and } a_l < d_i\} \tag{19}$$

After identifying set T_i, the overlapping time will be calculated. t_{iljx} is defined as the effective time to calculate the request t_l after completing the request of t_i on virtual machine j of resource provider x. The value of t_{iljx} depends on the speed of virtual machines, arrival time, deadline, and workload of t_i and t_j. t_{iljx} is calculated as follows:

$$t_{iljx} = \begin{cases} \min(D - U_{il}, d_l - a_l) & \text{if } a_l - a_i \geq \frac{w_i}{s_{jx}} \\ D - U_{il} & \text{if } a_l - a_i < \frac{w_i}{s_{jx}} \text{ and } d_l - a_i \geq D \\ d_l - (a_i + U_{il}) & \text{if } a_l - a_i < \frac{w_i}{s_{jx}} \text{ and } d_l - a_i < D \end{cases} \tag{20}$$

Therein $U_{il} = \frac{w_i}{s_{jx}} + \max(a_l - d_i, 0)$, s_{jx} the speed of virtual machine is mapped to request t_i

Mprofit algorithm
Input: S is the set of request which has been accepted by the SaaS provider, and the output of ACACO and ACPSO algorithms.
Output: An optimal schedule ST to map the request to virtual machine.

Description algorithm:

Sort all request in S accordingly to the provider, then all requests of the same provider will be in the same group;
FOR EACH provider x **in** S **DO**
  PUSH(t_i);// *Save t_i into the stack, t_i is the first request of the provider x*
  ST=ST+$\{t_i\}$; S=S-$\{t_i\}$;
  FOR EACH request t_i **in the provider** x **DO**
    t_i=POP(); // *Take t_i from stack*
    Find T_i and calculate t_{iljx} for the requests in T_i as in formula (19), (20);
    Find $max(t_{iljx})$, t_i has the largest overlapping time as the next request;
    Base on $max(t_{iljx})$ to find w_l reload all request status of t_i;
    PUSH(t_i);
    ST=ST+$\{t_i\}$; S=S-$\{t_i\}$;
  END FOR
END FOR
Base on ST to produce the mapped schedule onto the request of resources;

3.4 Correctness of the Algorithms

- Stutzle and Dorigo [3], Clerc [15] proved the convergence of the ACO and PSO algorithms which ensure the convergence of proposed algorithm ACACO and ACPSO.
- ACACO and ACPSO algorithms map the request i into the virtual machine j of the provider x based on the probability and fitness function as in formula (12) and (15). Therefore, the smaller the cost of virtual machine is, the greater the heuristic information and the greater fitness function are. This results in more probability of selecting low-cost VMs.
- The resource rental period is D-minute, therefore request t_i completes its task on itself with the lesser time than D-minute, but to pay the fee of D-minute. Mprofit algorithm takes advantage of this effective time interval to process the next request, which makes the costs of the entire system reduced, according to the objectives of SaaS providers

3.5 Simulation and Evaluation of the Algorithms

The algorithms are installed in NetBean 7.1.1, JDK 6, CloudSim 2.1 tools package
[16] with the following parameters: Use 4 Datacenter, 10 physical hosts, 150 virtual
machines. The parameters of users and resource providers are identified:

On the user side: the arrival time to be taken at random from 1 to 500, deadline is
generated randomly between (d_l, d_u) minutes and the different values of d_l and d_u
are limited from 10 to 1500, deadline must be greater than arrival time. Workload is
taken at random from 8×10^4 to 10^5 MI, based on the workload the required budget
is estimated, the remaining parameters are taken as implicit in CloudSim.

On the resource provider's side: the researchers simulate upon four resources
providers; each resource provider has a number of virtual machines, costs, speed,
and different bandwidth. In simulation installation, the Vm class of CloudSim is
inherited to create a virtual machines with the parameters of speed and cost are
defined as follows: speed is taken at random from 10^3 to 5×10^3 MIPS corresponding with the costs which are real numbers taken at random from 0.001 to 0.01,
other parameters of the virtual machine as a virtual machine initialization time,
bandwidth, etc. took default values of CloudSim.

The values in simulation is the result of the 5 tests and the average results are
obtained. In the simulation, we compare the total cost on the requests which were
accepted by the SaaS provider, these requests must be present in the output of the
algorithms. In contrast when compared to the benefit of providers, we calculate on
all requests are accepted.

3.5.1 Analyze the Total Cost and Total Profit as Fixed Requests

Figures 1 and 2 show the total cost and total profit of algorithms; using 150 VMs
and 1000 requests. Simulation results show that the total cost is often lower and

Fig. 1 The total cost of the
algorithms as fixed requests

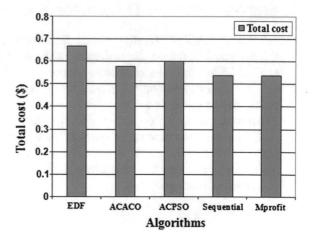

Fig. 2 The total profit of the
algorithms as fixed requests

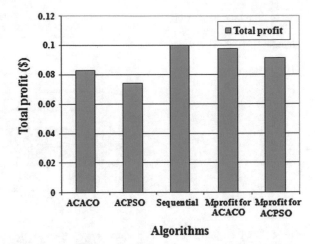

total profit of Mprofit algorithm is often higher than in remaining algorithms. The reason is that the ACACO and ACPSO algorithms are responsible for admission control, accepting or rejecting the user requests, the requests is accepted will be mapped to VMs which have low cost, then Mprofit algorithm continues taking advantage of the overlapping time of the requests in the same IaaS provider, which leads to the total of processing fee reduction as Fig. 1.

In contrast, sequential algorithm does not consider the overlapping period between requests, but uses exhaustive algorithm to find the resource, so there will be many cases that the request can't use all the rental time, this will make the cost of the sequential algorithm increase and take a huge amount of time to make a schedule. As EDF algorithms only consider the ratio used: $U = \sum_{i=1}^{m} \frac{C_i}{T_i} \leq 1$ (where C_i is the execution time and T_i corresponded deadline) [17, 18] to map the request to the resource, thus EDF algorithm only ensures the request to complete before the deadline, regardless of the cost of the request. So, we do not consider the EDF algorithm when comparing the benefits of SaaS providers.

The results in Fig. 2 shows the total profit given by SaaS providers of sequential algorithm and Mprofit algorithm are nearly equal. The sequential algorithm uses exhaustive algorithm to find the best resources, whereas Mprofit algorithm takes advantage of the period that is not used up to implement for the next request.

3.5.2 Analyze the Total Cost and Total Profit as Fixed Number of VMs and Changing Number of Requests

This section presents the results of the total cost, total profit when change the number of requests from 1000 to 5000 and also maintaining the fixed number of VMs as 150, as shown in Figs. 3 and 4. Sequential algorithm uses the exhaustive algorithm to find the resource, therefore the larger the requests are, the more time

Fig. 3 The total cost of the
algorithms when requests
change

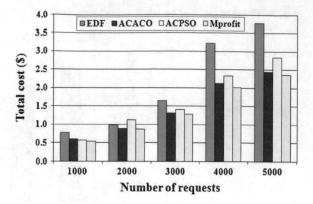

Fig. 4 The total profit of the
algorithms when requests
change

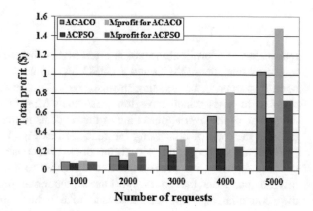

used for scheduling the complexity of algorithm will be exponential, so the
sequential algorithm is not considered in this section. When the number of requests
increases, it will have many requests that can't use all the rental time, while Mprofit
algorithm would be able to use all of such rental time. This will lead to the total cost
of Mprofit algorithm is much smaller than EDF, ACPSO and ACACO algorithms,
as shown in Fig. 3. Although the cost of ACPSO algorithm is higher than ACACO
algorithm but ACPSO load balancing and faster than ACACO. When the input data
of Mprofit algorithm is outputs of ACACO (Mprofit for ACACO) will give the
biggest benefit to the SaaS provider, as shown in Fig. 4.

4   Conclusions

The article focuses on researching the admission control algorithm and scheduling
for users' requests with QoS constraints, in each request, the researchers check all
factors such as arrive time, cost, deadline, budget, word-load, rates of penalization,

input and output file sizes; each virtual machine has speed, cost and different bandwidth. This paper proposes ACACO, ACPSO algorithms to admission control and find the resources with low cost in order to bring the lowest cost to users. In combining with Mprofit algorithm, the Mprofit algorithm is proposed to use up all the time that the requests rented for bringing the most profit to SaaS provider. According to the analysis and strategically experimental results of the same samples and using some CloudSim simulations show that ACACO, ACPSO and Mprofit algorithms have impressive improvement of cost compared to the sequential and EDF algorithms.

References

1. Li, K., Xu, G., Zhao, G., Dong, Y.: Cloud task scheduling based on load balancing ant colony optimization, In: 2011 Sixth Annual Sixth Annual Chinagrid, pp. 3–9. IEEE (2011)
2. Brucker, P.: Scheduling Algorithms, 5th edn., Springer, Berlin (2007)
3. Stutzle, T., Dorigo, M.: A Short convergence proof for a class of ant colony optimization algorithms. Trans. Evol. Comp, **6**, 358–365 (2002)
4. Kousalya, K., Balasubramanie, P.: An enhanced ant algorithm for grid scheduling problem. IJCSNS **8**, 262–271 (2008)
5. Kennedy, J., Eberhart, R.: Particle swarm optimization. In: Proceedings of IEEE International Conference on Neural Networks, pp. 1942–1948. IEEE (1995)
6. Bergh, F.V.D.: An analysis of particle swarm optimizers, Doctoral Dissertation, University of Pretoria Pretoria, South Africa, South Africa (2002)
7. Gomathi, B., Krishnasamy, K.: Task scheduling algorithm based on hybrid particle swarm optimization in cloud computing environment. JATIT **55**, 33–38 (2013)
8. Lee, Y.C., Wang, C., Zomaya, A.Y., Zhou, B.B: Profit-driven service request scheduling in clouds. In: Proceedings of the 2010 10th IEEE/ACM International Conference on Cluster, Cloud and Grid Computing, pp. 15–24. IEEE (2010)
9. Mao, M., Li, J.: Cloud auto-scaling with deadline and budget constraints. In: Grid Computing, 11th IEEE/ACM International Conference, pp. 41–48. IEEE (2010)
10. Kim, K.H., Beloglazov, A., Buyya, R.: Power-aware provisioning of cloud resources for real-time services. In: Proceedings of the 7th International Workshop on Middleware for Grids, Clouds and e-Science, pp. 1–6. ACM (2009)
11. Ramkumar, N., Nivethitha, S.: Efficient Resource utilization algorithm for service request scheduling in cloud. IJET **5**, 1321–1327 (2013)
12. Irugurala, S., Chatrapati, K.S.: Various scheduling algorithms for resource allocation in cloud computing. IJES **5**, 16–24 (2013)
13. Wu, L., Garg, S.K., Buyya, R.: SLA-based admission control for a Software-as-a-Service provider in cloud computing environments. JCSS **78**, 1280–1299 (2012)
14. Dorigo, M., Stutzle, T.: Ant Colony Optimization, A Bradford Book. The MIT Press, Cambridge, Massachusetts (2004)
15. Clerc, M.: The swarm and the queen: towards a deterministic and adaptive particle swarm. In: Proceedings of the 1999 Congress on Evolutionary Computation, CEC 99, vol 48, pp. 1951–1957. IEEE (1999)
16. Buyya, R., Ranjan, R.: Modeling and simulation of scalable cloud computing environments and the CloudSim toolkit: challenges and opportunities. In: Proceedings of the 7th High Performance Computing and Simulation, pp. 1–11. IEEE (2009)

17. Burns, A., Davis, R.I., Wang, P., Zhang, F.: Partitioned EDF scheduling for multiprocessors using a C = D scheme. Department of computer science. J. Real-Time Syst. **48**, 3–33 (2012)
18. Gupta, G., Kumawat, V., Laxmi, P.R., Singh, D., Jain, V., Singh, R.: A simulation of priority based earliest deadline first scheduling for cloud computing system. In: 2014 First International Conference on Networks & Soft Computing (ICNSC), pp. 35–39. IEEE (2014)

A Novel Clustering-Based 1-NN Classification of Time Series Based on MDL Principle

Vo Thanh Vinh and Duong Tuan Anh

Abstract In this work, we propose a clustering-based k-NN classifier for time series data. The classifier aims to select useful instances for the training set at the classifying step in order to reduce the training set and speed up the classification. First, we apply the MDL principle in selecting the core and peripheral sets for the clusters formed in the clustering step. Second, our classifier applies the Compression Rate Distance, a powerful distance measure for time series, which was proposed in our previous work. We conducted experiments over a vast majority number of time series datasets. The experimental results reveal that our proposed method can outperform the original Clustering-based k-NN and the two other methods INSIGHT, and Naïve Rank in most of the tested datasets. In comparison to the traditional k-NN method which use the whole original training set, our proposed method can run much faster while the accuracy rates decrease insignificantly (about 1.59 %) and in some data sets, the accuracy rates even increase.

Keywords Time series · MDL principle · Compression rate distance · Clustering-based classification · k-nearest neighbor · Instance selection

1 Introduction

Due to high dimensionality, a challenging property of time series data, there are just a few classification methods that can be applied in time series data such as: Artificial Neural Network, Bayesian Network, Decision Tree, etc. Among these

V.T. Vinh (✉)
Faculty of Information Technology, Ton Duc Thang University,
Ho Chi Minh City, Vietnam
e-mail: vtvinh@it.tdt.edu.vn

D.T. Anh
Faculty of Computer Science and Engineering, Ho Chi Minh City University
of Technology, Ho Chi Minh City, Vietnam
e-mail: dtanh@cse.hcmut.edu.vn

© Springer International Publishing Switzerland 2016
D. Król et al. (eds.), *Recent Developments in Intelligent Information
and Database Systems*, Studies in Computational Intelligence 642,
DOI 10.1007/978-3-319-31277-4_3

methods, the simple approach 1-Nearest Neighbor (1-NN) Classifier has been considered as very hard to beat for time series data since it can bring out the highest classification accuracy [1, 2]. There are some approaches to enhance the efficiency of 1-NN Classifier in time series such as using some index structures, reducing dimensions of time series, speed up the computational time of distance measure, and reducing the number of instances in the training set. For the last direction, one method was proposed by Xi et al. [1] which used an instance ranking function called Naïve Rank. In 2011, Buza et al. proposed another method based on graph theory [3], which also attempts to rank the instances in the training set.

Both methods, Naïve Rank [1] and INSIGHT [3] can reduce the training set to a specific percentage of instances based on an instance ranking function. Naïve Rank ranking formulas are as follows: for an instance x in the training set, $rank(x) = \sum_j k$, with $k = 1$ if $class(x) = class(x_j)$; otherwise, $k = -2$, where x_j is the instance having x as its nearest neighbor. When applying $rank(x)$ function, some time series can have the same ranking. Therefore, the authors proposed the second function $priority(x) = \sum_j 1/d(x, x_j)^2$ to break this tie, where $d(x, x_j)$ is the distance between instances x and x_j.

The INSIGHT (Instance Selection based on Graph-coverage and *Hubness* for Time series) was proposed by Buza et al. [3]. This method uses the score functions for an instance x based on *hubness* property of this instance. There are three score functions as follows: *good* 1-*occurrence* score $f_G(x) = f_G^1(x)$, relative score $f_R(x) = f_G^1(x)/(f_N^1(x) + 1)$, and Xi's score $f_{Xi}(x) = f_G^1(x) - 2f_B^1(x)$, where $f_G^1(x)$ is the number of instances in the training set having x as their 1-NN and x has the same label as them, $f_B^1(x)$ is the number of instances in the training set having x as their 1-NN and x has different label from them, $f_N^1(x)$ is the number of instances in the training set having x as their 1-NN.

In this work, for time series data, we proposed a novel clustering-based k-NN classifier which based on k-means clustering algorithm in its learning step. This kind of classifiers was firstly proposed by Hwang and Cho [4] for ordinary data and our work is the first attempt to apply their method for time series data. Our contributions in this work can be summarized as follows. First, we propose a systematical way based on the Minimum Description Length (MDL) principle to select the core and peripheral sets for the clusters formed in the clustering step of this classifier. Second, we apply our recently proposed distance measure in time series, the Compression Rate Distance (CRD) in our proposed classifier. The CRD is a powerful distance measure which was proved outperforming most of the widely used distance measure in time series data such as Euclidean Distance, Dynamic Time Warping Distance, Complexity-Invariant Distance. Finally, our experiments were conducted over a vast majority number of time series datasets. The experimental results reveal that our method outperforms the original Clustering-based k-NN [4], INSIGHT [3], and Naïve Rank [1] in most of the tested datasets. In comparison to the traditional method which uses the whole original training set, our proposed method can run remarkably faster while the accuracy rates decrease insignificantly (about 1.59 %) and in some datasets, the accuracy rates even increase.

2 Background and Related Works

2.1 Time Series and 1-Nearest Neighbor Classifier

A time series T is a sequence of real numbers collected at regular intervals over a period of time: $T = t_1, t_2,..., t_n$. Furthermore, a time series can be seen as an n-dimensional object in metric space. In 1-Nearest Neighbor Classifier (1-NN), the data object is classified the same class as its nearest object in the training set. The 1-NN has been considered hard to beat in classification of time series data among other methods such as Artificial Neural Network, Bayesian Network, etc. [1, 2].

2.2 Clustering-Based Reference Set Reduction for K-NN Classifier

The clustering-based k-NN classification method proposed by Hwang and Cho in [4] consist of two steps: preprocessing and classification. In the pre-processing step, the reference set is partitioned into clusters. K-means clustering is used because it is relatively fast. The instances assigned to each cluster are the split into "core set" and "peripheral set". The instances located within a certain distance from the cluster center are put into "core set" while the rest are put into "peripheral set". In classification step, first the distance from the new instance x_t to the cluster centers is calculated. The instances from the closest cluster and from the peripheral sets of adjacent clusters are put into the reference set. Finally, k-NN is performed with the reference set just obtained.

2.3 Minimum Description Length Principle: A Model Selection Method

The MDL principle is a formalization of Occam's razor in which the best hypothesis for a given set of data is the one that leads to the best compression of the data. MDL was introduced by Rissanen [5]. This principle is a crucial concept in information theory and computational learning theory. The MDL principle is a powerful tool which has been applied in many time series data mining tasks, such as discovery motif [6], criterion for clustering [7], semi-supervised classification of time series [8–10], discovery rules in time series [11], Compression Rate Distance measure for time series [12]. In this work, we use a version of MDL for time series which was applied in some previous studies such as in [7–12]. The MDL principle is described as follows:

Discrete Normalization Function: A discrete function *Dis_Norm* is the function to normalize a real-value subsequence T into b-bit discrete value of range [1, 2^b]. The maximum of the discrete range value 2^b is also called the *cardinality*:

$$Dis\_Norm(T) = round\left[(T-min)/(max - min) \times \left(2^b - 1\right)\right] + 1$$

where *min* and *max* are the minimum and maximum value in T, respectively. After casting the original real-valued data to discrete values, we are interested in determining how many bits are needed to store a particular time series T. It is called the *Description Length* of T.

Entropy: entropy can be understood as the average number of bits requires representing one element of a sequence

$$E(T) = - \sum_{t \text{ in } unique\,(T)} P(T = t)\log_2 P(T = t)$$

where $P(T = t) = frequency(t)/length(T)$ is the probability of value t in time series T, and *unique*(T) is a set of unique values in T. For example: $T = [2\ 1\ 2\ 1\ 1\ 2\ 2\ 3]$, $length(T) = 8$, $unique(T) = \{1, 2, 3\}$, $frequency(1) = 3$, $frequency(2) = 4$, $frequency$ (3) = 1. Therefore, $E(T) = -P(T = 1) \times \log_2 P(T = 1) - P(T = 2) \times \log_2 P(T = 2) - P(T = 3) \times \log_2 P(T = 3) = -(3/8) \times \log_2(3/8) - (4/8) \times \log_2(4/8) -(2/8) \times \log_2(2/8) = 1.5306$ bits. Note that time series T is a discrete time series which was transformed by the above *Dis_Norm* function.

Description Length: a description length DL of a time series T is the total number of bits required to represent it: $DL(T) = w \times E(T)$, where w is the length of T.

Hypothesis: a hypothesis H is a time series used to encode one or more time series of the same length. We are interested in how many bits are required to encode T given H. It is called the *Reduced Description Length* of T.

Reduced Description Length: a reduced description length of a time series T given hypothesis H is the sum of the number of bits required in order to encode T exploiting the information in H. i.e. $DL(T|H)$, and the number of bits required for H itself. The reduced description length is defined as follows:

$$DL(T,H) = DL(H) + DL(T|H)$$

One simple approach of encoding T using H is to store a difference vector between T and H. Therefore: $DL(T|H) = DL(T - H)$. The MDL is a method for selecting a good model. A model is considered good if there are some compressions obtained through encoding. For example: Given A and B, two time series of length 20:

$$A = [4\ 5\ 7\ 7\ 8\ 9\ 11\ 11\ 12\ 13\ 14\ 16\ 16\ 17\ 20\ 19\ 20\ 21\ 22\ 22]$$
$$B = [4\ 5\ 6\ 7\ 8\ 9\ 10\ 11\ 12\ 13\ 14\ 15\ 16\ 17\ 18\ 19\ 20\ 21\ 22\ 23]$$

Without encoding, the bit requirement to store A and B is $20 \times E(A) + 20 \times E$ $(B) = 20 \times 3.8219 + 20 \times 4.3219 = 162.876$ bits. The difference vector $A' = A - B = [0\ 0\ 1\ 0\ 0\ 0\ 1\ 0\ 0\ 0\ 0\ 1\ 0\ 0\ 2\ 0\ 0\ 0\ 0\ -1]$. The bit requirement is now just $20 \times E(A') + 20 \times E$ $(B) = 20 \times 1.154 + 20 \times 4.3219 = 109.518$ bits, which brings out a good data compression.

Now given B as above and $C = [23\ 18\ 13\ 22\ 8\ 13\ 5\ 10\ 11\ 9\ 6\ 21\ 19\ 20\ 4\ 6\ 9\ 6\ 10\ 11]$. We compute the bit requirement without encoding: $20 \times E(C) + 20 \times E$ $(B) = 20 \times 3.6842 + 20 \times 4.3219 = 160.122$ bits. With encoding, the bit requirement now is $20 \times E(C - B) + 20 \times E(B) = 20 \times 4.0219 + 20 \times 4.3219 = 166.876$ bits. This result illustrates that C is not similar to B because there is no compression when using encoding.

In this work, we apply the MDL principle to select the core sets for the clusters formed in the clustering-based classification which will be depicted in Sect. 3. Besides, the MDL principle is also a core spirit in the Compression Rate Distance which is described in the next subsection.

2.4   Compression Rate Distance

The Compression Rate Distance (CRD) was proposed in our previous work in 2015 [12]. In the work, we showed that the CRD can outperform Euclidean Distance, Complexity-Invariance Distance (CID) [13], and Dynamic Time Warping Distance (DTW) [14, 15] in classification problem using 1-NN. In addition, the CRD can satisfy many important properties such as Early Abandoning, Lower Bounding, and Relaxed Triangular Inequality in the *semi-metric space*. The spirit behind the CRD is based on the Minimum Description Length principle. If we can compress our time series when we use one time series as a hypothesis to encode the other one, their distance is reduced; otherwise, their distance increases. The CRD is defined as follows:

$$CRD(Q,C) = [CR(Q,C)^{\alpha} + 1] \times ED(Q,C)$$

where and CR is the *compression rate*, computed by

$$CR(Q,C) = E(Q-C)/[\min\{E(Q), E(C) + \varepsilon]$$

with $E(t)$ as the entropy of time series t, ED as the Euclidean Distance. For more information about the CRD, interested reader can refer to [12].

3 Proposed Method

The clustering-based classification method was firstly proposed by Hwang and Cho [4] for ordinary data, our work is the first attempt to apply this method for time series data. In the algorithm proposed by Hwang and Cho, in preprocessing step, the reference set is grouped into clusters. The patterns assigned to each cluster are split into "core set" and "peripheral set". The patterns located within a certain distance from the cluster center are put into "core set" while the rest are put into "peripheral set".

Different from Hwang and Cho's method which chooses the core set with the instances that have distance to their centroid lower than or equal to two times of the average distance to centroid of all instances in that cluster, we propose to select the core set by using the MDL principle. The spirit behind our method is that if we can compress the time series by using its centroid as hypothesis, we assign them to the core set of that cluster; otherwise, the time series is assigned into the peripheral set of that cluster. Our propose method is separated into two steps:

- *Step* 1: This step is the learning step. Firstly it bases on k-means clustering algorithm to cluster the training set; then, it selects the core set of each cluster bases on the MDL principle. The k-means is chosen for clustering here since its low complexity is very suitable for time series data which has very high dimensionality. Figure 1 illustrates the algorithm of this step. Note that Euclidean Distance is applied in k-means clustering only.
- *Step* 2: This step is the adaptive classification step. Firstly, it identifies which time series in the original training set should be used; then, it classifies the new time series with the selected patterns in training set by using 1-NN and CRD measure. CRD measure is used here since it is more accurate than Euclidean distance in classification and also much faster than DTW distance. In Fig. 2, we illustrate the algorithm of this step.

[*C, L, CoreSet, SurfaceSet*] = **Clustering-based_1-NN_Learning** (TRAIN, *k*)
// TRAIN: training set; *k*: number of clusters; *C*: centroids set, *L*: labels of each time series from *k*-means (different from real label of each time series in TRAIN).

1	[*C, L*] = *k*-means(TRAIN, *k*)
2	**for each** centroid *c* in *C*
3	**for each** time series *t* in the cluster with centroid *c*
4	**if** $CR(t, c) \leq 1$
5	*CoreSet(c)* = *CoreSet(c)* ∪ {*t*} // Add *t* to core set of cluster *c*
6	**else**
7	*PeripheralSet(c)* = *PeripheralSet(c)* ∪ {*t*} // Add *t* to peripheral set of cluster *c*
8	**end**
9	**end**
10	**end**

Fig. 1 Learning based on *k*-means algorithm

label = **1-NN_Classifier** (C, *CoreSet, PeripheralSet, t*)
// C: centroid set; *CoreSet*: core set of each cluster; *PeripheralSet*: peripheral set of each cluster; t: the time series to be classified; *label*: output label of t.

1	Find nearest centroid c_{nr} in C of time series t
2	**if** $CR(t, c_{nr}) \leq 1$
3	$\quad$ TRAIN = CoreSet(c_{nr})
4	**else**
5	$\quad$ TRAIN = CoreSet(c_{nr}) ∪ PeripheralSet(c_1) ∪ PeripheralSet(c_2) ∪...
	$\qquad\qquad\qquad\qquad$ ∪ PeripheralSet(c_k) // k is the number of clusters
7	**end**
8	*label* = 1-NN (TRAIN, t) // use the CRD measure

Fig. 2 Classification algorithm

Note that in line 4 of the algorithms in Fig. 1, and line 2 of the algorithm in Fig. 2, we use the compression rate to decide if the time series is assigned into the core set or the peripheral set. This compression rate bases on the MDL principle and is defined as in the CRD measure in Sect. 2.4. The rationale behind this core set selection method is that if we can compress the time series by using the centroid of its clusters as hypothesis, we assigned it into the core set of that cluster; otherwise, we assigned it into the peripheral set of that cluster. Recall that for the original method in [4], this criterion is $Distance(t, c) \leq 2 \times$ AVERAGE$\{Distance(t_i, c)|t_i$: time series in cluster with centroid $c\}$.

For the classification step, to select instances from the original training set to be used in the 1-NN classifier, there are two cases: If the time series to be classified is in the core set of its nearest cluster, the training set to be used is only the core set of that nearest cluster; otherwise, the training set to be used is all the time series of its nearest cluster and all the time series in the peripheral sets of the other clusters. Notice that in the classification step, CRD distance is used in the 1-NN procedure (Line 8 in Fig. 2).

4 Experimental Evaluation

We implemented our proposed method and the other methods in Matlab 2012 and conducted the experiments on the Intel Core i7-740QM 1.73 GHz, 4 GB RAM PC. In the experiments, we compare the accuracy rates of our method with those of Traditional method (use the whole original training set), Original Clustering-based [4], Naïve Rank [1], and INSIGHT [3]. The accuracy rate is the ratio of the number of correctly classified instances on tested data to the total number of tested instances. In general, the higher the accuracy rate is, the better the method is.

Our experiments are conducted by using 10-fold cross-validation over the datasets from the UCR Time Series Classification Archive [16]. There are 47 datasets used in these experiments, which is the before summer 2015 datasets; each has from 2 to 50 classes. The length of each time series in each datasets is from 60 to 1882. Due to space limit, we do not show details of the datasets. Interested reader can find more details about these datasets in [16]. A list of dataset names is also shown in Table 1 when we report main experimental results. Some initial parameters in the experiments are set as follows: α value of CRD is set to 2. Cardinality for the MDL principle (described in Sect. 2.3) is set to 8 (3-bit discrete values). For all the methods, we use CRD as distance measure. Euclidean Distance is applied only in k-means clustering in our methods (other parts of our method still use CRD). The number of clusters k in k-means algorithm is set to the number of classes of each dataset. The scoring function used in the INSIGHT method is *good* *1-occurrence* score.

4.1   Experimental Results

In Fig. 3, we show three scatter graphs of accuracy rates that compare the proposed method (Clustering-based with support of MDL) to the three methods: Original Clustering-based [4], Naïve Rank [1], and INSIGHT [3]. Each circle point in the scatter graphs represents a dataset. In general, the more circle points deviate on the lower triangle, the better our proposed method is. Figure 3a shows that our proposed method outperforms the Original Clustering-based in most of the datasets. That means the way we select the core set based on MDL principle brings better results. In details, our method outperform Original Clustering-based in 29 datasets, draw in 10 datasets, and lose in only 8 datasets. Due to limited space, we do not show the details of accuracy rates. In order to compare with Naïve Rank and INSIGHT, we first reduce the training set to the same size as in our proposed method and then we apply the k-NN classification. Figure 3b, c also shows that our method significantly outperform the two previous methods. The proposed method wins INSIGHT in 34 datasets, draws in 4 datasets and loses in 9 datasets; it wins Naïve Rank in 39 datasets, draws in 2 datasets and loses in 6 datasets. Moreover, in Table 1, we show details about the accuracy rates of the proposed method, Naïve Rank, and INSIGHT.

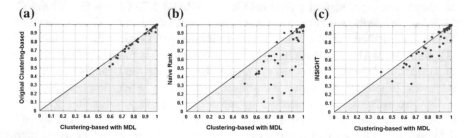

Fig. 3 Comparing the accurate rates of the proposed method with those of Original Clustering-based (**a**), Naïve Rank (**b**), and INSIGHT (**c**)

Table 1 Details of accuracy rate and running time (milliseconds) in 47 datasets

Datasets	Traditional k-NN		Naïve rank	INSIGHT	Clustering-based with MDL		
	(1) Accuracy ± STD	(2) Time (ms)	(3) Accuracy ± STD	(4) Accuracy ± STD	(5) Accuracy ± STD	(6) AVG TRAIN size and its percentage compared to original TRAIN size	(7) Time (ms) and Speed up compare to (2)
50words	0.780 ± 0.045	459	0.158 ± 0.057	0.505 ± 0.039	0.735 ± 0.026	26(3.1 %)	36(12.8)
Adiac	0.680 ± 0.042	338	0.115 ± 0.022	0.326 ± 0.035	0.661 ± 0.052	30(4.2 %)	29(11.7)
Beef	0.600 ± 0.189	28	0.440 ± 0.126	0.560 ± 0.207	0.600 ± 0.211	13(23.6 %)	10(2.8)
Car	0.792 ± 0.098	58	0.558 ± 0.111	0.733 ± 0.053	0.783 ± 0.119	32(29.6 %)	19(3.1)
CBF	1.000 ± 0.000	361	0.988 ± 0.023	0.997 ± 0.007	0.996 ± 0.010	290(34.6 %)	130(2.8)
Chlorine[1]	0.997 ± 0.003	2019	0.928 ± 0.011	0.824 ± 0.008	0.995 ± 0.004	1504(38.8 %)	684(3)
CinC[2]	0.999 ± 0.003	1102	0.997 ± 0.005	0.998 ± 0.005	0.987 ± 0.014	607(47.4 %)	512(2.2)
Coffee	1.000 ± 0.000	29	0.975 ± 0.079	1.000 ± 0.000	1.000 ± 0.000	30(57.7 %)	17(1.7)
Cricket_X	0.683 ± 0.083	405	0.464 ± 0.044	0.547 ± 0.054	0.619 ± 0.082	115(16.2 %)	68(6)
Cricket_Y	0.619 ± 0.048	405	0.431 ± 0.055	0.514 ± 0.071	0.592 ± 0.047	123(17.4 %)	74(5.5)
Cricket_Z	0.688 ± 0.061	392	0.439 ± 0.050	0.551 ± 0.061	0.618 ± 0.080	114(16.1 %)	70(5.6)
Diatom[3]	1.000 ± 0.000	158	0.983 ± 0.024	0.997 ± 0.011	1.000 ± 0.000	89(30.5 %)	50(3.2)
ECG200	0.911 ± 0.056	92	0.905 ± 0.085	0.916 ± 0.062	0.905 ± 0.060	111(61.3 %)	52(1.8)
ECG[4]	0.994 ± 0.008	411	0.976 ± 0.011	0.984 ± 0.011	0.990 ± 0.013	398(50 %)	186(2.2)
FaceAll	0.950 ± 0.014	1032	0.482 ± 0.067	0.751 ± 0.043	0.900 ± 0.037	196(9.6 %)	98(10.5)
FaceFour	0.944 ± 0.079	64	0.833 ± 0.094	0.867 ± 0.102	0.911 ± 0.102	29(28.2 %)	17(3.8)
FacesUCR	0.966 ± 0.011	1024	0.522 ± 0.049	0.787 ± 0.034	0.931 ± 0.015	198(9.7 %)	98(10.4)
FISH	0.829 ± 0.049	178	0.654 ± 0.065	0.711 ± 0.055	0.826 ± 0.039	59(18.7 %)	33(5.4)
Gun[5]	0.965 ± 0.047	96	0.950 ± 0.053	0.950 ± 0.047	0.940 ± 0.052	89(49.4 %)	42(2.3)
Haptics	0.407 ± 0.047	287	0.398 ± 0.109	0.462 ± 0.063	0.402 ± 0.055	103(24.6 %)	65(4.4)
Inline[6]	0.506 ± 0.058	507	0.323 ± 0.052	0.356 ± 0.037	0.497 ± 0.039	102(17.3 %)	84(6)
Italy[7]	0.970 ± 0.017	465	0.971 ± 0.020	0.969 ± 0.016	0.969 ± 0.018	689(69.7 %)	279(1.7)
Lighting2	0.782 ± 0.167	74	0.800 ± 0.094	0.809 ± 0.132	0.773 ± 0.162	70(63.6 %)	41(1.8)
Lighting7	0.750 ± 0.172	79	0.580 ± 0.114	0.650 ± 0.143	0.660 ± 0.171	29(21.8 %)	19(4.2)
MALLAT	0.985 ± 0.010	1526	0.626 ± 0.140	0.957 ± 0.008	0.984 ± 0.009	356(16.5 %)	214(7.1)

(continued)

Table 1 (continued)

Datasets	Traditional k-NN		Naïve rank	INSIGHT	Clustering-based with MDL		
	(1) Accuracy ± STD	(2) Time (ms)	(3) Accuracy ± STD	(4) Accuracy ± STD	(5) Accuracy ± STD	(6) AVG TRAIN size and its percentage compared to original TRAIN size	(7) Time (ms) and Speed up compare to (2)
Medical[8]	0.789 ± 0.026	517	0.593 ± 0.045	0.675 ± 0.035	0.755 ± 0.025	173(16.8 %)	86(6)
Mote[9]	0.947 ± 0.022	578	0.926 ± 0.022	0.939 ± 0.025	0.943 ± 0.019	578(50.4 %)	262(2.2)
Thorax1[10]	0.849 ± 0.017	2214	0.214 ± 0.029	0.641 ± 0.028	0.836 ± 0.021	104(3.1 %)	85(26)
Thorax2[11]	0.909 ± 0.009	2167	0.247 ± 0.048	0.647 ± 0.028	0.896 ± 0.016	98(2.9 %)	78(27.8)
OliveOil	0.950 ± 0.105	37	0.825 ± 0.169	0.950 ± 0.105	0.950 ± 0.105	21(37.5 %)	14(2.6)
OSULeaf	0.649 ± 0.050	264	0.495 ± 0.050	0.527 ± 0.076	0.646 ± 0.059	90(22.4 %)	50(5.3)
Plane	0.971 ± 0.025	113	0.914 ± 0.095	0.943 ± 0.044	0.967 ± 0.023	40(21.2 %)	22(5.1)
SonyII[12]	0.979 ± 0.015	490	0.976 ± 0.018	0.966 ± 0.023	0.978 ± 0.016	451(51.1 %)	205(2.4)
Sony[13]	0.987 ± 0.013	311	0.967 ± 0.027	0.977 ± 0.027	0.980 ± 0.017	307(54.8 %)	142(2.2)
Star[14]	0.895 ± 0.011	5402	0.925 ± 0.010	0.921 ± 0.010	0.892 ± 0.010	2941(35.4 %)	1810(3)
Swedish[15]	0.833 ± 0.027	457	0.409 ± 0.040	0.638 ± 0.034	0.821 ± 0.038	120(11.8 %)	58(7.9)
Symbols	0.961 ± 0.019	481	0.812 ± 0.061	0.960 ± 0.019	0.959 ± 0.017	205(22.3 %)	98(4.9)
Synthetic[16]	0.902 ± 0.039	252	0.907 ± 0.030	0.863 ± 0.043	0.835 ± 0.090	138(25.6 %)	63(4)
Trace	0.945 ± 0.064	91	0.850 ± 0.088	0.850 ± 0.058	0.940 ± 0.061	56(31.1 %)	30(3)
TwoLead[17]	0.995 ± 0.006	463	0.995 ± 0.008	0.992 ± 0.006	0.995 ± 0.006	523(50 %)	216(2.1)
Patterns[18]	0.995 ± 0.003	1946	0.982 ± 0.006	0.982 ± 0.008	0.986 ± 0.008	2303(51.2 %)	999(1.9)
uWave_X[19]	0.776 ± 0.020	1930	0.664 ± 0.041	0.745 ± 0.011	0.767 ± 0.017	560(13.9 %)	263(7.3)
uWave_Y[20]	0.730 ± 0.020	2020	0.641 ± 0.038	0.692 ± 0.018	0.723 ± 0.018	578(14.3 %)	269(7.5)
uWave_Z[21]	0.710 ± 0.010	2078	0.642 ± 0.011	0.674 ± 0.013	0.704 ± 0.009	580(14.4 %)	274(7.6)
wafer	0.999 ± 0.002	3007	0.997 ± 0.001	0.998 ± 0.002	0.998 ± 0.001	3564(55.3 %)	1523(2)
Words[22]	0.747 ± 0.053	502	0.311 ± 0.068	0.562 ± 0.037	0.723 ± 0.051	57(6.9 %)	39(12.9)
yoga	0.944 ± 0.011	1826	0.915 ± 0.020	0.923 ± 0.018	0.944 ± 0.011	1856(62.5 %)	890(2.1)

Some dataset names in this table are referred as follows:

[1]ChlorineConcentration, [2]CinC_ECG_torso, [3]DiatomSizeReduction, [4]ECGFiveDays, [5]Gun_Point, [6]InlineSkate, [7]ItalyPowerDemand, [8]MedicalImages, [9]MoteStrain, [10]NonInvasiveFatalECG_Thorax1, [11]NonInvasiveFatalECG_Thorax2 [12]SonyAIBORobotSurfaceII, [13]SonyAIBORobotSurface, [14]StarLightCurves, [15]SwedishLeaf, [16]synthetic_control, [17]TwoLeadECG, [18]Two_Patterns, [19]uWaveGestureLibrary_X, [20]uWaveGestureLibrary_Y, [21]uWaveGestureLibrary_Z, [22]WordsSynonyms

Note: AVG—Average, TRAIN—Training set, STD—Standard deviation, Speed up in (7) = Running time in (2)/Running time in (7)

In comparison to the Traditional k-NN which uses the whole original training set, the accuracy of our method decreases just a little, about 1.59 % in average. In some datasets, the accuracy rate does not change or even increases. Moreover, our proposed method just uses a small amount of time series in the training set, and it can run much faster than the Traditional k-NN method. Table 1 illustrates details about the speed up rate as well as the accuracy of our method compared to Traditional k-NN. For example, over CinC_ECG_torso dataset, the speed up rate is 2.2 (speed up rate is the ratio of the running time of Traditional k-NN to that of our method), and the accuracy rate only decrease 1.3 % (from 99.9 to 98.7 %). On NonInvasiveFatalECG_Thorax2 dataset, the speed up rate is 27.8, and the accuracy rate decrease only 1.3 % (from 90.9 to 89.6 %). On Yoga dataset, the accuracy does not change, and the speed up rate is 2.1. Many other good results of our method can be easily seen from Table 1.

In addition, we also compare our method using CRD measure against it using Euclidean Distance. The results show that CRD helps to bring out better results than Euclidean Distance. Due to limited space, we do not show these experiment results.

5 Conclusions and Future Works

Reducing the size of training set is very crucial in improving the accuracy and speed of k-NN classifier. In this work, our proposed method inherits some ideas from the clustering-based reference set reduction for k-NN classifier proposed by Hwang and Cho [4] and applies these ideas the first time in time series classification. We employed MDL principle in selecting the core sets of the clusters formed in clustering step and applied CRD measure in our time series 1-NN classifier. The experimental results reveal that our proposed method outperforms the two other methods in time series classification: Naïve Rank and INSIGHT. Besides, its accuracy rate is closely equal to the Traditional k-NN classifier, while the running time is remarkably better. We attribute the high performance of our method to the use of MDL principle in partitioning the clusters into core and peripheral sets and the use of CRD measure in 1-NN classification step.

As for future work, we intend to extend our method so that it can allow user to set a specific number of instances in the training set as in Naïve Rank and INSIGHT.

References

1. Xi, X., Keogh, E., Shelton, C., Wei, L., Ratanamahatana, C.A.: Fast time series classification using numerosity reduction. In Proceedings of the 23rd International Conference on Machine learning, ICML '06, pp. 1033–1040 (2006)
2. Ding, H., Trajcevski, G., Scheuermann, P., Wang, X., Keogh, E.: Querying and mining of time series data: experimental comparison of representations and distance measures. Proc VLDB Endowment 1(2), 1542–1552 (2008)

3. Buza, K., Nanopoulos, A., Schmidt-Thieme, L.: INSIGHT: efficient and effective instance selection for time-series classification. In: 15th Pacific-Asia Conference, PAKDD 2011, Proceedings, Part II, pp. 149–160, Shenzhen, China, 24–27 May 2011
4. Hwang, S, Cho, S.: Clustering-based reference set reduction for k-nearest neighbor. In: Proceedings of 4th International Symposium on Neural Networks, ISNN 2007, Part II, pp. 880–888. Nanjing, China, 3–7 June 2007
5. Rissanen, J.: Modeling by shortest data description. Automatica **14**(7), 465–471 (1978)
6. Tanaka, Y., Iwamoto, K., Uehara, K.: Discovery of time-series motif from multi-dimensional data based on MDL principle. Mach. Learn. **58**(2–3), 269–300 (2005)
7. Rakthanmanon, T., Keogh, E.J., Lonardi, S., Evans, S.: MDL-based time series clustering. Knowl. Inf. Syst. **33**(2), 371–399 (2012)
8. Vinh, V.T., Anh, D.T.: Some novel improvements for MDL-based semi-supervised classification of time series. In: Proceedings of Computational Collective Intelligence. Technologies and Applications, LNAI 8733, pp. 483–493. Springer, Berlin (2014)
9. Begum, N., Hu, B., Rakthanmanon, T., Keogh, E.: A minimum description length technique for semi-supervised time series classification. In: Integration of Reusable Systems Advances in Intelligent Systems and Computing, pp. 171–192 (2014)
10. Vinh, V.T., Anh, D.T.: Constraint-based MDL principle for semi-supervised classification of time series. In: Proceedings of 7th International Conference on Knowledge and System Engineering, pp. 43–48, Ho Chi Minh City, 8–10 Oct 2015
11. Shokoohi-Yekta, M., Chen, Y., Campana, B., Hu, B., Zakaria, J., Keogh, E.: Discovery of meaningful rules in time series. In: Proceedings of SIGKDD (2015)
12. Vinh, V.T., Anh, D.T.: Compression rate distance measure for time series. In: Proceedings of the 2015 IEEE International Conference on Data Science and Advance Analytics, Paris, 19–21 Oct 2015
13. Batista, G.E.A.P.A., Keogh, E.J., Tataw, O.M., Souza, V.M.A.D.: CID: an efficient complexity-invariant distance for time series. Data Min. Knowl. Disc. **8**(3), 634–669 (2014)
14. Keogh, E., Ratanamahatana, C.A.: Exact indexing of dynamic time warping. Knowl. Inf. Syst. **7**(3), 358–386 (2005)
15. Berndt, D., Clifford, J.: Using dynamic time warping to find patterns in time series. In: Proceedings of AAAI Workshop on Knowledge Discovery in Databases, KDD, pp. 359–370. Seattle, Washington, USA (1994)
16. Chen, Y., Keogh, E., Hu, B., Begum, N., Bagnall, A., Mueen, A., Batista, G.: The UCR time series classification archive (2015). www.cs.ucr.edu/~eamonn/time_series_data/

Efficient Subsequence Join Over Time Series Under Dynamic Time Warping

Vo Duc Vinh and Duong Tuan Anh

Abstract Joining two time series in their similar subsequences of arbitrary length provides useful information about the synchronization of the time series. In this work, we present an efficient method to subsequence join over time series based on segmentation and Dynamic Time Warping (DTW) measure. Our method consists of two steps: time series segmentation which employs important extreme points and subsequence matching which is a nested loop using sliding window and DTW measure to find all the matching subsequences in the two time series. Experimental results on ten benchmark datasets demonstrate the effectiveness and efficiency of our proposed method and also show that the method can approximately guarantee the commutative property of this join operation.

Keywords Time series · Subsequence join · Important extreme points · Nested loop join · Dynamic time warping

1 Introduction

Time series data occur in so many applications of various domains: from business, economy, medicine, environment, meteorology to engineering. A problem which has received a lot of attention in the last decade is the problem of similarity search in time series databases.

The similarity search problem is classified into two kinds: *whole sequence matching* and *subsequence matching*. In the whole sequence matching, it is supposed

V.D. Vinh (✉)
Faculty of Information Technology, Ton Duc Thang University,
Ho Chi Minh, Vietnam
e-mail: vdvinh@it.tdt.edu.vn

D.T. Anh
Faculty of Computer Science and Engineering, Ho Chi Minh City University,
Ho Chi Minh, Vietnam
e-mail: dtanh@cse.hcmut.edu.vn

© Springer International Publishing Switzerland 2016
D. Król et al. (eds.), *Recent Developments in Intelligent Information
and Database Systems*, Studies in Computational Intelligence 642,
DOI 10.1007/978-3-319-31277-4_4

41

that the time series to be compared have the same length. In the subsequence matching, the result of this problem is a consecutive subsequence within the longer time series data that best matches the query sequence.

Besides the two above-mentioned similarity search problems, we have the problem of *subsequence join* over time series. Subsequence join finds pairs of similar subsequences in two long time series. Two time series can be joined at any locations and for arbitrary length. Subsequence join is a symmetric operation and produces many-to-many matches. It is a generalization of both whole sequence matching and subsequence matching. Subsequence join can bring out useful information about the synchronization of the time series. For example, in the stock market, it is crucial to find correlation among two stocks so as to make trading decision timely. In this case, we can perform subsequence join over price curves of stocks in order to obtain their common patterns or trends. Figure 1 illustrates the results of subsequence join on two time series. This join operation is potentially useful in many other time series data mining tasks such as clustering, motif discovery and anomaly detection.

There have been surprisingly very few works on subsequence join over time series. Lin et al. in 2010 [5, 6] proposed a method for subsequence join which uses a non-uniform segmentation and finds joining segments based on a similarity function over a feature-set. This method is difficult-to-implement and suffers high computational complexity that makes it unsuitable for working in large time series data. Mueen et al. [7] proposed a different approach for subsequence join which is based on correlation among subsequences in two time series. This approach aims to maximizing the Pearson's correlation coefficient when finding the most correlated subsequences in two time series. Due to this specific definition of subsequence join, their proposed algorithm, Jocor, becomes very expensive computationally. Even though the authors incorporated several speeding-up techniques, the runtime of Jocor is still unacceptable even for many time series datasets with moderate size. These two previous works indicate that subsequence join problem is more challenging than subsequence matching problem.

In this work, we present an efficient method to subsequence join over time series based on segmentation and Dynamic Time Warping measure. Our method consists

Fig. 1 Two time series are joined to reveal some pairs of matching subsequences

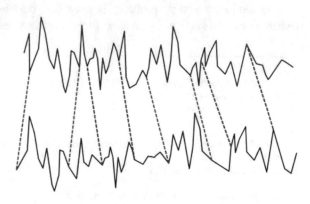

of two steps: (i) time series segmentation which employs important extreme points [1, 8] and (ii) subsequence matching which is a nested loop using sliding window and DTW measure to find all the matching subsequences in the two time series. DTW distance is used in our proposed method since so far it is the best measure for various kinds of time series data even though its computational complexity is high. We tested our method on ten pairs of benchmark datasets. Experimental results reveal the effectiveness and efficiency of our proposed method and also show that the method can approximately guarantee the commutative property of this join operation.

2 Related Work

2.1 Dynamic Time Warping

Euclidean distance is a commonly used measure in time series data mining. But one problem with time series data is the probable distortion along the time axis, making Euclidean distance unsuitable. This problem can be effectively handled by Dynamic Time Warping. This distance measure allows non-linear alignment between two time series to accommodate series that are similar in shape but out of phase. Given two time series Q and C, which have length m and length n, respectively, to compare these two time series using DTW, an n-by-n matrix is constructed, where the (ith, jth) element of the matrix is the cumulative distance which is defined as

$$\gamma(i, j) = d(q_i, c_j) + \min\{\gamma(i-1, j), \gamma(i, j-1), \gamma(i-1, j-1)\}$$

That means $\gamma(i, j)$ is the summation between $d(i, j) = (q_i - c_j)^2$, a square distance of q_i and c_j, and the minimum cumulative distance of the adjacent element to (i, j).

A warping path P is a contiguous set of matrix elements that defines a mapping between Q and C. The tth element of P is defined as $p_t = (i, j)_t$, so we have:

$$P = p_1, p_2, \ldots, p_t, \ldots, p_T \qquad \max(m, n) \leq T < m + n - 1$$

Next, we choose the optimal warping path which has minimum cumulative distance defined as

$$DTW(Q, C) = \min\left\{\sqrt{\sum_{t=1}^{T} p_t}\right\}$$

The DTW distance between two time series Q and C is the square root of the cumulative distance at cell (m, n).

Fig. 2 (*Left*) Two time series
Q and C which are similar in
shape but out of phase: Using
euclidean distance (*top left*)
and DTW distance (*bottom
left*). Note that Sakoe-Chiba
band with width R is used
to limit the warping path
(*right*) [9]

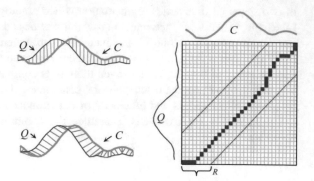

To accelerate the DTW distance calculation, we can constrain the warping path
by limiting how far it may stray from the diagonal. The subset of matrix that the
warping path is allowed to traverse is called *warping window*. Two of the most
commonly used global constraints in the literature are the Sakoe-Chiba band proposed by Sakoe and Chiba 1978 and Itakura Parallelogram proposed by Itakura
1975 [2]. Sakoe-Chiba band is the area defined by two straight lines in parallel with
the diagonal (see Fig. 2).

In this work, we use the Sakoe-Chiba Band with width R and assume that all the
time series subsequences in subsequence matching under DTW are of the same
length.

2.2 Important Extreme Points

Important extreme points in a time series contain important change points of the
time series. Based on these important extreme points we can segment the time series
into subsequences. The algorithm for identifying important extreme points was first
introduced by Pratt and Fink [8]. Fink and Gandhi [1] proposed the improved
variant of the algorithm for finding important extreme points. In [1] Fink and
Gandhi give the definition of strict, left, right and flat minima as follows.

Definition 1 (*Minima*) Given a time series $t_1, \ldots, t_n$ and t_i is its point such that
$1 < i < n$

- t_i is a *strict minimum* if $t_i < t_{i-1}$ and $t_i < t_{i+1}$.
- t_i is a *left minimum* if $t_i < t_{i-1}$ and there is an index *right* $> i$ such that
  $t_i = t_{i+1} = \ldots = t_{right} < t_{right+1}$.
- t_i is a *right minimum* if $t_i < t_{i+1}$ and there is an index *left* $< i$ such that
  $t_{left-1} > t_{left} = \ldots = t_{i-1} = t_i$.
- t_i is a *flat minimum* if there are indices *left* $< i$ and *right* $> i$ such that
  $t_{left-1} > t_{left} = \ldots = t_i = \ldots = t_{right} < t_{right+1}$

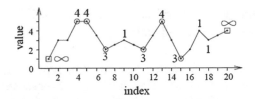

Fig. 3 Example of a time series and its extrema. The importance of each extreme point is marked with an integer greater than 1 [1]

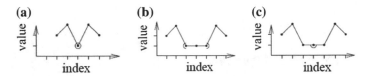

Fig. 4 Four types of important minima [1]. **a** Strict minimum. **b** Left and right minima. **c** Flat minimum

The definition for maxima is similar. Figure 3 shows an example of a time series and its extrema. Figure 4 illustrates the four types of minima.

We can control the compression rate with a positive parameter R; an increase of R leads to the selection of fewer points. Fink and Gandhi [1] proposed some functions to compute the distance between two data values a, b, in a time series. In this work we use the distance $dist(a, b) = |a - b|$.

For a given distance function $dist$ and a positive value R, Fink and Gandhi [1] gave the definition of important minimum as follows.

Definition 2 (*Important minimum*) For a given distance function $dist$ and positive value R, a point t_i of a series $t_1, \ldots, t_n$ is an *important minimum* if there are indices il and ir, where $il < i < ir$, such that t_i is a *minimum* among $t_{il}, \ldots, t_{ir}$, and $dist(t_i, t_{il}) \geq R$ and $dist(t_i, t_{ir}) \geq R$.

The definition for important maxima is similar.

Given a time series T of length n, starting at the beginning of the time series, all important minima and maxima of the time series are identified by using the algorithm given in [1]. The algorithm takes linear computational time and constant memory. The algorithm requires only one scan through the whole time series and, it can process new coming points as they arrive one by one, without storing the time series in memory.

After having all the extrema, we can select only *strict, left* and *right extrema* and discard all flat extrema.

In this work, we apply this algorithm for finding all important extrema from time series before extracting subsequences on them.

3   Subsequence Join Over Time Series Based on Segmentation and Matching

Our proposed method for subsequence join over time series is based on some main ideas as follows. Given two time series, first a list of subsequences is created for each time series by applying the algorithm for finding important extreme points in each time series [1]. Afterwards, we apply a nested loop join approach which uses a sliding window and DTW distance to find all the matching subsequences in the two time series. We call our subsequence join method EP-M (Extreme Points and Matching).

3.1   The Proposed Algorithm

The proposed algorithm EP-M for subsequence join over time series consists of the following steps.

Step 1:   We compute all important extreme points of the two time series T_1 and T_2. The results of this step are two lists of important extreme points $EP1 = (ep1_1, ep1_2, \ldots, ep1_{m1})$ and $EP2 = (ep2_2, ep2_2, \ldots, ep2_{m2})$ where $m1$ and $m2$ are the numbers of important extreme points in T_1 and T_2, respectively. In the next steps, when creating subsequences from a time series (T_1 or T_2), we extract the subsequence bounded by the two extreme points ep_i and ep_{i+2}

Step 2:   We keep time series T_1 fixed and for each subsequence s extracted from T_1 we find all its matching subsequences in T_2 by shifting a sliding window of the length equal to the length of s along T_2 one data point at a time. We store all the resulting subsequences in the result set S_1

Step 3:   We keep time series T_2 fixed and for each subsequence s extracted from T_2 we find all its matching subsequences in T_1 by shifting a sliding window of the length equal to the length of s along T_1 one data point at a time. We store all the resulting subsequences in the result set S_2

In Step 1, for the algorithm that identifies all the important extreme points in the time series, we have to determine the parameter R, the compression rate. According to Pratt and Fink [8], they only suggested that R should be greater than 1. In this work, we apply the following heuristic for selecting value for R. This heuristic is very intuitive. Let SD be the standard deviation of the time series. We set R to $1.5 * SD$.

The pseudo-code for describing Step 2 and Step 3 in EP-M is given as follows.

Algorithm Subsequence_Join ($T_1[1.. n_1]$, $T_2[1.. n_2]$)
Input: Two time series T_1 and T_2 which have been segmented using important extreme points
Output: Two result sets S_1 and S_2 which store matching pairs of subsequences
1.   For each subsequence s extracted from T_1 using the list of important extreme points *EP1*
      Find all the matching subsequences of s in T_2 by calling the procedure Subsequence_Matching (s, T_2)
      Store all the resulting pairs of subsequences in S_1
2.   For each subsequence s extracted from T_2 using the sequence of important extreme points *EP2*
      Find all the matching subsequences of s in T_1 by calling the procedure Subsequence_Matching (s, T_1)
      Store all the resulting pairs of subsequences in S_2

Procedure Subsequence_Matching($s[1.. m]$, $T[1.. n]$)
// T is a time series and s is a subsequence
1 **for** $i = 1$ to $n - m + 1$ **do**
2   *segment_of_T* = subsequence $T_{i, i+m-1}$
3   dtw_distance = DTW_EA(s, *segment_of_T*, *threshold*)
4   **if** (*dtw_distance* <= *threshold*) **then**
5   Store to the pair <s, *segment_of_T*> to the result set S
6 **endfor**
end Procedure

Notice that Procedure *Subsequence_Matching* invokes the procedure *DTW_EA* which computes DTW distance. The *DTW_EA* procedure applies Early Abandoning technique as mentioned in the following subsection.

3.2   Some Other Issues

- **Normalization**
  In order to make meaningful comparisons between two time series, both must be normalized. In our subsequence join method, before applying EP-M algorithm we use min-max normalization to transform the scale of one time series to the scale of the other time series based on the maximum and minimum values in each time series.

Min-max normalization can transform a value v of a numeric variable A to v' in the range [min_A, max_A] by computing:

$$v' = \frac{v - \min_A}{\max_A - \min_A}(new\_\max_A - new\_\min_A) + new\_\min_A$$

- **Speeding-up DTW distance computation**
  To accelerate the calculation of DTW distance, a number of approaches, such as using global constraints like Sakoe-Chiba band or Itakura parallelogram and using some lower bounding techniques have been proposed. Besides, Li and Wang [4] proposed the method which applies an early abandoning technique to accelerate the exact calculation of DTW. The main idea of this technique is described as follows. The method checks if values of the neighboring cells in the cumulative distance matrix exceed the tolerance, and if so, it will terminate the calculation of the related cell. Li and Wang 2009, called their proposed method EA_DTW (Early Abandon DTW). The details of the EA_DTW algorithm are given in [4].

4 Experimental Evaluation

We implemented all the comparative methods with Microsoft C#. We conducted the experiments on an Intel(R) Core(TM) i7-4790, 3.6 GHz, 8 GB RAM PC. The experiments aim to three goals. First, we compare the quality of EP-M algorithm in the two cases: using Euclid distance and using DTW distance. Second, we test the accuracy and time efficiency of EP-M algorithm. Finally, we check whether the EP-M algorithm can approximately guarantee the commutative property of the join operation. For simplicity, we always use equal size time series (i.e. $n_1 = n_2$) to join although our method is not limited to that.

4.1 Datasets

Our experiments were conducted over the datasets from the UCR Time Series Data Mining archive [3]. There are 10 datasets used in these experiments. The names and lengths of 10 datasets are as follows: Chromosome-1 (128 data points), Runoff (204 data points), Memory (6874 data points), Power-Demand-Italia (29931 data points), Power (35450 data points), Koski-ECG (144002 data points), Chromosome (999541 data points), Stock (2119415 data points), EEG (10957312 data points).

Since the subsequence join has to work on the two time series of some degree of similarity, we need a method to generate the synthetic dataset T_2 (inner time series in Step 2) for each test dataset T_1 (outer time series in Step 2). The synthetic dataset is generated by applying the following rule:

$$x_i = x_{i-1} \pm |x_{i-1} - \varepsilon| \qquad \text{where } \varepsilon = \frac{\sum_{i=1}^{6} x_i}{6}$$

In the above formula, + or − is determined by a random process. The synthetic dataset is created after the correspondent dataset has been normalized, therefore, the noisy values does not affect to the generated dataset. Specifically for Runoff dataset (from The Be River in Southern Vietnam) we use the real datasets (monthly data from 1977 to 1993) at two measurement stations Phuoc Long and Phuoc Hoa. The dataset at Phuoc Long Station will play the role of time series T_1 and the dataset at Phuoc Hoa Station will be time series T_2.

4.2   DTW Versus Euclidean Distance in EP-M Algorithm

In time series data mining, distance measure is crucial factor which has a huge impact on the performance of the data mining task. Here we need to know between DTW and Euclidean distance which measure is more suitable for our proposed subsequence join method. We conducted the experiment to compare the performance of EP-M algorithm using each of the two distance measures in terms of the number of pairs of matching subsequences found by the algorithm. Table 1 reports the experimental results over 8 datasets. The experimental results show that with Euclidean distance, EP-M can miss several pairs of matching subsequences in the two time series (see the last column). That means EP-M with Euclidean Distance can not work well in subsequence join as EP-M with DTW. The experiment recommends that we should use DTW distance in the EP-M algorithm even though DTW incurs much more computational time than Euclidean distance.

4.3   Accuracy of EP-M Algorithm

Now we turn our discussion to the accuracy of EP-M algorithm. Following the tradition established in previous works, the accuracy of a similarity search algorithm is basically based on human analysis of the matching subsequences discovered by that algorithm. That means through human inspection we can check if the matching subsequences of a query in question identified by the proposed algorithm in a given time series have almost the same shape as the query pattern. If the check result is positive in most of the test datasets, we can conclude that the proposed algorithm brings out the accurate similarity search.

Table 1 Experimental results in comparing DTW to euclidean distance

Dataset	Length	ED	DTW	ED omissions
Chromosome-1	128	180	277	97
Runoff	204	387	1502	1115
Memory	1000	4040	4101	61
Power-Demand-Italia	1000	5139	23289	18150
Power	1000	8926	10523	6067
Koski-ECG	1000	6562	6562	0
Stock	1000	3236	4531	1295
EEG	1000	13550	35366	21786

Through our experiment, we found out that the result sets provided by our EP-M algorithm are quite accurate over all the test datasets.

4.4   Time Efficiency of EP-M

The names and runtimes (in minutes) of 9 datasets by EP-M are as follows: Chromosome-1 (0.003), Runoff (0.007), Memory (3.227), Power-Demand-Italia (10.8), Power (82), Koski-ECG (106.2), Chromosome (64.4), Stock (34.7), EEG (30.45).

The maximum length of these datasets is 30000. The execution time of EP-M on the Koski-ECG dataset with 30,000 data points is about 1 h 46 min. So, the runtimes reported are acceptable in practice.

4.5   The Commutative Property of Join Operation by EP-M

The reader may wonder if subsequence join achieved by EP-M can satisfy the commutative property of join *operation*, i.e. $T_1 \bowtie T_2 = T_2 \bowtie T_1$. We can check this theoretic property through experiments over 10 datasets with the maximum length 5000. For $T_1 \bowtie T_2$, time series T_1 is considered in the outer loop and time series T_2 is considered in the inner loop of Step 2 in EP-M. For $T_2 \bowtie T_1$, the order of computation is in the opposite. Table 2 reports the experimental results on commutative property of join operation by EP-M in terms of the number of pairs of matching subsequences found by the algorithm in two cases $T_1 \bowtie T_2$ and $T_2 \bowtie T_1$. The experimental results reveal that the difference between two cases is small enough to confirm that the EP-M method can approximately guarantee the commutative property of this join operation.

Table 2 Experimental results on commutative property of join operation by EP-M

Dataset	$T_1 \bowtie T_2$	$T_2 \bowtie T_1$	Error (%)
Chromosome-1	277	272	2
Runoff	1502	1591	6
Memory	4101	4073	0.007
Power-Demand-Italia	511092	524256	3
Power	309576	287576	7.4
Koski-ECG	6562	6366	3
Chromosome	7376	7766	5
Stock	4531	4817	6
EEG	35336	32072	9.6

5 Conclusions

Subsequence join on time series data can be used as a useful analysis tool in many domains such as finance, health care monitoring, environment monitoring. In this paper, we have introduced an efficient and effective algorithm to perform subsequence join on two time series based on segmentation and Dynamic Time Warping measure.

As for future works, we intend to extend our subsequence join method for the context of streaming time series. Besides, we plan to adapt our EP-M method with Compression Rate Distance [10], a new distance measure that is better than DTW, but with $O(n)$ computational complexity. Future work also includes considering the utility of EP-M algorithm for the problem of time series clustering, classification, motif discovery and anomaly detection.

References

1. Fink, E., Gandhi, H.S.: Important extrema of time series. In: Proceedings of IEEE international conference on system, Man and cybernetics, pp. 366–372. Montreal, Canada (2007)
2. Keogh, E.: Exact indexing of dynamic time warping. In: Proceedings of 28th international conference on very large databases (VLDB) (2002)
3. Keogh, E.: The UCR time series classification/clustering homepage. http://www.cs.ucr.edu/~eamonn/time_series_data/ (2015)
4. Li, J., Wang, Y.: Early abandon to accelerate exact dynamic time warping. Int. Arab J. Inf. Technol. **6**(2), 144–152 (2009)
5. Lin, Y., McCool, M.D.: Subseries join: A similarity-based time series match approach. In: Proceedings of PAKDD 2012, Part I, LNAI 6118, pp. 238–245 (2010)
6. Lin, Y., McCool, D., Ghorbani, A.: Time series motif discovery and anomaly detection based on subseries join. Int. J. Comput. Sci. **37**(3) (2002)
7. Mueen, A., Hamooni, H., Estrada, T.: Time series join on subsequence correlation. Proc. ICDM **2014**, 450–459 (2014)
8. Pratt, K.B., Fink, E.: Search for patterns in compressed time series. Int. J. Image Graphics **2**(1), 89–106 (2002)

9. Rakthanmanon, T., Campana, B., Mueen, A., Batista, G., Westover, B., Zhu, Q., Zakaria, J., Keogh, E.: Searching and mining trillions of time series subsequences under dynamic time warping. In: Proceedings. of SIGKDD (2012)
10. Vinh, V.T., Anh, D.T.: Compression rate distance measure for time series. In: Proceedings of 2015 IEEE international conference on data science and advanced analytics, Paris, France, 19–22 October (2015)

Multi-Instrumental Deep Learning for Automatic Genre Recognition

Mariusz Kleć

Abstract The experiments described in this paper utilize songs in the MIDI format to train Deep Neural Networks (DNNs) for the Automatic Genre Recognition (AGR) problem. The MIDI songs were decomposed into separate instrument groups and converted to audio. Restricted Boltzmann Machines (RBMs) were trained with the individual groups of instruments as a method of pre-training of the final DNN models. The Scattering Wavelet Transform (SWT) was used for signal representation. The paper explains the basics of RBMs and the SWT, followed by a review of DNN pre-training methods that use separate instrument audio. Experiments show that this approach allows building better discriminating models than those that were trained using whole songs.

Keywords RBM · Deep neural network · Automatic genre recognition · Unsupervised Pre-training · Neural networks · Music information retrieval

1 Introduction

Deep Learning (DL) is used in multiple application domains such as speech recognition, language processing, recommendation systems and computer vision [2]. In particular, Music Information Retrieval (MIR) found several potentials for DL in the areas of music representation and classification [8, 14, 15, 23]. Deep models are often constructed from several Restricted Boltzmann Machines (RBMs) stacked one on top of another. They are trained one layer at a time in an unsupervised manner. RBMs primarily discover statistical correlation occurring in the data. They operate as features detectors learned from that data [3]. In the case of training a DNN to recognize

M. Kleć (✉)
Multimedia Department, Polish-Japanese Academy of Information Technology,
Warsaw, Poland
e-mail: mklec@pjwstk.edu.pl

© Springer International Publishing Switzerland 2016
D. Król et al. (eds.), *Recent Developments in Intelligent Information and Database Systems*, Studies in Computational Intelligence 642,
DOI 10.1007/978-3-319-31277-4_5

handwritten digits[1] from the image, first RBM learns to recognise "pen strokes". The next one learns segments of the image formed by these strokes, e.g. simple shapes, circles, squares, crosses. Finally the model learns to recognize whole digits on the last layer [4]. This technique was inspired by the biology and the ability of human brain for decomposing visual and auditory signals. Although, the brain signal processing is far beyond the scope of this paper, most would agree that a musically trained person can recognize different instruments playing simultaneously in music. But at the same time, the number of simultaneous sounds and the interference they cause between each other are the reasons why it is still a challenge for MIR to deal with the instruments recognition computationally. The techniques that have been proposed so far are either limited to a slightly defined set of signals, or have low accuracy [16].

1.1 Recent Work

All the recent research on Automatic Genre Recognition (AGR) relied on training different models with the representation of overall songs computed by MFCC or Scattering Wavelets Transform (SWT) [1, 6, 21]. According to the survey described in [22], 23 % of papers about AGR use GTZAN[2] dataset for their experiments. The second widely used benchmark is ISMIR2004[3] which is 17 % of all papers about AGR. Many researchers (58 %) construct their private datasets, that unfortunately prevent them from being compared with others. As the GTZAN is most popular, the following survey and described experiments are also based on it.

In [21] the authors achieved 83 % accuracy using spectral features on GTZAN, dividing it into 50, 25 and 25 % of songs for training, validation and evaluation. In [1] the authors utilized SWT as an signal representation for training SVM classifier. They achieved 89.3 % accuracy in 10-fold cross-validation test, which is much better compared to 82 % achieved using MFCC in the same test. Hence, SWT has turned out to be very effective signal representation for automatic music classification. The power of SWT was utilized in [19] together with sparse representation-based classifiers. They achieved 92.4 % accuracy, which is the state of the art according to the author's knowledge.

1.2 Insights into the Role of Musical Instruments

In a musical band, each instrument has a specific role. Percussion provides rhythm and tempo. Bass gives a base for underlying musical harmony. Some instruments

---

[1]http://yann.lecun.com/exdb/mnist.
[2]http://marsyasweb.appspot.com/download/data_sets.
[3]http://ismir2004.ismir.net/genre_contest.

may play backgrounds whilst other the first-plan parts of the song. The musical genre is determined by a well-defined set of possible instruments. It is also determined by the type of instruments, their interaction and its consonance. The instruments "communicate" with each other, playing a composition that makes the final sound of given song. The problem of polyphonic musical signal decomposition into separate instrument streams is still a largely unsolved due to many harmonic dependencies and the signal complexity [10].

As one instrument depends on other instruments, the features representation of them in hidden layers of a DNN seems to be a promising approach. But obtaining audio files for training with separate instrument audio is a very difficult task. For this reason, this paper uses a simulation of real instruments using the sound of virtual instruments driven by MIDI messages.

This paper is organized as follows: in the next two sections Restricted Boltzmann Machine and Scattering Wavelet Transformation will be described. Following are sections describing the dataset and finally, some conclusions are presented.

2 Restricted Boltzmann Machine

A Restricted Boltzmann Machine (RBM) is a generative stochastic neural network that consists of two layers of visible $V = (v_1, \ldots, v_j)$ and hidden $H = (h_1, \ldots, h_i)$ units. In addition, each layer has bias units associated to each unit in that layer. The connections between V and H are symmetrically weighted and undirected. The word "restricted" refers to the structure of the model where direct interaction between units in the same layer is not allowed. The visible units correspond to the states of observations (data). Hidden units are treated as feature detectors that learn a probability distribution over the inputs [5]. The parameters of an RBM model can be estimated by maximum likelihood (ML) estimation of weights in an unsupervised training. The RBM model maximizes the product of likelihood given the training data, $P(data)$ [5]. An RBM is characterized by the energy function $E(V, H)$ which denotes the joint probability distribution over all units assigned by the network. It is given by $P(V,H) = Z^{-1} exp(-E(V,H))$, where $Z = \sum_{V,H} exp(-E(V,H))$ is a normalizing factor obtained by summing over all possible pairs of visible and hidden vectors. The basic RBM model has the following energy function:

$$E(V,H) = -\sum_{j \in V} a_j v_j - \sum_{i \in H} b_i h_i - \sum_{j,i} v_j h_i w_{ji} \tag{1}$$

where v_j, h_i are the binary states of visible and hidden units, a_j, b_i are their biases, and w_{ji} is the weight between them [11]. This energy function refers to an RBM with binary units (called Bernoulli units). An unsupervised training algorithm for RBM, has been proposed by Hinton and is called Contrastive Divergence (CD) [12]. The algorithm performs data reconstruction (called Gibbs sampling) from the hidden probability distribution of the data. The algorithm is described in detail in [12]. In addition, many practical hints about training RBMs are given in [11].

RBMs have been used before with different types of data, including images [12], mel-cepstral coeficients (MFCC) for music and speech representation [18], bags of words in documents [13], and user ratings of movies [20]. But RMBs owe their popularity to their ability to be stacked on top of each other, forming Deep Belief Networks (DBN) [12] that can subsequently be fine-tuned using back-propagation [2–4].

3   The Scattering Wavelet Transform

The SWT uses a sequence of Wavelet Transforms to compute the modulation spectrum coefficients of multiple orders [17]. It works by computing a series of Wavelet decompositions iteratively (the output of one decomposition is decomposed again), producing a transformation which is both transformation invariant (like the MFCC) and experiences no information loss (proven by producing an inverse transform—something which cannot be done using MFCC without loss). The WT begins by defining a family of dilated signals known as wavelets. A single mother wavelet $\psi(t)$ is expanded to a dictionary of wavelets $\psi_{u,s}$, translated to u and scaled by s, using the formula

$$\psi_{u,s}(t) = \frac{1}{\sqrt{s}} \psi\left(\frac{t-u}{s}\right) \tag{2}$$

These wavelets are then used to decompose the input signal by using a convolution operator (denoted by $\langle \cdot \rangle$):

$$Wf(u,s) = \langle f, \psi_{u,s} \rangle = \int f(t) \frac{1}{\sqrt{s}} \psi\left(\frac{t-u}{s}\right) dt \tag{3}$$

Unlike the Fourier transform (which decomposes the signal into sinusoidal waves of infinite length), the Wavelet Transform (WT) encodes the exact location of the individual components. The Fourier transform encodes the same information as the phase component, but this is usually discarded in the standard MFCC feature set. This means that Fourier-based methods are very good at modeling harmonic signals, but are very weak at modeling sudden changes or short-term instabilities of the signal, something that the WT seems to deal with very well [17].

4   Multi-Instrumental Dataset Preparation

GTZAN consists of 1000 musical pieces (30 s. lengths) organized in 10 genres, distributed uniformly as 100 songs per each genre. In this work, 20 MIDI files belonging to 4 genres were selected: disco, reggae, metal, rock. The particular songs were also a part of the GTZAN. The MIDI files were processed in Apple Logic Pro X software.[4] First, the songs were trimmed to 30 s in length (omitting the first 30 s intro). Next, instrument tracks were grouped into 3 categories: drums (D), bass (B) and others (O). The latter group consists of all other instruments that were not a percussion nor a bass but also not a vocal part (which is usually imitated by the sound of an arbitrary instrument in the MIDI files). The EXS24 sampler in Logic Pro X was controlled by the MIDI events generating real sounding instrumental parts. Finally, these parts were grouped to one of the three categories and saved as audio files.

However, the task of processing MIDI files manually is very time consuming. The author is developing an automatic method for such processing which should allow him to extend the SIDB dataset into a much larger dataset to be used in the future.

The SWT transform was computed to a depth of 2 using the ScatNet toolbox.[5] The first layer contained 8 wavelets per octave of the Gabor kind and the second had 2 wavelets per octave of the Morlet type. The window length was set to 370 ms. Although in [1] 740 ms was pointed out as the more effective for AGR, the author decided to choose a shorter window in order to obtain more data for unsupervised training. The reason was the limited number of songs in SIDB. The GTZAN was also prepared according to the same genre taxonomy as in the SIDB, containing the 4 genres (4-GTZAN). This allowed to perform an experiment with 400 songs and obtain the baseline (see Sect. 5.2). The goal of this paper is to investigate a new approach and compare it with the baseline established using common window length settings.

The data was standardized to have zero mean and unit variance. The 4-GTZAN songs were randomly shuffled and divided into 2 equal subsets (200 songs for training and 200 for evaluation). Each subset had exactly 50 songs per genre. These subsets were used to perform a 2-fold cross validation test.

The core of the code was developed entirely in MATLAB. The implementation of RBM comes from the Deep Learning Toolbox for Matlab [6]

5 The Experiments

Training a traditional neural network (NN) involves initializing its parameters randomly [7, 9]. While training a DNN, there are two phases: unsupervised pre-trainig followed by supervised fine-tuning using labeled data. SIDB was used for training the RBMs during the pre-training phase. Final results were obtained on 4-GTZAN during fine-tuning. In this paper, SIDB pre-training (see Sect. 5.3) was compared with two established baselines (see Sect. 5.2): training a DNN with 4-GTZAN, and learning standard NNs on the same dataset.

---

[4]http://www.apple.com/pl/logic-pro.

[5]http://www.di.ens.fr/data/software/scatnet.

[6]http://www.mathworks.com/matlabcentral/fileexchange/38310-deep-learning-toolbox.

5.1 Evaluation

All models were evaluated by performing 2-fold cross validation (CV) on the
4-GTZAN. The frame-level maximum voting on the test set was used for scoring.
This means that the most correctly classified frames within the song indicate its final
genre. The supervised fine-tuning was performed through 1000 iterations. The
accuracy was recorded throughout the training. The CV tests were performed three
times. Reported outcomes present an average accuracy over three trials of training.

5.2 Baseline

The first baseline used standard NNs with different number of hidden layers on the
4-GTZAN database. There were 4 standard NN models trained. Each layer had 603
units—using the same dimensionality as that of the input. The last layer had 4 units
corresponding to each of the 4 genres. The second baseline was with a DNN, where
GTZAN was used for both, pre-training and fine-tuning. This model had 3 hidden
layers. The final results are presented in Fig. 1.

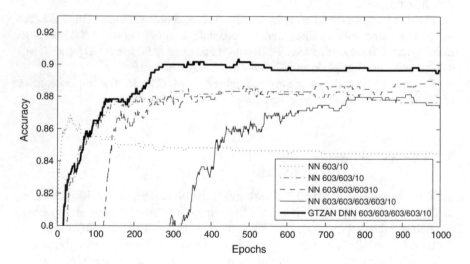

Fig. 1 The figure presents the two baselines. It plots accuracy versus iterations in a 2-fold cross
validation test. The *lines* refer to different topologies of the models and indicate the average
accuracy values over the three trials of training the same model. Hidden units had sigmoid
activation functions. The standard NN models were trained with back-propagation using
cross-entropy cost function, 0.01 learning rate, no momentum and the 1e-4 value of regularization.
DNN in turn, was pre-trained on 4-GTZAN using RBMs through 350 iterations

5.3 Multi-Intrumental Learning

There were 3 RBMs trained separately with 3 different instrument groups in the SIDB dataset. In order to train them, the CD-1 algorithm was used for 350 iterations. Momentum was set to 0.5 for only the first 10 epochs. After the 10th epoch, the momentum was increased to 0.9. During the first 10 epochs, the initial progress in the reduction of reconstruction error relatively settled down. This technique was recommended by Hinton in [11]. The training was performed with a mini-batch data size equal to 30 frames.

The RBMs had 603 units in both their visible and hidden layers. Their parameters were used to initialize the final DNN model. Equal number of units in the visible and hidden layers allowed to examine the 6 combinations of pre-training of the final models. This was possible to achieve by combining the order in which the RBMs took the place in the individual layers of final DNN. The author assumed that such approach is allowed, due to the fact that all RBMs learned the features for individual instrument groups and altogether they can form the features for the whole songs.

There were 6 different DNN models studied. They used RBMs in the following orders (D-drums, B-bass, O-other): O/B/D, O/D/B, B/O/D, B/D/O, D/O/B and D/B/O. The models were fine-tuned with back-propagation using 0.0007 learning rate (it was very low in order not to blur the pre-trained parameters too much) using the cross-entropy cost function. The value of regularization was decreased during fine-tuning. As mentioned in [4], unsupervised pre-training acts as strong regularization so it was decreased from 1e-4 to 1e-7. Described experiments allowed to observe how the order of particular RBM influences the genre classification accuracy. The results are presented in Fig. 2.

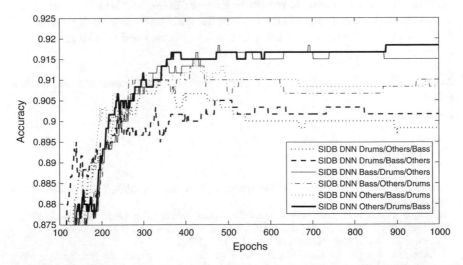

Fig. 2 The figure plots accuracy versus iterations in DNNs after pre-training their weights with SIDB. The legend indicates the group of instruments being used for pre-training of the subsequent hidden layer: drums (D), bass (B), other (O)

6 Conclusions

In this paper, Restricted Boltzmann Machines and the Scattering Wavelet Transform were used together in the problem of Automatic Genre Recognition. Deep Neural Networks were trained to recognize 4 genres in the 4-GTZAN database (reduced from the original GTZAN database). The research took advantage of features recognized by the three RBMs. They were trained with the three groups of instruments: drums, bass and others.

Particular attributes of music can be better recognized when analysed separately. One might not hear characteristic notes for jazz in the double bass part, while hearing all instruments playing simultaneously. As the experiments showed, analysing them separately can bring advantages that help training DNNs for AGR. The intuition about how human perceive the music was one of the inspiration for writing this paper.

From the research that was carried out, it is possible to conclude that the order of pre-trained weights plays a significant role (see Fig. 2). Every case gives at least a minor improvement in accuracy compared to training the DNN only with GTZAN (see Fig. 1). The order in which the first layer was pre-trained by others, next by drums and finally by bass, can boost the results by about 2 % in the simplified 4 genre recognition task. Also the order of *B/D/O* performs similarly very well. Using this approach, it was possible to obtain 92 % accuracy (see Fig. 2) which is better than the 89 % and 90.5 % obtained in the baselines (see Fig. 1).

According to the author's knowledge, there were no publications about training RBMs with separate musical instruments for this task. More data and additional experiments are required to validate if this method performs the same with more genres. The author hopes that this paper is a modest contribution to the ongoing discussions about how human recognize genre of music and how it can be translated into machine learning. On the basis of the quite promising findings presented in this paper, work on the remaining issues is continuing and will be presented in future papers.

Acknowledgments I would like to thank Danijel Koržinek and Krzysztof Marasek, for their support.

References

1. Andén, J., Mallat, S.: Deep scattering spectrum. Sign. Process, IEEE Trans. **62**(16), 4114–4128 (2014)
2. Bengio, Y.: Learning deep architectures for ai. Found. trends Mach. Learn. **2**(1), 1–127 (2009)
3. Bengio, Y., Lamblin, P., Popovici, D., Larochelle, H.: Greedy layer-wise training of deep networks. Adv. neural inf. process. syst. **19**, 153 (2007)
4. Erhan, D., Bengio, Y., Courville, A., Manzagol, P.A., Vincent, P., Bengio, S.: Why does unsupervised pre-training help deep learning? J. Mach. Learn. Res. **11**, 625–660 (2010)
5. Fischer, A., Igel, C.: An introduction to restricted boltzmann machines. Progress in Pattern Recognition, Image Analysis, Computer Vision, and Applications, pp. 14–36. Springer (2012)

6. Fu, Z., Lu, G., Ting, K.M., Zhang, D.: A survey of audio-based music classification and annotation. Multimedia, IEEE Trans. **13**(2), 303–319 (2011)
7. Glorot, X., Bengio, Y.: Understanding the difficulty of training deep feedforward neural networks. International conference on artificial intelligence and statistics, pp. 249–256 (2010)
8. Hamel, P., Eck, D.: Learning features from music audio with deep belief networks. ISMIR, pp. 339–344. Utrecht, The Netherlands (2010)
9. He, K., Zhang, X., Ren, S., Sun, J.: Delving deep into rectifiers: Surpassing human-level performance on imagenet classification. arXiv preprint arXiv:1502.01852 (2015)
10. Herrera-Boyer, P., Klapuri, A., Davy, M.: Automatic classification of pitched musical instrument sounds. Signal Processing Methods For Music Transcription, pp. 163–200. Springer (2006)
11. Hinton, G.: A practical guide to training restricted boltzmann machines. Momentum **9**(1), 926 (2010)
12. Hinton, G., Osindero, S., Teh, Y.W.: A fast learning algorithm for deep belief nets. Neural Comput. **18**(7), 1527–1554 (2006)
13. Hinton, G.E., Salakhutdinov, R.R.: Replicated softmax: an undirected topic model. Advances in Neural Information Processing Systems, pp. 1607–1614 (2009)
14. Humphrey, E.J., Bello, J.P., LeCun, Y.: Moving beyond feature design: Deep architectures and automatic feature learning in music informatics. ISMIR, pp. 403–408. Citeseer (2012)
15. Kleć, M., Koržinek, D.: Pre-trained deep neural network using sparse autoencoders and scattering wavelet transform for musical genre recognition. Comput. Sci. **16**(2), 133–144 (2015)
16. Kostek, B.: Intelligent musical instrument sound classification. Perception-Based Data Processing in Acoustics, pp. 39–186. Springer (2005)
17. Mallat, S.: Group invariant scattering. Commun. Pure Appl. Math. **65**(10), 1331–1398 (2012)
18. Mohamed, A.r., Sainath, T.N., Dahl, G., Ramabhadran, B., Hinton, G.E., Picheny, M.A.: Deep belief networks using discriminative features for phone recognition. 2011 IEEE international conference on acoustics, speech and signal processing (ICASSP), pp. 5060–5063 (2011)
19. Panagakis, Y., Kotropoulos, C., Arce, G.R.: Music genre classification using locality preserving non-negative tensor factorization and sparse representations. ISMIR, pp. 249–254 (2009)
20. Salakhutdinov, R., Mnih, A., Hinton, G.: Restricted boltzmann machines for collaborative filtering. Proceedings of the 24th international conference on machine learning, ACM, pp. 791–798. (2007)
21. Sigtia, S., Dixon, S.: Improved music feature learning with deep neural networks. 2014 IEEE international conference on acoustics, speech and signal processing (ICASSP), pp. 6959–6963 (2014)
22. Sturm, B.L.: A survey of evaluation in music genre recognition. Adaptive Multimedia Retrieval: Semantics, Context, and Adaptation, pp. 29–66. Springer (2014)
23. Yu, D., Deng, L.: Deep learning and its applications to signal and information processing [exploratory dsp]. Sig. Process. Mag. IEEE **28**(1), 145–154 (2011)

A Multiple Kernels Interval Type-2 Possibilistic C-Means

Minh Ngoc Vu and Long Thanh Ngo

Abstract In this paper, we propose multiple kernels-based interval type-2 possi-bilistic c-Means (MKIT2PCM) by using the kernel approach to possibilistic clus-tering. Kernel-based fuzzy clustering has exhibited quality of clustering results in comparison with "routine" fuzzy clustering algorithms like fuzzy c-Means (FCM) or possibilistic c-Means (PCM) not only noisy data sets but also overlap-ping between prototypes. Gaussian kernels are suitable for these cases. Interval type-2 fuzzy sets have shown the advantages in handling uncertainty. In this study, multiple kernel method are combined into interval type-2 possibilistic c-Means (IT2PCM) to produce a variant of IT2PCM, called multiple kernels interval type-2 possibilistic c-Means (MKIT2PCM). Experiments on various data-sets with validity indexes show the performance of the proposed algorithms.

Keywords Type-2 fuzzy sets · Fuzzy clustering · Multiple kernels · PCM

1 Introduction

Clustering is one of the most important technique in machine intelligence, approach to unsupervised learning for pattern recognition. The fuzzy c-Means (FCM) algorithm [1] has been applied successfully to solve a large number of problems such as patterns recognition (fingerprint, photo) [2], image processing (color separated clustering, image segmentation) [3]. However, the drawbacks of FCM algorithm are sensitive to noise, loss of information in process and overlap clusters of different volume. To handle this uncertainty, Krishnapuram et al. used the compulsion of memberships in FCM, called Possibilistic c-Means (PCM) algorithm [4]. The PCM algorithm solves

M.N. Vu (✉) · L.T. Ngo
Department of Information Systems, Le Quy Don Technical University,
236 Hoang Quoc Viet, Hanoi, Vietnam
e-mail: ngoc114@gmail.com

L.T. Ngo
e-mail: ngotlong@mta.edu.vn

© Springer International Publishing Switzerland 2016   63
D. Król et al. (eds.), *Recent Developments in Intelligent Information and Database Systems*, Studies in Computational Intelligence 642,
DOI 10.1007/978-3-319-31277-4_6

so well some problems of actual data input which may be uncertainty, contradictory and vaguenesses that FCM algorithm cannot handle correctly. Though, PCM still exists some drawbacks such as: depending deeply on initializing centroid or sensitivity to noise data [5, 6]. To settle these weakness associated with PCM and improve accuracy in data processing, multiple kernel method based clustering is used. The main idea of kernel method by mapping the original data inputs into a higher dimensional space by transform function. However, different input features may impact on the results, the multiple kernel method can be combined from various kernels.

Type-2 fuzzy sets, which are the extension of fuzzy sets of type-1 [7], have shown advantages when solving uncertainty, have been researched and exploited to many different applications [8, 9], including fuzzy clustering. The interval type-2 Fuzzy Sets (IT2FS) and the kernel method [10, 11] considered separately have shown potential for the clustering problems. This combination of IT2FS and multiple kernels gives the extended interval type-2 possibilistic c-Means [12] in the kernel space. In this paper, we propose a multiple kernels interval type-2 possibilistic c-Means (MKIT2PCM) which is the kernel method based on the possibilistic approach to clustering. Because data-sets could come from many sources with different characteristics, the multiple kernels approach to IT2PCM clustering could be to handle the ambiguousness of data. In which, the final kernel function is combined from component kernel functions by sum of weight average. The algorithm is experimented with various data to measure the validity indexes.

The paper is organized as follows: Sect. 2 brings a brief background of type-2 fuzzy sets and kernel method. Section 3 describes the MKIT2PCM algorithm. Section 4 shows some experimental results and Sect. 5 concludes the paper.

2   Preliminaries

2.1   Interval Type-2 Fuzzy Sets

A type-2 fuzzy set [9] defined in X denoted $\tilde{A}$, comes with a membership function of the form $\mu_{\tilde{A}}(x, u), u \in J_x \subseteq [0, 1]$, which is a type-1 fuzzy set in $[0, 1]$. The elements of field of $\mu_{\tilde{A}}(x, u)$ are called primary membership grades of x in $\tilde{A}$ and membership grades of primary membership grades in $\mu_{\tilde{A}}(x, u)$ are called secondary ones.

Definition 1 A type-2 fuzzy set, denoted $\tilde{A}$, is described by a type-2 membership function $\mu_{\tilde{A}}(x, u)$ where $x \in X$ and $u \in J_x \subseteq [0, 1]$, i.e.

$$\tilde{A} = \{((x, u), \mu_{\tilde{A}}(x, u)) | \forall x \in X, \forall u \in J_x \subseteq [0, 1]\} \tag{1}$$

or

$$\tilde{A} = \int_{x\in X} \int_{u\in J_x} \mu_{\tilde{A}}(x,u))/(x,u), J_x\subseteq[0,1] \qquad (2)$$

in which $0 \le \mu_{\tilde{A}}(x,u) \le 1$

At each value of x, say $x = x'$, the 2-D plane whose axes are u and $\mu_{\tilde{A}}(x',u)$ is called a vertical slice of $\mu_{\tilde{A}}(x,u)$. A secondary membership function is a vertical slice of $\mu_{\tilde{A}}(x,u)$. It is $\mu_{\tilde{A}}(x = x',u)$ for $x \in X$ and $\forall u \in J_{x'}\subseteq[0,1]$, i.e.

$$\mu_{\tilde{A}}(x = x',u) \equiv \mu_{\tilde{A}}(x') = \int_{u\in J_{x'}} f_{x'}(u)/u, J_{x'}\subseteq[0,1] \qquad (3)$$

in which $0 \le f_{x'}(u) \le 1$ (Fig. 1).

A type-2 fuzzy set is called an interval type-2 fuzzy set if its secondary membership function $f_{x'}(u) = 1 \ \forall u \in J_x$ i.e. such constructs are defined as follows:

Definition 2 An interval type-2 fuzzy set $\tilde{A}$ is characterized by an interval type-2 membership function $\mu_{\tilde{A}}(x,u) = 1$ where $x \in X$ and $u \in J_x\subseteq[0,1]$, i.e.,

$$\tilde{A} = \{((x,u),1)|\forall x \in X, \forall u \in J_x\subseteq[0,1]\} \qquad (4)$$

Footprint of uncertainty (FOU) of a type-2 fuzzy set is union of primary functions i.e. $FOU(\tilde{A}) = \bigcup_{x\in X} J_x$. Upper/lower bounds of membership function (UMF/LMF) are denoted by $\bar{\mu}_{\tilde{A}}(x)$ and $\underline{u}_{\tilde{A}}(x)$.

Fig. 1 Visualization of a type-2 fuzzy set

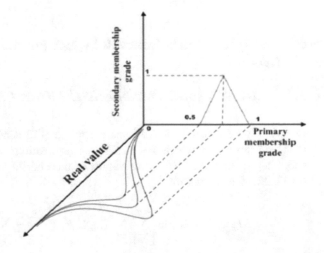

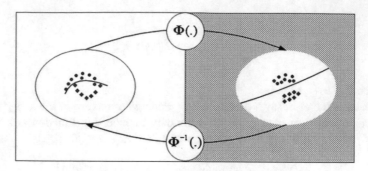

Fig. 2 Expression of the mapping from feature space to kernel space

2.2 Kernel Method

Kernel method gives an exclusive approach which maps all data ϕ from the basic dimensional feature space R^p to a different higher possibly dimensionality vector space (kernel space). The kernel space could be of infinite dimensionality.

Definition 3 K is a kernel function on X space if

$$K(x, y) = \phi(x).\phi(y) \tag{5}$$

where ϕ is a nonlinear map from X to a scalar space F: $\phi : R^p \to F$; $x, y \in X$. Figure 2 [13] represents the mapping from the original space to kernel space

The function $k : X \times X \to F$ is called Mercer kernel if $t \in I$ and $x = (x_1, x_2 \ldots, x_t) \in X^t$ the $t \times t$ matrix K is positive semi definite.

3 Multiple Kernels Interval Type-2 Possibilistic C-Means Clustering

3.1 Interval Type-2 Possibilistic C-Means Clustering

Interval type-2 possibilistic c-Means clustering (IT2PCM) is an completion of the PCM clustering in which we use two fuzzification coefficients m_1, m_2 corresponding to the upper and lower values of membership. Objective functions used for IT2PCM is shown in:

$$J_{m_1}(U, v) = \min \left[\sum_{i=1}^{c} \sum_{j=1}^{n} (u_{ij})^{m_1} d_{ij}^2 + \sum_{i=1}^{c} \eta_i \sum_{j=1}^{n} (1 - u_{ij})^{m_1} \right] \tag{6}$$

$$J_{m_2}(U, v) = \min \left[\sum_{i=1}^{c} \sum_{j=1}^{n} (u_{ij})^{m_2} d_{ij}^2 + \sum_{i=1}^{c} \eta_i \sum_{j=1}^{n} (1 - u_{ij})^{m_2} \right] \qquad (7)$$

where $d_{ij} = \parallel x_j - v_i \parallel$ is the Euclidean distance between the jth data and the ith cluster centroid, c is the number of clusters and n is number of patterns. m_1, m_2 is the degree of fuzziness and η_i is a suitable positive number. Upper/lower degrees of membership $\bar{u}_{ij}$ and $\underline{u}_{ij}$ are formed by involving two fuzzification coefficients m_1, m_2 ($m_1 < m_2$) as follows:

$$\bar{u}_{ij} = \begin{cases} \dfrac{1}{1 + \sum_{j=1}^{c} \left(\frac{d_{ij}^2}{\eta_i} \right)^{1/(m_1-1)}} & if \quad \dfrac{1}{1 + \sum_{j=1}^{c} \left(\frac{d_{ij}^2}{\eta_i} \right)^{1/(m_1-1)}} \geq \dfrac{1}{1 + \sum_{j=1}^{c} \left(\frac{d_{ij}^2}{\eta_i} \right)^{1/(m_2-1)}} \\ \dfrac{1}{1 + \sum_{j=1}^{c} \left(\frac{d_{ij}^2}{\eta_i} \right)^{1/(m_2-1)}} & otherwise \end{cases} \qquad (8)$$

$$\underline{u}_{ij} = \begin{cases} \dfrac{1}{1 + \sum_{j=1}^{c} \left(\frac{d_{ij}^2}{\eta_i} \right)^{1/(m_1-1)}} & if \quad \dfrac{1}{1 + \sum_{j=1}^{c} \left(\frac{d_{ij}^2}{\eta_i} \right)^{1/(m_1-1)}} \leq \dfrac{1}{1 + \sum_{j=1}^{c} \left(\frac{d_{ij}^2}{\eta_i} \right)^{1/(m_2-1)}} \\ \dfrac{1}{1 + \sum_{j=1}^{c} \left(\frac{d_{ij}^2}{\eta_i} \right)^{1/(m_2-1)}} & otherwise \end{cases} \qquad (9)$$

$$i = 1, 2, \ldots, c; j = 1, 2, \ldots, n$$

where $\bar{\eta}_{ij}$ and $\underline{\eta}_{ij}$ are upper/lower degrees of typicality function calculated by

$$\bar{\eta}_{ij} = \frac{\sum_{j=1}^{n} \bar{u}_{ij} d_{ij}^2}{\sum_{j=1}^{n} \bar{u}_{ij}} \qquad (10)$$

and

$$\underline{\eta}_{ij} = \frac{\sum_{j=1}^{n} \underline{u}_{ij} d_{ij}^2}{\sum_{j=1}^{n} \underline{u}_{ij}} \qquad (11)$$

Because each pattern comes with the membership interval as the bounds $\bar{u}$ and $\underline{u}$. Cluster prototypes are computed as

$$v_i = \frac{\sum_{j=1}^{n} (u_{ij})^m x_j}{\sum_{j=1}^{n} (u_{ij})^m} \qquad (12)$$

In which $i = 1, 2, \ldots, c$

Algorithm 1: Determining Centroids.

Step 1: Find $\bar{u}_{ij}, \underline{u}_{ij}$ using (8), (9)
Step 2: Set $m \geq 1$

Compute $v'_j = (v'_{j1}, \ldots, v'_{jP})$ using (12) with $u_{ij} = \frac{(\bar{u}_{ij} + \underline{u}_{ij})}{2}$.

Sort n patterns on each of P features in ascending order.

Step 3: Find index k where: $x_{kl} \leq v'_{jl} \leq x_{(k+1)l}$ with $k = 1, .., N$ and $l = 1, .., P$

Update u_{ij}: if $i \leq k$ then do $u_{ij} = \underline{u}_{ij}$ else $u_{ij} = \bar{u}_{ij}$

Step 4: Compute v''_j by (12) and compare v'_{jl} with v''_{jl}

If $v'_{jl} = v''_{jl}$ then do

$v_R = v'_j$.

  Else

  Set $v'_{jl} = v''_{jl}$.

  Back to Step 3.

For example we compute v^L

  In step 3

  Update u_{ij}:

  If $i \leq k$ then do

   $u_{ij} = \bar{u}_{ij}$.

  Else

   $u_{ij} = \underline{u}_{ij}$.

  In Step 4 change V_R with v_L.

  After having v_i^R, v_i^L, the prototypes is computed as follows

$$v_i = \left(\frac{v_i^R + v_i^L}{2} \right) \tag{13}$$

For the membership grades we obtain

$$u_i(x_k) = (u_i^R(x_k) + u_i^L(x_k))/2, j = 1, \ldots, c \tag{14}$$

where

$$u_i^L = \sum_{l=1}^{M} u_{il}/P, u_{il} = \begin{cases} \bar{u}_i(x_k) & \text{if } x_{il} \text{ uses } \bar{u}_i(x_k) \text{ for } v_i^L \\ \underline{u}_i(x_k) & \text{otherwise} \end{cases} \tag{15}$$

$$u_i^R = \sum_{l=1}^{M} u_{il}/P, u_{il} = \begin{cases} \bar{u}_i(x_k) & \text{if } x_{il} \text{ uses } \bar{u}_i(x_k) \text{ for } v_i^R \\ \underline{u}_i(x_k) & \text{otherwise} \end{cases} \tag{16}$$

Next, defuzzification follows the ruleif $u_i(x_k) > u_j(x_k)$ for $j = 1, \ldots, c$ and $i \neq j$ then x_k is assigned to cluster i

3.2   Multiple Kernel Interval Type-2 Possibilistic C-Means Algorithm

Here, we introduce MKIT2PCM algorithm which can use a combination of different kernels. The pattern $x_j, j = 1, \ldots n$ and prototype $v_i, i = 1, \ldots c$ are mapped to the kernel space through a kernel function ϕ. The objective functions used for MKIT2PCM is shown in

$$Q_{m_1}(U, V) = \sum_{i=1}^{c} \sum_{j=1}^{n} u_{ij}^{m_1} \parallel \phi_{com}(x_j) - \phi_{com}(v_i) \parallel^2 + \sum_{i=1}^{c} \eta_i \sum_{j=1}^{n} (1 - u_{ij})^{m_1} \quad (17)$$

$$Q_{m_2}(U, V) = \sum_{i=1}^{c} \sum_{j=1}^{n} u_{ij}^{m_2} \parallel \phi_{com}(x_j) - \phi_{com}(v_i) \parallel^2 + \sum_{i=1}^{c} \eta_i \sum_{j=1}^{n} (1 - u_{ij})^{m_2} \quad (18)$$

where $\parallel \phi_{com}(x_j) - \phi_{com}(v_i) \parallel = d_{\phi ij}$ is the Euclidean distance from the prototype v_i to the pattern x_j which is computed in the kernel space.

Where ϕ_{com} is the transformation defined by the combined kernels

$$k_{com}(x, y) = \; <\phi_{com}(x), \phi_{com}(y)> \quad (19)$$

The kernel k_{com} is the function that is combined from multiple kernels. A commonly used the combination is a linearly summary that reach as follows:

$$k_{com} = w_1^b k_1 + w_2^b k_2 + \cdots + w_l^b k_l \quad (20)$$

where $b > 1$ is a certain coefficient and satisfying the constraint $\sum_{i=1}^{l} w_i = 1$.

The typical kernels defined on $\mathbb{R}^p \times \mathbb{R}^p$ are used: Gaussian kernel $k(x, y) = \exp\left(\frac{-x-y^2}{2\sigma^2}\right)$ and polynomial kernel $k(x_i, x_j) = (x_i * x_j + d)^2$

The next is how to compute $\bar{u}_{ij}$ and $\underline{u}_{ij}$ which can be reached by:

$$\bar{u}_{ij} = \begin{cases} \dfrac{1}{1 + \sum_{j=1}^{c} \left(\frac{d_{\phi com}(x_j, v_i)^2}{\eta_i}\right)^{1/(m_1-1)}} & if \; \dfrac{1}{1 + \sum_{j=1}^{c} \left(\frac{d_{\phi com}(x_j, v_i)^2}{\eta_i}\right)^{1/(m_1-1)}} \geq \dfrac{1}{1 + \sum_{j=1}^{c} \left(\frac{d_{\phi com}(x_j, v_i)^2}{\eta_i}\right)^{1/(m_2-1)}} \\[4mm] \dfrac{1}{1 + \sum_{j=1}^{c} \left(\frac{d_{\phi com}(x_j, v_i)^2}{\eta_i}\right)^{1/(m_2-1)}} & otherwise \end{cases}$$

$$(21)$$

$$\underline{u}_{ij} = \begin{cases} \dfrac{1}{1 + \sum_{j=1}^{c} \left(\frac{d_{\phi com}(x_j, v_i)^2}{\eta_i}\right)^{1/(m_1-1)}} & if \; \dfrac{1}{1 + \sum_{j=1}^{c} \left(\frac{d_{\phi com}(x_j, v_i)^2}{\eta_i}\right)^{1/(m_1-1)}} \leq \dfrac{1}{1 + \sum_{j=1}^{c} \left(\frac{d_{\phi com}(x_j, v_i)^2}{\eta_i}\right)^{1/(m_2-1)}} \\[4mm] \dfrac{1}{1 + \sum_{j=1}^{c} \left(\frac{d_{\phi com}(x_j, v_i)^2}{\eta_i}\right)^{1/(m_2-1)}} & otherwise \end{cases}$$

$$(22)$$

Respectively, where $d_{\phi_{com}}(x_j, v_i)^2 = \| \phi_{com}(x_j) - v_i \|^2$

$$= k_{com}(x_j, x_j) - \frac{2\sum_{h=1}^{n}(u_{ih})^m k_{com}(x_j, x_h)}{\sum_{h=1}^{n}(u_{ih})^m} + \frac{\sum_{h=1}^{n}\sum_{l=1}^{n}(u_{ih})^m (u_{il})^m k_{com}(x_h, x_l)^2}{\left(\sum_{h=1}^{n}(u_{ih})^m\right)}$$

$$(23)$$

where $\overline{\eta}_{ij}$ and $\underline{\eta}_{ij}$ are calculated by

$$\overline{\eta}_{ij} = \frac{\sum_{j=1}^{n} \overline{u}_{ij}\phi_{com}(x_j) - v_i^2}{\sum_{j=1}^{n} \overline{u}_{ij}} \qquad (24)$$

$$\underline{\eta}_{ij} = \frac{\sum_{j=1}^{n} \underline{u}_{ij}\phi_{com}(x_j) - v_i^2}{\sum_{j=1}^{n} \underline{u}_{ij}} \qquad (25)$$

Algorithm 2. Multiple kernel interval type 2 possibilistic c-Means clustering

Input: n patterns $X = \{x_i\}_{i=1}^{n}$, kernel functions $\{k_i\}_{i=1}^{l}$, fuzzifiers m_1, m_2 and c clusters.

Output: a membership matrix $U = \{u_{ij}\}_{i=1,..,n}^{j=1,..,c}$ and weights $\{w_i\}_{i=1}^{l}$ for the kernels.

Step 1: Initialization of membership grades

1. Initialize centroid matrix $V^0 = \{v_i\}_{i=1}^{C}$ is randomly chosen from input data set
2. The membership matrix U^0 can be computed by:

$$u_{ij} = \frac{1}{\sum_{k=1}^{c}\left(\frac{d_{ij}}{d_{lj}}\right)^{\frac{2}{m-1}}} \qquad (26)$$

where $m > 1$ and $d_{ij} = d(x_j - v_i) = \|x_j - v_i\|$.

Step 2: Repeat

1. Compute distance $d(x_j, v_i)^2$ using formula (23)
2. Compute $\overline{\eta}_{ij}$ and $\underline{\eta}_{ij}$ using formulas (24), (25)
3. Calculate the upper and lower memberships $\overline{u}_{ij}$ and $\underline{u}_{ij}$ using formulas (21), (22)
4. Use **Algorithm 1** for finding centroids and using formula (12) to update the centroids.
5. Update the membership matrix using formula (14)

Step 3: Conclusion criteria satisfied or maximum iterations reached:

1. Return U and V
2. Assign data x_j to cluster c_i if data $(u_j(x_i) > u_k(x_i))$, $k = 1,\ldots,c$ and $j \neq k$.
Else
Back to step 2.

4 Experimental Results

In this paper, the experiment is implemented on remote sensing images with validity index comparisons. The vector of pixels is acquired from various sensors with different spectral. So that, we can identify different kernels for different spectrals and use the combined kernel in a multiple-kernel method.

The data inputs involve Gaussian kernel k_1 of pixel intensities and Gaussian kernel k_2 of spatial information. The spatial function is defined as follows [14]

$$p_{ij} = \sum_{k \in NB(x_j)} u_{ik} \qquad (27)$$

where $NB(x_j)$ represents a squared neighboured 5×5 window with the center on the pixel x_j. The spatial function p_{ij} express membership grade that pixel x_j belongs to the cluster i.

The kernel algorithm produces Gaussian kernel k of pixel intensities as data inputs. We use Lansat-7 images data in this experiment with study data of Cloud Landsat TM at IIS, U-Tokyo, (SceneCenterLatitude: 11.563043 S SceneCenterLongitude: 15.782159 W).

The results are shown in Fig. 3 where (a) is for VR channel, (b) is for NIR channel, (c) is for SWIR channel and (d), (e) and (f) are image results of the

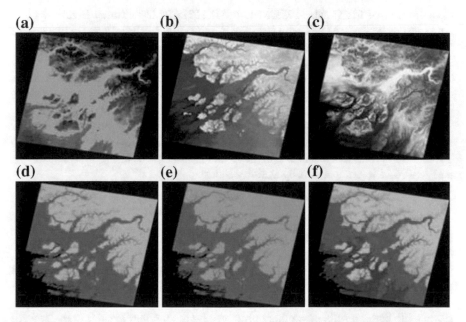

Fig. 3 Data for test: land cover classification. **a** VR channel image; **b** NIR channel image; **c** SWIR channel image; **d** MKPCM classification; **e** IT2PCM classification; **f** MKIT2PCM classification

Fig. 4 Comparisons between
the result of MKPCM,
IT2PCM, MKIT2PCM

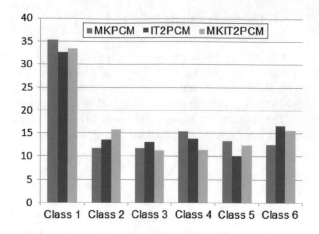

Table 1 The various validity
indexes

Validity Index	MKPCM	IT2PCM	MKIT2PCM
PC-I	0.334	0.225	**0.187**
CE-I	1.421	1.645	**1.736**
S-I	0.019	0.171	**0.0608**
DB-I	2.42	2.051	**1.711**

classification of MKPCM, IT2PCM and MKIT2PCM algorithms. Figure 4 compares results between MKPCM, IT2PCM-F, MKIT2PCM algorithms (in percentage %).

We consulted the various validity indexes -which maybe the Bezdek's partition coefficient (PC-I), the Classification Entropy index (CE-I), the Separation index (S-I) and the Davies-Bouldin's index (DB-I) [15]. We can see the results of these indexes in the Table 1

The value of validity indexes coming from MKIT2PCM algorithm has smaller values of PC-I, S-I, DB-I and larger value only of CE-I. So that, the results in Table 1 show that the MKIT2PCM have better quality clustering than KIT2PCM and MKPCM in comparison by validity indexes.

5   Conclusion

In this paper, we applied the multiple kernel method to interval type-2 PCM clustering algorithm which enhances the efficiency of the clustering results. The experiments completed for remote sensing images dataset that the proposed algorithm quite better results than others produced by the familiar clustering methods of PCM. We demonstrated that the proposed algorithm can determine proper clusters, and they can realize the advantages of the possibilistic approach. Some future studies may be focused on other possible extensions can be incorporated interval type 2 fuzzy sets to hybrid clustering approach to data classification use multiple kernel fuzzy possibilistic c-Means clustering (FPCM).

References

1. Bezdek, J.C., Ehrlich, R., Full, W.: The fuzzy c-means clustering algorithm. Comput. Geosci. **10**(2–3), 191–203 (1984)
2. Das, S.: Pattern recognition using the fuzzy c-means technique. Int. J. Energy Inf. Commun **4** (1), February (2013)
3. Patil, A., Lalitha, Y.S.: Classification of crops using FCM segmentation and texture, color feature. IJARCCE **1**(6), (2012)
4. Krishnapuram, R., Keller, J.: A possibilistic approach to clustering. IEEE Trans. Fuzzy Syst. **1**, 98–110 (1993)
5. Kanzawa, Y.: Sequential cluster extraction using power-regularized possibilistic c-means. JACIII **19**(1),67–73 (2015)
6. Krishnapuram, R., Keller, J.M.: The possibilistic c-means: insights and recommendations. IEEE Trans. Fuzzy Syst. **4**, 385–393 (1996)
7. Sanchez, M.A., Castillo, O., Castro, J.R., Melin, P.: Fuzzy granular gravitational clustering algorithm for multivariate data. Inf. Sci. **279**, 498–511 (2014)
8. Rubio, E., Castillo, O.: Designing type-2 fuzzy systems using the interval type-2 fuzzy c-means algorithm. Recent Adv. Hybrid Approaches Des. Intell. Syst., 37–50 (2014)
9. Nguyen, D.D., Ngo, L.T., Pham, L.T., Pedrycz, W.: Towards hybrid clustering approach to data classification: multiple kernels based-interval type-2 fuzzy C-means algorithms. J. Fuzzy Sets Syst. **279**, 17–39 (2015)
10. Zhao, B., Kwok, J., Zhang, C.: Multiple kernel clustering. In: Proceedings of 9th SIAM International Conference Data Mining, pp. 638–649, 2009
11. Filippone, M., Masulli, F., Rovetta, S.: Applying the possibilistic C-means algorithm in Kernel-induced spaces. IEEE Trans. Fuzzy Syst. **18**(3), 572–584 (2010)
12. Rubio, E., Castillo, O.: Interval type-2 fuzzy clustering for membership function generation. HIMA, 13–18, (2013)
13. Raza, M.A., Rhee, F.C.H.: Interval type-2 approach to Kernel possibilistic C-means clustering. In: IEEE International Conference on Fuzzy Systems, 2012
14. Chuang, K.S., Tzeng, H.L., Chen, S., Wua, J., Chen, T.-J.: Fuzzy c-means clustering with spatial information for image segmentation. Comput. Med. Imaging Graph. **30**(1), 9–15 (2006)
15. Wang, W., Zhang, Y.: On fuzzy cluster validity indices. Fuzzy Sets Syst. **158**, 2095–2117 (2007)

Application of Artificial Neural Networks in Condition Based Predictive Maintenance

Jiri Krenek, Kamil Kuca, Pavel Blazek, Ondrej Krejcar
and Daniel Jun

Abstract This paper reviews different techniques of maintenance, artificial neural networks (ANN) and their various applications in fault risk assessment and an early fault detection analysis. The predictive maintenance is in focus of production facilities supplying in long supplier chains of automotive industry to ensure the reliable and continuous production and on-time deliveries. ANN offer a powerful tool to evaluate machine data and parameters which can learn from process data of fault simulation. Finally there are reviewed examples of usage of ANN in specific predictive maintenance cases.

Keywords Artificial neural networks · Predictive maintenance · Condition based maintenance · Simulations · Defect detection · Classification

J. Krenek (✉) · K. Kuca · P. Blazek · O. Krejcar
Center for Basic and Applied Research Faculty of Informatics and Management,
University of Hradec Kralove, Hradec Kralove, Czech Republic
e-mail: jiri.krenek@uhk.cz

K. Kuca
e-mail: kamil.kuca@uhk.cz

P. Blazek
e-mail: pavel.blazek@uhk.cz

O. Krejcar
e-mail: ondrej.krejcar@uhk.cz

P. Blazek · D. Jun
Faculty of Military Health Sciences,
University of Defence Hradec Kralove, Hradec Kralove, Czech Republic
e-mail: daniel.jun@unob.cz

© Springer International Publishing Switzerland 2016   75
D. Król et al. (eds.), *Recent Developments in Intelligent Information
and Database Systems*, Studies in Computational Intelligence 642,
DOI 10.1007/978-3-319-31277-4_7

1   Introduction

Contemporary worldwide industrial environment with high demands of customers requires a robust, reliable and lean manufacturing processes. This is mainly caused by requirements like Kanban system of production and Just-In-Time strategy [1]. It basically means that any overproduction is evaluated as a waste, the storage capacity at the customer is minimized to essential stock to cover the few following days or even hours of production.

Due to fact that any overproduction is undesirable and the stock of finished goods at customer and supplier is limited to necessary minimum, the high demands are focused on production processes from incoming goods input till the final product [2]. Overall equipment effectiveness (OEE) [3] is one of the most important production facilities key performance indicators and shows how effectively are manufacturing operations used.

This paper is focused on OEE supporting process, namely the maintenance and its strategies that are described in second chapter. The third chapter sums up the artificial neural networks and the areas of their usage. In the follow-up fourth part comes the review of application of artificial neural network in predictive maintenance. The last, fifth chapter discusses the used applications and their advantages and/or disadvantages.

2   Maintenance

Maintenance is an activity required to keep the assets close to their original condition to make sure the asset can fulfil the function it was developed or even purchased for. There exist three strategies of maintenance: Corrective, Preventive and Predictive.

2.1   Corrective Maintenance

Corrective maintenance, known as a run-to-failure is the one that is activated in case of machine failure [4]. Such an activity is a must and leads to fast recovery of running process. In case the problem occurs during production the machine failure has to be solved immediately. The biggest disadvantage of such a case is that it often happens without notice of any warning signals and comes in the less appropriate time causing unexpected interruption of manufacturing process. It is directly related to additional production costs. This kind of maintenance is extensively time consuming and a cost challenging.

2.2 Preventive Maintenance

Preventive maintenance is also called condition based maintenance. It is based on the knowledge from machine manufacturer, experience of maintenance engineers and a lessons learnt from past service incidents and means a set of activities to be done within certain period on a set regular frequencies (daily, weekly, monthly and an annually). Daily activities could be simple checks done by machine operators like visual inspection of hydraulic oil level or air leakage in selected areas, inspection of machine noises, control of machine gauges with record of observed measurements, defined clean-up activities etc. Weekly and monthly activities could be very similar to daily maintenance tasks but could be more detailed and provided by specially trained person. Annual preventive maintenance is mainly done by dedicated engineer if not directly by manufacturer. Such work could also contain replacement of some machine parts that are identified by manufacturer, different runtime counters, or by any other definition. The advantage of this maintenance strategy is that it can be well planned and causes no interruption to manufacturing processes due to machine idle or during downtime periods. It is also counterbalanced with a disadvantage where there could be replaced costly ok parts "just in case" even if they would do a lot of further job. A value added to the condition based maintenance is a fact that it leads to system optimization and improvement of productivity [5].

2.3 Predictive Maintenance

The predictive maintenance brings the new look and a philosophy on the maintenance strategies to achieve the maximum lifetime of machines while the risk of machine failure is minimized. Some of the failures being fixed within corrective maintenance could have been identified earlier based on different and specific signs like uncommon noises, shorter cycle time or obvious change in machine run like motion speed and fluency, baseless stopping of machinery, overheating parts, etc. Some of these signs cannot be properly measured and therefore are dependent on human observation. The other signs might progress over the longer time period and are not identifiable for human on day-to-day basis. Today's manufacturers can supply machines with additional diagnostics software that can capture the machine system data and parameters or there exist variety of different sensors and techniques to monitor the noises, heat distribution, torque etc. and store them for further analysis.

When there are used sophisticated methods of data capturing, the datasets normally contain tens of thousands records. On the other side the implementation of preventive maintenance might be very expensive as there is a need of use at least statistical tools or even better an advanced tools for data processing like data mining or artificial neural network, additional sensory systems and therefore needs a good evaluation whether it is valuable to install these tools.

3   Artificial Neural Networks

Artificial neural network (ANN) is a method of artificial intelligence inspired by biological structure—a brain [6]. Similarly to this structure the artificial neural network is built of body called processing element, inputs and outputs. The significance of each input is multiplied by weight and together with bias they get to cell body, the processing element. In the first step the multiplied inputs are summed by summation function and in the second step they are propagated by transfer function to an output.

Artificial neural networks are mostly used to solve the following tasks: association, classification, clustering, pattern recognition, image processing, control, optimisation and modelling. The basic design of artificial neuron is very simple (Fig. 1). It consists of three basic types of layers: an input layer, a hidden layer and an output layer. The hidden layer can have one or more layers, commonly up to three based on complexity of solved problem. Weighted signal goes to hidden layers where there is done most of the computations. Finally it is followed to the output layer.

3.1   Designs and Topologies

ANNs have different topologies suitable for solving of specific problems. It can be from simple feed-forward network where the signal flows directly from inputs to outputs in one direction only till the recurrent or back-propagation networks where

Fig. 1 Layers of artificial neural network

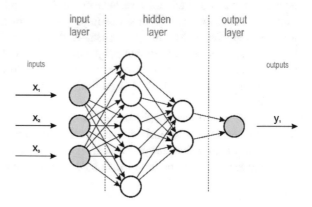

the signals beside the direct flow can make the loops to a previous layer or even within the same layer [7].

A very specific neural network design are the self-organising maps (SOM) alternatively called Kohonen maps. These topologies use unique architecture with rectangular two dimensional grids up to three dimensional arrays and employ a characteristic neighbourhood function (Fig. 2). The SOMs are often used for image processing.

3.2 Learning Algorithms

The process of learning is basically the manipulation of input signal weights to ensure the best computational results are achieved. During the ANN design and the testing phase the dataset should be divided into three sets: a training set, a validation set and a test set. The training set is used for the basic training of the network. The validation set is used for tuning the algorithm to ensure the best results and the test

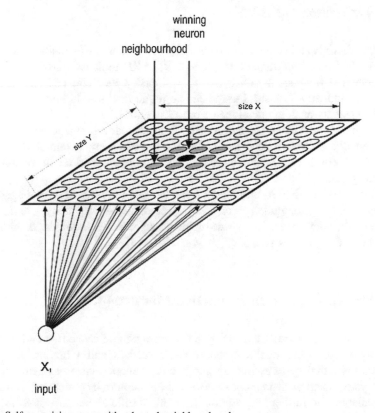

Fig. 2 Self-organising map with adopted neighbourhood

set is there for evaluation of finally designed network with the right algorithm. It is important, that each of the set contain different data.

There exist two basic types of learning: supervised and unsupervised. The target of unsupervised training is the system optimization and in practice it can be seen as maximizing the profit, minimizing the costs or energy, etc.

3.3 Software Computational Tools

Work with artificial neural networks is always connected with processing of high volume of data which requires advanced commercial and open-source software tools [8]. Hardware requirements are mostly limited to minimum requirement from developers. Any computer not older than 8 years will be able to run the SW tools. The biggest impact on software performance has the number of CPU cores able to run parallel calculations, RAM and bus speed.

3.4 Applications of ANN

Well trained ANN can cover noisy or incomplete data or is even suitable to predict unknown data based on defined parameters [9]. ANN tools are massively used for predictions in economics, weather forecast, chemistry, medicine and pharmacy and of course in industry. Industrial usage can be found in manifold intelligent sensory or visual systems or in applications of process control.

In pharmacy the ANN are used for drug design and development where based on experimentally collected data the ANN can predict new potential drugs with expected attributes, higher efficiency or lower toxicity based on in vitro-in vivo relationship [10].

Applications of ANN in biomedicine are focused on early disease diagnosis in radiology, CT colonography or EEG (electro-encefalography) and EMG (electro-myography) [11]. The wide applications can be found in diagnosis of cancer in its early stages [12, 13].

4 ANN Application in Predictive Maintenance

There are available various sensors systems, from simple capacitive and inductive proximity sensors and photoelectric or laser sensors and vibration sensors to advanced industrial systems like thermographic cameras, vision systems, process control systems and measurement sensors that can capture large amount of machine data available for further elaboration [14–16]. Artificial neural networks show promising results as a robust tool for evaluation these data in order to support predictive maintenance activities. There exist a lot of papers focused on application of ANN in maintenance. Mainly multi-layer perceptrons (MLP) are used for fault diagnosis of bearings, induction motors, non-destructive evaluation of check valve performance and degradation and in robotic systems [17]. Sun et al. [18] has developed self-organising map for monitoring multi-equipment health management system with fault detection and a real-time monitoring.

4.1 Mechanical Damage and Crack Detection

Stiffness and damping data based on engine load of large diesel engine crankshafts were used for early detection of cracks by Villanueva et al. [19]. Their method was capable to evaluate risk of forming cracks using radial basis function neural network (RBF) and suggest an implementation possibility of on-line monitoring. Castejón et al. [20] applied artificial neural networks beside the multiresolution analysis to diagnose damage of rolling bearings which are a frequent cause of machinery breakdowns and added radial load data to their model. The data about the crankshaft rotation came from accelerometer, phototachometer and the simulations were done with the new bearing and with bearings combinating artificial damage of inner race, outer race and a ball. For fault classification, multi-layer perceptron with the conformance of 80 % was successfully used.

Saxena et al. [21] used combination of accelerometers and an acoustic emission sensor for the diagnosis of mechanical bearings and compared results of genetic algorithm (GA) supported ANNs. Pacheco et al. [22] used for bearing fault detection a method of acoustic measurement with conformity rate 95 % and higher, Unal et al. [23] with the same results above 95 %. On the top of the above mentioned defects, Rao et al. [24] defines the additional faults and defects of bearings that are cage fault, surface irregularities, misaligned races, cracks, pits and spalls on rolling surfaces. Tian [25] suggests that the late in the bearing life is the prediction accuracy more important as the aging could affect decision making about whether the bearing should be replaced or not and uses for computations feed forward ANN (FF ANN).

4.2 Early Detection of Faulty Electrical Devices

Infrared thermography in non-corrective maintenance is used for detection of heat distribution of the machine parts. This method profits from the phenomenon that every object emits infrared radiation. In industrial maintenance, the application of infrared thermography, is suitable for example for detection of heat losses in detection of heat of the electrical equipment, transformer load, etc. [26, 27].

Thermal snapshot of machine electrical control cabinet can quickly and easily display invisible temperature irregularities of electrical devices such as circuit breakers, transformers, fuses, relays etc. Abnormal heat concentration on these devices is caused by faulty, oxidised or loosen contacts, damaged or cracked insulations or broken surge arrestors [28].

The evaluation could be done by simple look on picture with the temperature scale finding the places coloured above the set temperature limits. Huda et al. [29] used multi-layer perceptron with one hidden layer and 5-fold cross validation method for evaluation of temperature differences from ambient temperature or a component with the similarly expected temperature value.

The temperatures were classified in three classes: no defect with temperature difference less than 5 °C; probable deficiency with difference above 5 °C and a defect with major discrepancy with temperature difference above 15 °C. This procedure advances from the fact, that it is robust against the differences of subject emissivity on measuring its temperature. Normalisation of the images was done by conversion to greyscale bitmaps where focused regions were selected for detailed analysis.

Huda et al. [29] used comparison of hotspot features like highest grey value, kurtosis, standard deviation, entropy, contrast, energy against the reference areas. For feature selection they used a discriminant analysis to reduce noisy data and applied further various tests to evaluate the features performance. They achieved up to 83 % of sensitivity rate on defect classification while parallel discriminant analysis prediction produced better results on specificity and accuracy.

4.3 Detection of Faults on Pneumatic Systems

Pneumatic systems consisting from various pneumatic cylinders are used as a cheap and reliable instruments for machine driven parts where there is no requirement for precise positioning. System of the pneumatic cylinder and an air valves block was in focus of Demetgul et al. [30]. There was used a station with testing set of linear potentiometer and proximity sensors (optical, capacitive and inductive) and for training process used simulated modes like empty magazine, dropped work piece, no pressure at valve, low pressure with a conformity rate from 93 till 98 %.

Some of the selected sensors used were vacuum analogue pressure sensor, material handling arm pressure sensor where the data were classified into two classes, signal below or over the 3 V. The type of ANN used was adaptive resonance theory (ART) with back propagation topology.

Although the fault modes used in this study were set to 11, authors recommend to reduce the number of fault modes below 5 to improve ANN classification reliability due to experience of misidentification of some faults.

4.4 Monitoring of Robotic Manipulator

Robotic manipulators are used for precise positioning of manufactured parts during transfer to the following process step or they can be equipped as an automated welding machines. These robots have numerous joints with rotational axis for wide movement and positioning flexibility. Due to complicated design and implementation in fully automated process, these have to be well maintained and monitored mainly for early risk of fault analysis.

Eski et al. [31] deployed radial basis function network and a Kohonen self-organising maps for analysis of joint accelerometer data and compared performance of these two methods with the approach of vibration prediction on industrial robots. The employed SOM shown differences on acceleration variations data at the peak values independent on the joint running speed. Root mean square error in this case was from 0.04 till 0.06. On the other side, radial basis function demonstrated superior performance on accelerations variation descriptions for different running speeds with root mean square error below 4×10^{-4}.

5 Discussion

From the above reviewed methods the often used is multi-layer perceptron ANN (see Table 1). The advantage of MLP is in its simplicity. It is suitable for simple classification of faults or machine (instrument) condition. In case of implementation of MLP for fault detection caused by mechanical damage, it would be convenient to use boundary samples in training phase.

The usage of artificially produced faults might lead to phenomenon, where an undiscovered fault in an early stage could cause irreversible damage on other parts. This could lead to an accelerated wear of the machine if the first damage is not

Table 1 Overview of used ANN designs with achieved results

Predictive maintenance task	ANN design	Achieved results	References
Crankshaft crack	RBF	96 %	[19]
Bearing condition	MLP	80 %	[20]
Bearing condition	GA	93 %	[21]
Bearing condition	MLP	95 %	[22]
Bearing condition	MLP	95 %	[23]
Bearing condition	FF ANN	89 %	[25]
Bearing condition	FF ANN	87 %	[4]
Electrical fault	MLP	83 %	[29]
Pneumatic system	ART ANN	93 %	[30]
Robotic manipulator	SOM	RMSE 0.0004	[31]
Robotic manipulator	RBF	RMSE 0.04	[31]

recognised correctly and at the right time. This approach is useful in cases where there is expected as a fault a destroyed part or machine.

In case the focus of predictive maintenance is on the extensively worn machine parts it is more helpful to use continuous data as an output of the ANN, that would for example give an estimation of remaining useful life of the parts as suggested by study [25].

The application of artificial neural networks is useful for processing of high volume data and a prediction. Analysis of thermal images of electrical devices from the above mentioned study [29] still needs of manual selection of regions of interest, conversion to greyscale. Beside that there is a problem that thermal images are commonly adjusted to the range of detected temperatures, i.e. there exist no reference to the absolute temperature scale and as a reference there serves the adopted scale for each picture and therefore application of automated and unsupervised methods seems to be complicated. This would need a referenced pictures with exactly the same range or an algorithm for automated referencing from the attached scales.

6 Conclusions

Artificial neural networks show strong potential in industrial applications, especially in predictive maintenance tasks. Due to need of well-trained person for ANN computations, necessity of special software and availability of sensors able to capture and systematically store the collected data it has to be properly evaluated whether the use in real production will be meaningful and profitable. The advantage of these methods is the potential of effective equipment failure prevention and the increase of the OEE performance results.

Beside the research of different ANN applications the authors focus on applications of ANN in automotive industry chain. The future research will be focused on injection moulding machines for rubber compounds, mainly on predictive maintenance activities based on machine process data and evaluation of the injection moulding process parameters.

Acknowledgements This work and the contribution were also supported by project "Smart Solutions for Ubiquitous Computing Environments" FIM, University of Hradec Kralove, Czech Republic (under ID: UHK-FIM-SP-2016).

References

1. Hines, P., Holweg, M., Rich, N.: Learning to evolve: a review of contemporary lean thinking. Int. J. Oper. Prod. Manag. **24**(10), 994–1011 (2004)
2. Naylor, J.B., Naim, M.M., Berry, D.: Leagility: integrating the lean and agile manufacturing paradigms in the total supply chain. Int. J. Prod. Econ. **62**(1), 107–118 (1999)

3. Abdulmalek, F.A., Rajgopal, J.: Analyzing the benefits of lean manufacturing and value stream mapping via simulation: a process sector case study. Int. J. Prod. Econ. **107**, 223–236 (2007)
4. Wu, S.J., Gebraeel, N., Lawley, M.A., Yih, Y.: A neural network integrated decision support system for condition-based optimal predictive maintenance policy. Man Cybern. Part A: Syst. Hum. IEEE Trans. Syst. **37**(2), 226–236 (2007)
5. Shin, J.H., Jun, H.B.: On condition based maintenance policy. J. Comput. Des. Eng. (2015)
6. Yegnanarayana, B.: Artificial neural networks. PHI Learning Pvt, Delhi (2009)
7. Maltarollo, V.G., da Silva, A.B.F., Honório, K.M.: Applications of Artificial Neural Networks in Chemical Problems. INTECH Open Access Publisher (2013)
8. Krenek, J., Kuca, K., Krejcar, O., Maresova, P., Sobeslav, V., Blazek, P.: Artificial neural network tools for computerised data modelling and processing. In: 15th IEEE International Symposium on Computational Intelligence And Informatics, pp. 255–260 (2014)
9. Cheng, F., Sutariya, V.: Applications of artificial neural network modeling in drug discovery. Clin. Exp. Pharmacol. **2**(3), 1–2 (2012)
10. Mendyk, A., Tuszyński, P.K., Polak, S., Jachowicz, R.: Generalized in vitro-in vivo relationship (IVIVR) model based on artificial neural networks. Drug Des. Dev. Ther. **7**, 223 (2013)
11. Krenek, J., Kuca, K., Bartuskova, A., Krejcar, O., Maresova, P., Sobeslav, V.: Artificial neural networks in biomedicine applications. In: Proceedings of the 4th International Conference on Computer Engineering and Networks, Springer, pp. 133–139 (2015)
12. Motalleb, G.: Artificial neural network analysis in preclinical breast cancer. Cell J. (Yakhteh) **15**(4), 324 (2014)
13. Samanta, R.K., Mitra, M.: A Neural Network Based Intelligent System for Breast Cancer Diagnosis (2013)
14. Jancikova, Z., Zimny, O., Kostial, P.: Prediction of metal corrosion by neural networks. Metalurgija **3**(52), 379–381 (2013)
15. Krejcar, O., Frischer, R.: Non destructive defect detection by spectral density analysis. Sensors **11**(3), 2334–2346 (2011)
16. Machacek, Z., Hajovsky, R.: Modeling of temperature field distribution in mine dumps with spread prediction. Elektron. Ir Elektrotechnika **19**(7), 53–56 (2013)
17. Meireles, M.R.G., Almeida, P.E.M., Simoes, M.G.: A comprehensive review for industrial applicability of artificial neural networks. IEEE Trans. Industr. Electron. **50**(3), 585–601 (2003)
18. Sun, F., Gao, L., Zou, J., Wu, T., Li, J.: Study on multi-equipment failure prediction based on system network. Sens. Transducers **158**(11), 427–435 (2013)
19. Villanueva, J.B., Espadafor, F.J., Cruz-Peragon, F., García, M.T.: A methodology for cracks identification in large crankshafts. Mech. Syst. Sig. Process. **25**(8), 3168–3185 (2011)
20. Castejón, C., Lara, O., García-Prada, J.C.: Automated diagnosis of rolling bearings using MRA and neural networks. Mech. Syst. Signal Process. **24**, 289–299 (2009)
21. Saxena, A., Saad, A.: Evolving an artificial neural network classifier for condition monitoring of rotating mechanical systems. Appl. Soft Comput. **7**(1), 441–454 (2007)
22. Pacheco, W.S., Castro Pinto, F.A.N.: Classifier based on artificial neural networks and beamforming technique for bearing fault detection. In: ABCM, 22nd International Congress of Mechanical Engineering COBEM, Brazil (2013)
23. Unal, M., Demetgul, M., Onat, M., Kucuk, H.: Fault diagnosis of rolling bearing based on feature extraction and neural network algorithm. Recent Adv. Electr. Eng. Ser. **10** (2013)
24. Rao, B.K.N., Srinivasa, P., Nagabhushana, T.N.: Failure diagnosis and prognosis of rolling— element bearings using artificial neural networks: a critical overview. J. Phys.: Conf. ser. **364**, 1–28 (2012)
25. Tian, Z.: An artificial neural network method for remaining useful life prediction of equipment subject to condition monitoring. J. Intell. Manuf. **23**(2), 227–237 (2012)
26. Al-Kassir, A.R., Fernandez, J., Tinaut, F.V., Castro, F.: Thermographic study of energetic installations. Appl. Therm. Eng. **25**(2), 183–190 (2005)

27. Meola, C., Carlomagno, G.M.: Recent advances in the use of infrared thermography. Meas. Sci. Technol. **15**(9), R27 (2004)
28. Lizák, F., Kolcun, M.: Improving reliability and decreasing losses of electrical system with infrared thermography. Acta Electrotechnica et Informatica **8**(1), 60–63 (2008)
29. Nazmul Huda, A.S., Taib, S.: Application of infrared thermography for predictive/preventive maintenance of thermal defect in electrical equipment. Appl. Therm. Eng. **61**, 220–227 (2013)
30. Demetgul, M., Tansel, I.N., Taskin, S.: Fault diagnosis of pneumatic systems with artificial neural network algorithms. Experts Syst. Appl. **36**, 10512–10519 (2009)
31. Eski, I., Erkaya, S., Setrac, S., Yildirim, S.: Fault detection on robot manipulators using artificial neural networks. Robot. Comput-Integr. Manufact. **27**, 115–123 (2011)

Long Term Analysis of the Localization Model Based on Wi-Fi Network

Rafał Górak and Marcin Luckner

Abstract The paper presents the analysis of long term accuracy of the localization solution based on Wi-Fi signals. The localization model is built using random forest algorithm and it was tested using data collected between years 2012–2014 inside of a six floor building.

1 Introduction

While there are effective localization solutions available outdoor [2] such as Global Positioning System (GPS), finding a similar solution for the indoor environment is more difficult. The lack of GPS signal as well as the fact that greater accuracy is required, makes the problem of indoor localization more challenging. In our paper we describe the solution that is based on the existing Wi-Fi infrastructure of a building. This particular approach brings attention of many researchers however solutions as popular as GPS positioning are still not available. One of the indoor localization systems that is used commercially is Ekhau© Real-Time Location System. It uses readings of RSSI (Received Signal Strength Indication) vectors. For the calibration of the system a database of fingerprints (measurements) is required. The independent tests [3] showed that the average error was about 7 m.

In our paper we present a localization solution that also requires a collection of fingerprints and the calibration process. The model is built using Random Forest© algorithm [1]. It is worth mentioning that different learning methods were

The research is supported by the National Centre for Research and Development, grant No. PBS2/B3/24/2014, application No. 208921.

R. Górak · M. Luckner (✉)
Faculty of Mathematics and Information Science, Warsaw University of Technology,
ul. Koszykowa 75, 00-662 Warszawa, Poland
e-mail: M.Luckner@mini.pw.edu.pl

R. Górak
e-mail: R.Gorak@mini.pw.edu.pl

© Springer International Publishing Switzerland 2016
D. Król et al. (eds.), *Recent Developments in Intelligent Information and Database Systems*, Studies in Computational Intelligence 642,
DOI 10.1007/978-3-319-31277-4_8

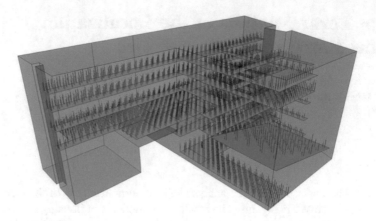

Fig. 1 The building with measurement points marked

previously applied. In [7] multilayer perceptron was used and in [6] k Nearest Neighbours was tested. While preparing this paper we tested all three methods. It appeared that random forest approach gave us as good results as multilayer perceptron and far better than kNN algorithm (several natural metrics were considered). Different methods can be found in [9–11]. At the same time creating and applying the random forest model is the fastest. However, the aim of this paper is not to present the comparison of different machine learning methods. We focus on long term changes in accuracy of the localization solution based on random forests. We analyse data collected between years 2012–2014 and check if there is any decrease in accuracy. This is very important issue from a perspective of future implementations since building a localization model as well as updating it can be quite expensive and time consuming.

All the tests were held inside the building of Faculty of Mathematics and Information Science of Technical University of Warsaw. This is a 6 floor building of irregular shape (see Fig. 1) that fits inside the rectangle of dimensions 50 m by 70 m. In order to build and test the model we used the data gathered in five different series, each one covering the same region of the building (see Fig. 1).

The data were collected using the Android application. For more details about the application see [8].

2　Precise Definition of the Localization Problem

Definition 1

(i) $\mathcal{TS}$ is the set of all transceiver stations used for the localization model.
(ii) $\mathcal{F} = \mathbf{R}^3 \times \mathbf{R} \times \mathbf{R}^n$ is the set of all measurements (fingerprints).

(a) $n = \sharp \mathcal{TS}$.
   For $f \in \mathcal{F}$

(b) first two coordinates $f_1[m], f_2[m]$ denote a horizontal position of a place where measurement was taken and $f_3 \in \{0, 1, \ldots\}$ is the floor number;

(c) $f_4[s]$ is the time of the measurement;

(d) $f_k[dBm]$, where $4 \leq k \leq n+4$, is the RSSI from kth source from $\mathcal{TS}$. If there is no signal from kth transceiver station than $f_{k+4} = noSignal$, where $noSignal$ is a special unique value. In our considerations we define $noSignal[dBm]$ as the minimal signal strength ever recorded minus 2 (our devices reports only odd values). The minimal value that was ever reported was -115 hence we set $noSignal = -117$.

(iii) We call a set of fingerprints $S \subset \mathcal{F}$ a measuring series. Usually S is collected during a few days in a particular building.

(iv) $\mathcal{L} = (\mathcal{L}_x, \mathcal{L}_y, \mathcal{L}_f) : \mathcal{F} \mapsto \mathbf{R}^3$ is the projection onto the first three coordinates of the set $\mathcal{F}$ and $\pi : \mathcal{F} \mapsto \mathbf{R}^n$ is the projection onto the last n coordinates.

(v) For $v \in \mathbf{R}^n$ we denote $supp(v) = \{i : v_i \neq noSignal\}$ which is just the set of visible transceiver stations in the point where the measurement v (RSS vector) was taken.

The localization problem is to construct a function $\widehat{\mathcal{L}} : \mathbf{R}^n \mapsto \mathbf{R}^3$ such that given a RSSI vector $v \in \mathbf{R}^n$, $\widehat{\mathcal{L}}(v)$ predicts a localization of the point where the measurement v was taken.

Definition 2 Let S be a measuring series. For an element $s \in S$ we introduce the following notions:

horizontal error

$$\mathcal{E}_h(s) = \sqrt{(\hat{x} - x)^2 + (\hat{y} - y)^2};$$

and the floor error:

$$\mathcal{E}_f(s) = |\hat{f} - f|;$$

where $\widehat{\mathcal{L}}(\pi(s)) = (\hat{x}, \hat{y}, \hat{f})$ and $\mathcal{L}(s) = (x, y, f)$.

In order to evaluate how well $\widehat{\mathcal{L}}$ predicts a localization for a given testing series S, we introduce the following:

Definition 3 For a testing series S let us define:

(i) Mean horizontal error

$$\mu \mathcal{E}_h(S) = mean\{\mathcal{E}_h(s) \colon s \in S\};$$

(ii) Median horizontal error

$$m \mathcal{E}_h(S) = median\{\mathcal{E}_h(s) \colon s \in S\};$$

(iii) Percentile horizontal error

$$p_{80} \mathcal{E}_h(S) = Perc_{80}\{\mathcal{E}_h(s) \colon s \in S\};$$

(iv) Accuracy of floor's prediction

$$\alpha_f(S) = \frac{\sharp\{s \in S \colon \mathcal{E}_f(s) = 0\}}{\sharp S};$$

Obviously the main goal in the localization problem is to make $\mu \mathcal{E}_h, m \mathcal{E}_h, p_{80} \mathcal{E}_h$ small and to obtain α_f as close to 1 as possible. We will not consider any separate central tendency measures as we will provide graphs of cumulative distributions of $\mathcal{E}_h(S_i)$ instead.

3   Data Collection

Let us denote by S_i where $i \in \{1, 2, 3, 4, 5\}$ the data set of fingerprints collected during ith series. Here series S_1, S_2, S_3 were collected in 2012 within three months period and data sets S_4, S_5 were collected during two series in 2014 within two months period. In each of five series fingerprints were taken in a 1.5 × 1.5 m grid. There were 40 fingerprints taken in every point and each grid covers the same region. Creating a model we considered only the academic net that consists of 46 APs. Hence, following the introduced notation $\sharp TS = 46$. At this point it is worth to mention that there were far more than 46 Access Points visible inside the building (almost 600) but we decided not to take them into account while building a model. It is because we had control only over the academic net and we can guarantee no changes in its infrastructure (location of APs, device changes etc.)

4   The Localization Model

In this report we apply a random forest method to create five localization models $\widehat{\mathcal{L}}^i$, $i \in \{1, 2, 3, 4, 5\}$, where $\widehat{\mathcal{L}}^i$ is created using S_i for training purposes. More precisely, we create $\widehat{\mathcal{L}}^i_x$, $\widehat{\mathcal{L}}^i_y$ and $\widehat{\mathcal{L}}^i_f$ where $i \in \{1, 2, 3, 4, 5\}$. In order to create $\widehat{\mathcal{L}}^i_x$, $\widehat{\mathcal{L}}^i_y$ and $\widehat{\mathcal{L}}^i_f$ we apply a random forest algorithm described in [1] where the training set is $\pi(S_i)$ with responses $\mathcal{L}_x(S_i)$, $\mathcal{L}_y(S_i)$ and $\mathcal{L}_f(S_i)$, respectively. Obviously for creating $\widehat{\mathcal{L}}_x^i$ and $\widehat{\mathcal{L}}_y^i$ we grow regression trees and for $\widehat{\mathcal{L}}_f^i$ the decision trees are grown. We select a number of grown trees to be 30 as we checked that growing more does not improve the accuracy of the localization algorithm. Finally we may define $\widehat{\mathcal{L}}^i(v) = (\widehat{\mathcal{L}}_x^i(v), \widehat{\mathcal{L}}_y^i(v), \widehat{\mathcal{L}}_f^i(v))$.

5 Accuracy of Indoor Localization Model

In order to revel any changes in the long term performance of the localization model, we test each $\widehat{\mathcal{L}}^i$, $1 \leq i \leq 5$ against the series S_j for every $1 \leq j \leq 5$. Let us recall that there were no changes inside the net infrastructure that is used for the localization purposes, that is no changes of devices (APs) as well as their location. We also have not observed any major changes in the building construction during this two years period. Four Tables 1, 2, 3 and 4 represent the measures of accuracy introduced in Definition 3. Rows correspond to the results of a model built on the appropriate data set (see the header of each row). The header of each column is the name of the testing series.

5.1 Floor's Prediction

Let us first analyse the floor's prediction. The results are presented in Table 1. While looking at the accuracy of floor's prediction we can see that we lose very

Table 1 Accuracy of floor's prediction (α_f)

	Series 1	Series 2	Series 3	Series 4	Series 5
Training series 1 (2012)	0.99	0.97	0.98	0.92	0.92
Training series 2 (2012)	0.90	0.99	0.95	0.92	0.92
Training series 3 (2012)	0.97	0.97	1.00	0.91	0.91
Training series 4 (2014)	0.89	0.94	0.93	1.00	0.95
Training series 5 (2014)	0.89	0.94	0.93	0.94	0.99

Table 2 Horizontal mean error ($\mu\mathcal{E}_h$[m])

	Series 1	Series 2	Series 3	Series 4	Series 5
Training series 1 (2012)	1.32	3.61	3.52	4.39	4.52
Training series 2 (2012)	4.18	0.69	3.50	4.71	4.71
Training series 3 (2012)	4.25	3.63	0.53	4.40	4.54
Training series 4 (2014)	5.14	4.80	4.40	0.40	3.20
Training series 5 (2014)	5.23	4.80	4.36	3.04	0.35

Table 3 Horizontal median error ($m\mathcal{E}_h[m]$)

	Series 1	Series 2	Series 3	Series 4	Series 5
Training series 1 (2012)	0.18	2.85	2.95	3.70	3.79
Training series 2 (2012)	2.93	0.15	3.00	4.03	4.04
Training series 3 (2012)	3.11	2.96	0.15	3.73	3.82
Training series 4 (2014)	3.78	3.94	3.74	0.12	2.68
Training series 5 (2014)	3.75	3.78	3.58	2.52	0.14

Table 4 Horizontal 80th percentile error ($p_{80}\mathcal{E}_h[m]$)

	Series 1	Series 2	Series 3	Series 4	Series 5
Training series 1 (2012)	1.19	5.20	5.09	6.49	6.48
Training series 2 (2012)	5.64	0.72	4.99	6.91	6.86
Training series 3 (2012)	5.54	5.05	0.62	6.37	6.49
Training series 4 (2014)	7.48	7.08	6.36	0.38	4.59
Training Series 5 (2014)	7.53	6.74	6.18	4.28	0.42

little with the time passing. In Table 1 we can see that the worst possible situation occurs when we teach the model using series from 2014 and we test it on series 1 from 2012. The accuracy of these models can be as high as 95 % when testing on different series from 2014. It drops to 89 % when testing on series 1 which is still considerably good result taking into account that the localization is on spot that means it is based on just one measurement. While Series 1 serves as the worst testing series it should be observed that Series 2 from 2012 serves as the best testing series when checking floor's prediction. On the other hand, when we look at the performance of the model built on Series 1 we see that we obtain very good results of floor's prediction throughout the whole two years period.

5.2 Horizontal Error

While looking at the horizontal accuracy we can see from Tables 2, 3 and 4 that we have a decrease of all the accuracy measures we proposed ($\mu\mathcal{E}_h$, $m\mathcal{E}_h$ and $p_{80}\mathcal{E}_h$) by not more than 1.5 m which is a very good result in such a big building. Obviously we exclude from that the values on the diagonals which are just tests of the models performed on their training data sets. It appears that when looking at the horizontal accuracy the models built on Series 4 and 5 and tested on Series 1 give the worst results.

In fact Series 1 serves as an example of a testing series with the worst $\mu\mathcal{E}_h$ for all the five models (see Table 2). This time we can expect such results because the horizontal error of the model built on Series 1 (2012) and tested on the same series gives us the mean horizontal error of about 1.5 m which is considerably more when

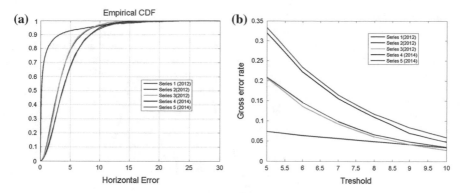

Fig. 2 Horizontal error analysis of the model created using S_1 as the learning series. **a** Horizontal error cumulative distribution. **b** Gross error rate

considered a similar situation for other models built on other series where a similar error is less than 0.5 m (see the diagonal of Table 2). Tables 3 and 4 shed some light on the problem since we can easily spot that while median error is more or less the same on the diagonal for all the five models, while looking at $p_{80}\mathcal{E}_h$ we see that the model built on Series 1 performs the worst. Hence, the gross errors were responsible for high $\mu\mathcal{E}_h$. On the other hand when looking at α_f, $\mu\mathcal{E}_h$, $m\mathcal{E}_h$ and $p_{80}\mathcal{E}_h$ for the model built on Series 1 tested on Series 2 to 5 (first row) we see that the performance is as good as for the other four remaining models or better. Thus we can draw a conclusion that the data set collected during Series 1 contains many outliers that were misclassified due to the pruning process when $\widehat{\mathcal{L}}^1$ was tested against S_1. This explains why the model $\widehat{\mathcal{L}}^1$ performs as good as the others but worse when tested on the training data S_1.

Let us focus now on the performance of the $\widehat{\mathcal{L}}^1$ model. Figure 1 presents an empirical cumulative distribution of $\mathcal{E}_h(S_i)$ for all $i \in \{1, 2, \ldots, 5\}$. Figure 1 presents the gross error rate for different Thresholds that defines the gross error. More precisely the gross error rate is the value $\frac{\#\{\mathcal{E}_h(s) \geq \text{Threshold} : s \in S_i\}}{\#S_i}$. Figure 2 shows the performance's decrease with time of the localization model $\widehat{\mathcal{L}}^1$ even more visible.

5.3 Visible APs

The above analysis suggests that the time influences the accuracy of the localization algorithm. It is reasonable to check if there are any characteristics of the data sets $\{S_i\}_{i \in \{0,1,\ldots,5\}}$ that changed with time and hence can be somehow related to the accuracy of the proposed solution. It appears that one of such characteristics is the function $\tau_i \colon \{0, 1, 2, \ldots\} \mapsto \mathbf{R}$ given by the formula $\tau_i(k) = \frac{\#\{s \in S_i : \#supp(s) = k\}}{\#S_i}$. The

Fig. 3 Graphs of τ_i

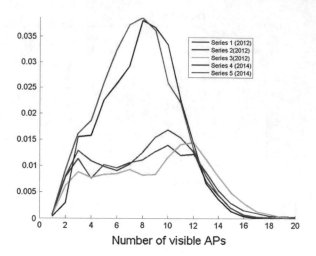

Number of visible APs

graphs of $\{\tau_i\}_{1 \le i \le 5}$ are presented in Fig. 3. The value $\tau_i(k)$ describes the rate of fingerprints taken in S_i for which there is k APs visible.

We can clearly see the difference between data sets from 2012 and 2014 when looking not only at the performance of the localization models but also at the characteristics τ_i. Let us recall that each set $\mathcal{L}(S_i)$ covers the same region of the building. More precisely each time it is a 1.5 m by 1.5 m grid where there is a 40 fingerprints in S_i for every point from $\mathcal{L}(S_i)$. We are also aware that no revolutionary changes took place in the building's infrastructure between years 2012 and 2014. Most probably this was caused by the accumulation of minor changes in the infrastructure of building such as furniture replacements etc. All the series were collected during academic holidays with minimal building occupancy.

6   Summary and Directions for Future Research

The analysis presented in this short note provide a very important results concerning localization based on Wi-Fi networks. Our analysis indicates that although performance of the model that was once built changes with time, its accuracy is still quite good. For instance, a horizontal mean error is about 5 m (the worst case—see Table 2) inside of a big building and α_f around 90 % (the worst case—see Table 1) in a six floor building. These are very promising results. Especially that it could be a good starting point for building a localization system that predicts the localization not on just one reading but takes into account several previous readings as well, which could give us further accuracy improvements.

In order to obtain localization solution that are stable in the long term one could consider frequent updates of the model by just collecting a new data set of fingerprints for learning purposes. Although most probably it would solve the problem, the process of fingerprints collection with their exact locations is time consuming and hence expensive. This suggests the following:

Problem 1 Create a localization system that determines the necessity of collecting additional series of fingerprints for training purposes.

To avoid collecting fingerprints inside the whole building it would be reasonable to create a system which determines an area of the building that requires additional measurements. At this point it is worth mentioning an interesting paper [4] that suggests an adaptive localization system. However the proposed solution still requires a real life tests for the bigger areas or even the whole buildings. Another self adaptive solution was proposed by authors in [5]. However this one deals with the problem of changes inside the Wi-Fi infrastructure (missing APs etc.) which is slightly different problem to the one we consider in this paper.

From our analysis we can see that the gross errors of $\mathcal{E}_h$ are also an important issue one should address in further investigations. We can observe in Fig. 1 that for the localization model $\widehat{\mathcal{L}}^1$ we can obtain $\mathcal{E}_h$ greater than 7 m for about 10 % of fingerprints or even more depending on the testing series. It suggests the following:

Problem 2 Create a localization system that detects readings for which we obtain a gross localization error.

Readings for which the localization system produces a gross error could be excluded from the decision (localization) process providing more reliable localization solution.

References

1. Breiman, L.: Random forests. Mach. Learn. **45**(1), 5–32 (2001)
2. Enge, P., Misra, P.: Special issue on global positioning system. Proc. IEEE **87**(1), 3–15 (1999)
3. Gallagher, T., Tan, Y.K., Li, B., Dempster, A.G.: Trials of commercial Wi-Fi positioning systems for indoor and urban canyons. In: IGNSS Symposium on GPS/GNSS, Gold Coast, Australia (2009)
4. Garcia-Valverde, T., Garcia-Sola, A., Gomez-Skarmeta, A., Botia, J., Hagras, H., Dooley, J., Callaghan, V.: An adaptive learning fuzzy logic system for indoor local-isation using wi-fi in ambient intelligent environments. In: 2012 IEEE International Conference on Fuzzy Systems (FUZZ-IEEE), pp. 1–8 (2012)
5. Górak, R., Luckner, M.: Malfunction immune wi-fi localisation method. In: Nez, M., Nguyen, N., Camacho, D., Trawiski, B. (eds.) Computational Collective Intelligence, Lecture Notes in Computer Science, vol. 9329, pp. 328–337. Springer International Publishing (2015). http://dx.doi.org/10.1007/978-3-319-24069-5-31
6. Grzenda, M.: On the prediction of floor identification credibility in RSS-based positioning techniques. In: Ali, M., Bosse, T., Hindriks, K.V., Hoogendoorn, M., Jonker, C.M., Treur, J. (eds.) 26th International Conference on Industrial, Engineering and Other Applications of Applied Intelligent Systems, IEA/AIE 2013 Recent Trends in Applied Artificial Intelligence, Amsterdam, The Netherlands, June 17–21, 2013. Proceedings. Lecture Notes in Computer Science, vol. 7906, pp. 610–619. Springer (2013). http://dx.doi.org/10.1007/978-3-642-38577-3_63

7. Karwowski, J., Okulewicz, M., Legierski, J.: Application of particle swarm optimization algorithm to neural network training process in the localization of the mobile terminal. In: Iliadis, L.S., Papadopoulos, H., Jayne, C. (eds.) Engineering Applications of Neural Networks—14th International Conference, EANN 2013, Halkidiki, Greece, September 13–16, 2013 Proceedings, Part I. Communications in Computer and Information Science, vol. 383, pp. 122–131. Springer (2013). http://dx.doi.org/10.1007/978-3-642-41013-0_13

8. Korbel, P., Wawrzyniak, P., Grabowski, S., Krasinska, D.: Locfusion API—programming interface for accurate multi-source mobile terminal positioning. In: 2013 Federated Conference on Computer Science and Information Systems (FedCSIS), pp. 819–823 (2013)

9. Papapostolou, A., Chaouchi, H.: Scene analysis indoor positioning enhancements. Ann. Télécommun. **66**, 519–533 (2011)

10. Roos, T., Myllymaki, P., Tirri, H., Misikangas, P., Sievanen, J.: A probabilistic approach to WLAN user location estimation. Int. J. Wireless Inf. Networks **9**(3), 155–164 (2002)

11. Xiang, Z., Song, S., Chen, J., Wang, H., Huang, J., Gao, X.G.: A wireless LAN-based indoor positioning technology. IBM J. Res. Dev. **48**(5–6), 617–626 (2004)

Part II
Ontologies, Social Networks and Recommendation Systems

Specific Behavior Recognition Based on Behavior Ontology

Ngoc Q. Ly, Anh M. Truong and Hanh V. Nguyen

Abstract In this paper, we propose a novel method based on ontology for human behavior recognition. The state-of-art behavior recognition methods based on low visual features or high visual features have still achieved low recognition accuracy rate in reality. The two most important challenges of this problem are still remaining such as the semantic gap and the variety of appearance of human behaviors in reality. By using prior knowledge, our system could completely detect a behavior without training data of entire process and could be reused in other cases. In experimental results, our method have achieved the encouraging performance on PETS 2006 and PETS 2007 datasets. These results have proved the good prospect of the human behavior recognition system based on behavior ontology.

Keywords Behavior ontology · Human behavior recognition · Specific behavior

1 Introduction

Human Behavior consisted of *Gestures*, *Actions*, *Activities* and *Expressions*. In real life, Human Behavior could be expressed as a sentence with subjects, verbs and objects. Moreover, actions, activities or expressions could be considered as the verbs of that sentence. In this work, we only focus on *Activity* aspects of *human behaviors*. There are many studies of general human activity recognition; however, the result of them are not high on real data. In addition, there are many applications

N.Q. Ly · A.M. Truong (✉) · H.V. Nguyen
Faculty of Information Technology, University of Science,
VNU-HCMC, 227 Nguyen Van Cu Street, Ho Chi Minh City,
Viet Nam
e-mail: tmanh93@outlook.com

N.Q. Ly
e-mail: lqngoc@fit.hcmus.edu.vn

H.V. Nguyen
e-mail: nguyenvanhanh.cs@gmail.com

© Springer International Publishing Switzerland 2016
D. Król et al. (eds.), *Recent Developments in Intelligent Information and Database Systems*, Studies in Computational Intelligence 642,
DOI 10.1007/978-3-319-31277-4_9

such as surveillance system need the good precision result on some specific behaviors. Because of that reason, in this paper, we only focus on the approach to improve the recognition result on the specific behaviors.

In the field of behavior recognition, there are many current studies based on using low-level visual features such as local space-time features (e.g. [1]), dense point trajectories (e.g. [2]) or learning features (e. g. [3]). The main drawback of this approach is lack of semantic information; therefore, they have limited ability to deal with complex data.

Another approach of behavior recognition tackles the lack of information by using semantic units such as action poses, action attributes or action hierarchical (e.g. [4]) and scene terms (e.g. [5, 6]). The use of semantic units is reducing the gap between low-level features and behavior terms. However, they still have lack in relations and constraints between defined semantic units.

The next approach addresses the semantic gap problem by defining an ontology. For example, Vu [7] and Akdemir [8] proposed the method to describe the ontology for human activities as the set of entities, environments and interactions between them. In general, this approach could break behavior recognition issue into sub-problems and resolve semantic gap problem better because it not only uses semantic units but also defines in detail about the constraints, relationships between those semantic units. However, not all of behaviors can be modeled because some of them are too complex and diverse such as fighting, abnormal behavior.

In this paper, we proposed a method to recognize specific behavior based on behavior ontology by using common concepts from previous works. The basic ideas is using the strategy "Divide and Conquer" to break the recognition problem into sub-problems by the guidelines was create from behavior ontology. We also create the basic ontology for left luggage and loitering at the corridor behaviors. To evaluate the system and behavior ontology, the experiments on PETS 2006 and PETS 2007 are performed and received the encouraging result.

The rest of this paper is structured as follows: In Sect. 2, we review some related works. The the general frame work, model of behaviors and how the system work are detailed in Sects. 3, 4 and 5 respectively. Then, the experiment was show in Sect. 6. Finally, we conclude our work in Sect. 7.

2   Related Works

Ontology is the term which was borrowed from the philosophy. In computer science, this word usually means the specification of a conceptualization. In Gruber [9] also proposed the very first guideline to design and develop an Ontology for Artificial Intelligent System with five important criteria to evaluate the ontology design:

- **Clarity**: An ontology should explicitly express the meaning of defined terms.
- **Coherence**: An ontology should be logical, so they do not have any defined concepts which are conflicting or incompatible.

- **Extendibility**: The design of the ontology should be easily to upgrade in the future by adding more terms based on the original definitions of the system.
- **Minimal Encoding Bias**: in other words, it means the ontology need to specify at the general knowledge level without depending on a particular situation.
- **Minimal Ontological Commitment**: this evaluation criterion is used for reducing as much as possible the redundant terms. So it help the ontology become simple and general.

To address the recognition activity based on ontology approach, there are many researches such as Vu [7], Akdemir [8], Gómez-Romero [10]. In [7, 8], the ontology was created in the same way based on the scenario term to describe the plot of the human behavior and the relationships between the objects. In Gómez-Romero [10], applied the ontology for improving tracking objects result for surveillance system.

3 Framework Outline

Our framework is divided into three levels are low, medium and high. The high-level system is the main processing area of behavior inference. At this level, the system has two tasks: requesting the behavior inference input from the medium-level and making recognition decision. The missions of medium level are managing the low-level processing method and executing the requirement of the high-level. Finally, at the low level, the fundamental processing method such as preprocessing, features detection and selection was implemented. And the input of this level is an image sequence or a video was recorded from cameras (Fig. 1).

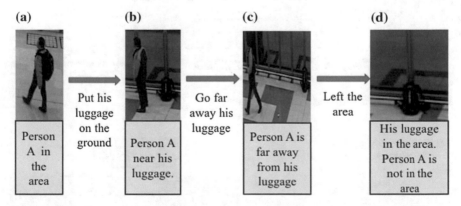

Fig. 1 The example regarding the scenes of left luggage behavior. There are 4 scenes: **a** person A in the area, **b** person A is near his luggage placed on the ground, **c** person A is far away from his luggage was placed on the ground, **d** person A is not in the area and his luggage is in the area

4   Modeling Behavior

In general, a specific behavior can be acted in various ways however they still share **general plot** (e.g. the left luggage behavior has the general plot as follow: the person enters the area then puts the luggage on the ground. Finally, he leaves the area). To express the general plot of a specific behavior, we use the concept *scene* (Figs. 2 and 3).

4.1   Object Modeling

The objects, in our work, are everything that appear in the video. We use same the idea with Vu [7] to describe an object by four types of attribute:

- Position: including position, size of an object
- Appearance: describing the identifying characteristics of an object such as color, texture, shape and visual features [11]
- Component: describing the local appearance of an object by its component and relationship of those component. There are also have 2 special types of component are parts (e.g. legs, arms, etc.) and belonging objects (The belonging objects are things owned by one people, e.g. luggage, shirt, etc.)
- Name: the name of object in real life (e.g. luggage as "Luggage") (Fig. 4).

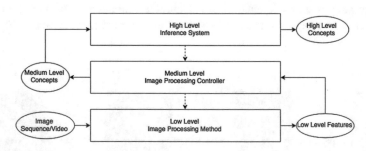

Fig. 2 The framework of our recognition system

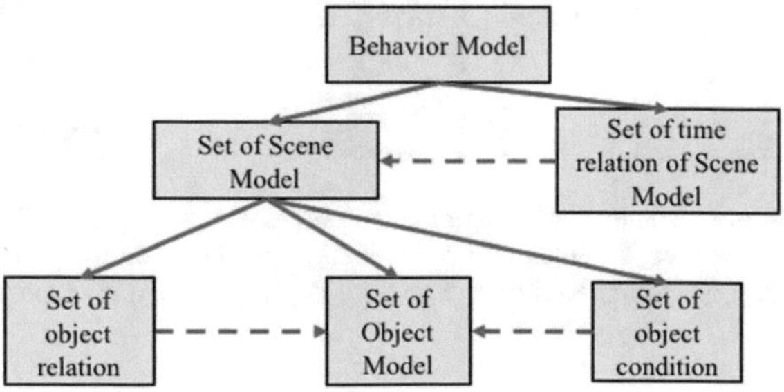

Fig. 3 The modeling of a specific behavior

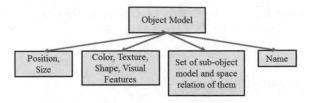

Fig. 4 The modeling of an object

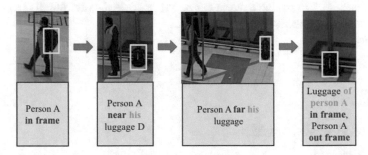

Fig. 5 The example interested objects and their relations, conditions in left luggage behavior

4.2 Relation and Condition Modeling

In our work, we model all relationships of objects by three kinds of relation and two types of condition (Fig. 5).

- Relations:
  - Space Relation: this relation expresses the correlation of position of an object with other via the terms are used in daily. Maillot in [11] introduced the ontology to model various types of space relation.
  - Space-Time Relation: this relation was built based on space relation to express the change of space relation by the time.
  - Action Relation: this relation is used for expressing the action or sub-behavior interaction of objects in one behavior

- Conditions:
  - Position condition: this condition expresses the constraint regarding position of objects that should be satisfied.
  - Action condition: this condition is used for expressing the action or sub-behavior constraint of objects in one behavior (the action or sub-behavior in action condition is not interaction with other objects)

We also applied three kinds of space relation (*near, far and tangential*) and two kinds of position condition (*in frame* and *out frame*).

4.3   Time Relation Modeling

In our work, we use the time relation to express the constraint between the scenes and Allen's interval algebra for the time operator [12]. In addition, we define the time operator *before* for the experiment on PETS 2006 dataset, for all scenes S_1, S_2 happen at t_1, t_2 respectively.

Before relation:

$$S_1 \textbf{ before } S_2 \cong t_1 < t_2 \tag{1}$$

5   Behavior Recognition

5.1   Recognition Process

Behavior recognition process
Input: Images sequence/Video
Output: List of new behavior
Step 1: The high-level system send the request to medium-level to get objects and their relations, conditions for the inference.
Step 2: The medium-level system send the request to low-level to get required features corresponding to scenes detection.
Step 3: The low-level system perform the processing method corresponding to medium-level's request.
Step 4: The medium-level system perform detecting objects and their relations, conditions.
Step 5: The high-level system detect scenes and their relations based on the scene ontology.
Step 6: The high-level system deduce specific behaviors based on behavior ontology.

5.2   Ontology for Left Luggage Behavior

In general, left luggage behavior starts with the person A brings his luggage appeared in the frame. Then, he puts his luggage on the ground (the person no longer carries the luggage). Finally, he goes far away his luggage (maybe out the frame or in frame but the distance between him and luggage is far enough). Therefore, we define two model of left luggage behavior for those situation (Fig. 6).

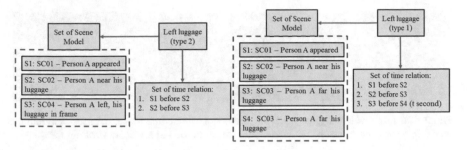

Fig. 6 The modeling of left luggage type 1 and 2

The details of all scenes are shown as follow:

SC01: *person A appeared* [*Space condition: person A in frame*]	**SC02**: *person A near his luggage* [***Space condition***: *person A in frame*] [***Space condition***: *luggage D in frame*] [***Space relation***: *luggage D near person A*] [***Belonging Objects***: *luggage D*]
SC03: *person A far his luggage* [***Space condition***: *person A in frame*] [***Space condition***: *luggage D in frame*] [***Space relation***: *luggage D far person A*] [***Belonging Objects***: *luggage D belong to person A*]	**SC04**: *person A near his luggage* [***Space condition***: *person A out frame*] [***Space condition***: *luggage D in frame*] [***Belonging Objects***: *luggage D belong to person A*]

5.3 Evaluating Behavior Ontology

Evaluating the specific behavior model: the specific behavior ontology is based on fundamental terms such as scene, object and relationship in space, space-time, time, action. Therefore, our specific behavior model is able to ensure the quality of simplicity and generality to satisfy the criteria of the ontology.

Evaluating the left luggage behavior model: According to observation, the left luggage behavior was created. It's not only suitable for normal situation but also special case. In reality, there are many special situation such as obstruction or the owner throws away his luggage. The explanations in detail of those cases are below:

- Obstruction case: In complex situation, the tracking module may have errors because of the tracking technical limitation. In this case, the information from other cameras in surveillance system can be used to resolve this problem.
- Throwing luggage case: in this case, the owner throw away the luggage in the scene. Our left luggage behavior ontology still has ability to address this

Fig. 7 The result of left luggage behavior recognition. In the first frame, the man went into the scene. In the second frame, he put his luggage down. In the last scene, he went out the scene

situation because the changes of the relation between owner and luggage are the same as our defined left luggage behavior ontology. (First, the luggage moves far away the owner, then, the owner leaves the scene or keeps the long distance with the luggage in a long time)

- Reusing case: we presented the ontology for left luggage behavior with four scenes related the space relation between the owner and his luggage. The system can be reused the defined scenes and relationships. In case of the specific behaviors are related to luggage such as swapping luggage or stealing luggage, the system could be extend by adding some scenes involved to bringing baggage farther the owner. While other kind of specific behaviors may need some different modifications (Fig. 7).

6 Experiments

6.1 PETS 2006 Dataset

This datasets contains seven frame sequences about left luggage behavior are recorded from four surveillance cameras in different views. In PETS 2006 dataset, luggage is defined to include all kinds of hand baggage as bag, valise, suitcase, briefcase, backpack, rucksack, ski gear carrier.

6.2 PETS 2007 Dataset

PETS 2007 dataset includes 8 sequences contain 3 specific behaviors: loitering, theft luggage, left luggage; 1 sequence for background training and 1 sequence does not have any specific behavior that was recorded from surveillance cameras (Fig. 8).

Fig. 8 The example scene from 4 camera of PETS 2007 dataset

6.3 Left Luggage Behavior Recognition

The goal of this experiment is checking the efficient of our framework when applying on a real specific behavior. Table 1 shows the comparison our proposed method with other researches regarding detecting unattended luggage in PETS 2006 dataset. As we can see, our system has acceptable results and the false alarm errors are smaller than others. Because the methods used in [13, 14] only detect unattended luggage without concerning with the behavior. While our method considers the unattended luggage is an object in left luggage behavior and detects its behavior by an ontology based on the space, time relationships (Table 2).

Though this method has good precision in unattended luggage detection, the accuracy when recognizing left luggage behavior is lower. The main reason of these errors is the wrong result in tracking module. If luggage is obstructed in a long time while being tracking, the system may make the wrong tracking result or owner detection (Table 3).

Table 1 The comparison between the best result of our proposed method and other methods on detecting unattended luggage from sequence S1 to S7 in PETS 2006 dataset

Method	S1	S2	S3	S4	S5	S6	S7
Proposed method	1/0/1	1/0/1	0/0/0	1/0/1	1/0/1	1/0/1	1/0/1
[13]	1/0/1	1/0/1	0/0/0	1/0/1	1/13/1	1/0/1	1/3/1
[14]	1/0/1	1/0/1	0/0/0				1/1/1

The result of each sequence shows true positive, false positive and total number of unattended luggage in the image sequence respectively

Table 2 The result of recognition left luggage behavior of our proposed method

No	Detail	Dataset	Precision (%)	Recall (%)
1	Left luggage type 1 (size filter)	PETS 2006	71.4	71.4
2	Left luggage type 1 (no size filter)	PETS 2006	71.4	71.4
3	Left luggage type 2 (size filter)	PETS 2006	50	71.4
4	Left luggage type 2 (no size filter)	PETS 2006	38.46	71.4
5	Left luggage type 2 (no size filter)	PETS 2007	50	50

Table 3 The comparison between our proposed method and other methods on loitering behavior recognition in PETS 2007 dataset

Sequence	Our proposed method		[15]		Total loitering person
	TP	FP	TP	FP	
S1	1	0	1	0	1
S2	1	0	1	0	1

TP True positive, *FP* False positive. The result shows true positive and false positive number of loitering people was detected by the system

7 Conclusion

In this paper, we presented the framework to recognize specific behavior by using behavior ontology to recognize the specific behavior without training process. The experiment in PETS 2006 and PETS 2007 datasets show that our proposed method have a good precision. It also can be flexibly reused and easily extended to other specific behaviors such as stealing or swapping luggage by adding some scenes.

In future work, we have planned to improve the behavior ontology system and apply it to more complicated specific behavior. In this work, the tracking and detecting module of our system could have better results by using ontology.

Acknowledgment This research is funded by Vietnam National University, Ho Chi Minh City (VNU-HCM) under grant number B2014-18-02.

References

1. Tran, Q.D., Ly, N.Q: Sparse spatio-temporal representation of joint shape-motion cues for human action recognition in depth sequences. In: Proceedings of the 2013 IEEE-RIVF International Conference on Computing and Communication Technologies, pp. 253–258 (2011)
2. Wang, H., Klaser, A., Schmid, C., Liu, C.: Action recognition by dense trajectories. In: Proceedings of the 2011 IEEE Conference on Computer Vision and Pattern Recognition, pp. 3169–3176 (2011)
3. Le, Q.V., Zou, W.Y., Yeung, S.Y., Ng, A.Y.: Learning hierarchical invariant spatio-temporal features for action recognition with independent subspace analysis. In: Proceedings of the 2011 IEEE Conference on Computer Vision and Pattern Recognition, pp. 3361–3368 (2011)
4. Wang, C., Wang, Y., Yuille, A.L.: An approach to pose-based action recognition. In: Proceedings of the 2013 IEEE Conference on Computer Vision and Pattern Recognition, pp. 915–922 (2013)
5. Oh, S., McCloskey, S., Kim, I.: Multimedia event detection with multimodal feature fusion and temporal concept localization. J. Mach. Vis. Appl. **25**, 49–69 (2014)
6. Pham, T.T., Trinh, T.T., Vo, V.H., Ly, N.Q., Duong, D.A.: Event retrieval in soccer video from coarse to fine based on multi-modal approach. In: Proceedings of the 2010 IEEE-RIVF International Conference on Computing and Communication Technologies, pp. 1–6 (2010)
7. Vu, V.-T., Bremond, F., Thonnat, M.: Automatic video interpretation: a novel algorithm for temporal scenario recognition. In: Proceedings of the 18th International Joint Conference on Artificial Intelligence, pp. 1295–1300. Morgan Kaufmann Publishers Inc., San Francisco, CA, USA (2003)
8. Akdemir, U., Turaga, P., Chellappa, R.: An ontology based approach for activity recognition from video. In: Proceedings of the 16th ACM International Conference on Multimedia, pp. 709–712 (2008)
9. Gruber, T.R.: Toward principles for the design of ontologies used for knowledge sharing. J. Int. J. Hum.-Comput. Stud. **43**, 907–928 (1995)
10. Gómez-Romero, J., Patricio, M.A., García, J., Molina, J.M.: Ontology-based context representation and reasoning for object tracking and scene interpretation in video. J. Expert Syst. Appl. **38**, 7494–7510 (2011)
11. Maillot, N., Thonnat, M., Boucher, A.: Towards ontology-based cognitive vision. J. Mach. Vis. Appl. **16**(1), 33–40 (2004)
12. Allen, J.F.: An interval-based representation of temporal knowledge. In: Proceedings of the 7th International Joint Conference on Artificial Intelligence, vol. 1, pp. 221–226 (1981)
13. Auvinet, E., Grossmann, E., Rougier, C., Dahmane, Meunier, J.: Left-luggage detection using homographies and simple heuristics. In: Proceedings of the 9th IEEE International Workshop on Performance Evaluation of Tracking and Surveillance, pp. 51–58 (2006)
14. Martínez-del-Rincón, J., Herrero-Jaraba, E., Gómez, J.R., Orrite-Uruñuela, C.: Automatic left luggage detection and tracking using multi-camera UKF. In: Proceedings of the 9th IEEE International Workshop on Performance Evaluation of Tracking and Surveillance, pp. 59–66 (2006)
15. Arsic, D., Hofmann, M., Schuller, B., Rigoll, G.: Multi-camera person tracking and left luggage detection applying homographic transformation. In: Proceedings 10th IEEE International Workshop on Performance Evaluation of Tracking and Surveillance, PETS 2007, IEEE, Rio de Janeiro, pp. 55–62 (2007)

A New Method for Transforming TimeER Model-Based Specification into OWL Ontology

Toan Van Nguyen, Hoang Lien Minh Vo, Quang Hoang
and Hanh Huu Hoang

Abstract Building ontologies in Semantic Web from the Entity—Relationship model (ER model) is one of the approaches which attracts interests from the content theory research community. In this paper, we propose a method to transform specifications in TimeER model into OWL ontology by transforming the entities and relationships in a TimeER model to an OWL 2 ontology. Our transformation approach is based on the improvement of the transformation methods from ER model to OWL ontology, and adding the transforming rules for the temporal components of the TimeER model.

Keywords Semantic web · OWL · Ontology · TimeER · Data model

1 Introduction

Ontology has become a well-known term in the field of computer science and has different significance from its original meaning. Ontologies help people and machines communicate, and help the machine "understand" and so that they are enable to process the information efficiently.

The current Web are usually designed from ER or extended ER models, while the Semantic Webs is mainly built with ontologies serialized in OWL. OWL is a language describing the classes, properties and relationships between these objects

T. Van Nguyen (✉) · H.L.M. Vo (✉) · Q. Hoang (✉)
Hue University of Sciences, Hue University, 77 Nguyen Hue, Hue City, Vietnam
e-mail: toan.fiit@gmail.com

H.L.M. Vo
e-mail: minhvhl@gmail.com

Q. Hoang
e-mail: hquang@hueuni.edu.vn

H.H. Hoang (✉)
Hue University, 03 Le Loi, Hue City, Vietnam
e-mail: hhhanh@hueuni.edu.vn

© Springer International Publishing Switzerland 2016   111
D. Król et al. (eds.), *Recent Developments in Intelligent Information
and Database Systems*, Studies in Computational Intelligence 642,
DOI 10.1007/978-3-319-31277-4_10

in a way that machines can understand Web contents. OWL 2 is an extension of OWL DL by adding the expression syntax but still not loses the deciding ability [1]. Particularly, the OWL 2 is able to present the key of the class similar to key of entities in the ER model.

TimeER model is an ER model that supports temporal attributes. Comparing with other ER models that support temporal attributes, the TimeER model is fully supported to the temporal attributes [2]. Currently, there are no studies about the transformation method for TimeER model into OWL ontology, although it is considered as the expansion of the transformation method from ER model into OWL ontology allowing the design of OWL ontology with the temporal attributes.

The W3C vision of Semantic Web is to transform the current Web contents into a Semantic Web contents in which reuse of the previous legacy system. Many previous legacy systems was based on the ER model. Therefore, the upgrading and transformation of the ER model into Ontology to reduce cost is necessary. A majority of this research has been done [3, 4, 5] but no study has yet to mention the transformation of the TimeER to OWL ontology. Our transformation rules is useful for integrating heterogeneous data sources as well as upgrade the legacy systems by providing a method for transformating ER model into OWL ontologies equivalent.

In this paper, a systematic approach for an automatic transformation method to the OWL ontology from an ER model is presented, and this result allows preserving the information of ER models, including the entity with key constraints, attributes and the relationship. In addition, the mapping method also performed the time factor of TimeER model in OWL ontology. Accordingly, this paper is organized as follows: in Sect. 2 provides an overview of related work. Section 3 we show some existing works on the research field, and add additional rules key constraints to OWL ontology resulting ensure key constraints. Section 4 we are going to propose a method for transforming TimeER to OWL ontology. This method inherits the transformation methods in Sect. 3 and simultaneously adds the rules to convert the temporal attributes in TimeER model to OWL ontology. Section 5 concludes and points out fields for future work.

2 Related Work

There are several approaches and frameworks on building ontologies for Semantic Web by mapping ER model to OWL ontology.

Fahad [3] presented the rule to transform ERD to OWL-DL in details. The framework provides OWL ontology for Semantic Web fundamental. This framework helps software engineers in upgrading the structured analysis and design of ERD artifacts, to components of Semantic Web. Myroshnichenko and Murphy [4] introduced a solution to a notable special case of the automatic and semantically equivalent transformation of well-formed ER schemas to OWL-Lite ontologies. They presented a set of mapping rules that fully capture the ER schema semantics, along with an overview of an implementation of the complete mapping algorithm integrated into the SFSU ER Design Tools software. Chujai et al. [5] proposed an approach of building OWL ontology from relational database based on ER model using the ontology editor Protégé. Their rules can transform each part of an ER model: entities, relationship and attributes into each part of OWL ontology.

However, mentioned transforming approaches remain following limitations: the results of transformations were presented as OWL 1, a language has not been supported in key constraints as in ER model. Therefore, the results have not satisfied the key constraints, the rules have not been formed or have already formed apparently, just for orientation, multi-valued attribute and complexes have not been mentioned.

3 Transformation EER Model into OWL Ontology

3.1 Transformation of Entity Types and Object-Oriented Components

3.1.1 Transformation of Entity Types

Rule EER1: Transform each entity type into a class with the same name in OWL ontology [6].

3.1.2 Transformation of an Inheritance Relationship

Rule EER2: For each inheritance relationship, add the subclasses constraint (inheritance) for the corresponding class of set of entity subclasses [6] is shown in Fig. 1.

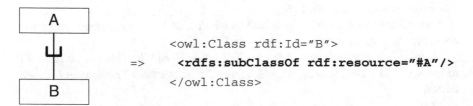

```
<owl:Class rdf:Id="B">
    <rdfs:subClassOf rdf:resource="#A"/>
</owl:Class>
```

Fig. 1 Transformation of an entity type and an inheritance relationship

3.1.3   Transformation of Disjoint Specializations

Rule EER3: With each entity of a subclass in disjoint specializations, use owl:
disjointWith to add to the corresponding class of constraints which do not intersect
with the corresponding class of other entity set in subclasses inheriting from that
disjoint subclass [6].

3.1.4   Transformation of Overlap Specializations

Rule EER4: For each overlap specializations, add to the corresponding class which
is the entity type which has this overlap specializations the union constraint
including the classes corresponding with the inherited entity type to that overlap
subclass [6].

3.1.5   Transformation of Union Specializations

Rule EER5: For each union specializations, add the union constraint of disjoint
specializations for class corresponding with this entity type of that union specializations which the parameters are the corresponding with this entity type in union
specializations.

3.2   Transformation of Attributes

The composite and single-valued attribute can be represented by its single and
single-valued attributes. By considering the multivalued composite attributes of an
entity type as an weak entity type of this owner entity, so we can convert each
multivalued composite attribute in a similar way as the transformation rules of weak
entity types.

   Rule EER6: With each attribute *attA* of entity *E*, add a datatype property *attA*
with *E* as domains and range is the actual datatype of that attribute [6].

   Rule EER7: With each single valued attributes *attA*, set functional characteristics for datatype attribute *A* [6].

   Rule EER8: With each attribute *attA* (not Null), we can set minimum cardinality
constraint to one for datatype property *attA* [4].

   Rule EER9: For each key attribute *KE* of entity *E*, add the datatype property *KE*
to the set of key attributes of class *E*. Figure 2 demonstrates EER6, EER7, EER8,
EER9.

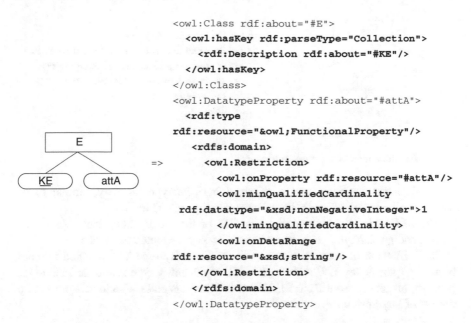

```
<owl:Class rdf:about="#E">
  <owl:hasKey rdf:parseType="Collection">
    <rdf:Description rdf:about="#KE"/>
  </owl:hasKey>
</owl:Class>
<owl:DatatypeProperty rdf:about="#attA">
  <rdf:type
rdf:resource="&owl;FunctionalProperty"/>
    <rdfs:domain>
      <owl:Restriction>
        <owl:onProperty rdf:resource="#attA"/>
        <owl:minQualifiedCardinality
rdf:datatype="&xsd;nonNegativeInteger">1
        </owl:minQualifiedCardinality>
        <owl:onDataRange
rdf:resource="&xsd;string"/>
      </owl:Restriction>
    </rdfs:domain>
</owl:DatatypeProperty>
```

Fig. 2 Transformation of attributes and set functional characteristics, min cardinality constraint for property

3.3 Transformation of the Weak Entity Type

Let's consider that W is a weak entity type of the identifying relationship R and the owner entity type E. Suppose W with partial key KW, and E with key KE. A weak entity always participates in the identifying relationship with cardinality constraint (1,1). In general, because a key of the weak entity type W is created by combining the partial key KW with the key KE of the entity type E, therefore, we have the key constraint KEY(KW, KE).

Rules EER10: Transform each weak entity type W of the owner entity type E into a class W, and add two object properties which show the relationship between class E and class W. They have identifier, domain, and range shown in Table 1. Depending on the cardinality constraint of the identifying relationship, add the corresponding min/max constraint to the object property just added with domain is the class with corresponding entity type which has pairs cardinalities.

Table 1 Properties added when transformation of weak entity type

Identity	Domain	Range
EhasW	E	W
WOfE	W	E

Note If *W* is a weak entity type of many owner entity types, for each owner entity type, add two inverse object properties which show relationship between class *W* and class of the owner entity type

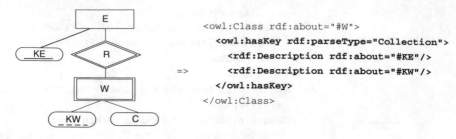

Fig. 3 Transformation of the weak entity types

Weak entity attributes are transformed into datatype property, similar to the transformation rules of attributes of strong entity type (Fig. 3).

Rule EER11: For each partial key *KW* of the weak entity type *W*, add the corresponding datatype property to the set of key attributes of class *W*.

Rule EER12: In general, a key of the weak entity type W is created by combining the partial key KW with the key KE of the entity type E and for each object property originally from Rule EER10 with domain is class *W*, add this property to the set of key attributes of class *W*.

3.4   Transformation of Relationship

3.4.1   Transformation of Relationship Without Attributes

Rule EER13: For each binary relationship without attributes between entity *A* and *B*, add two inverse object properties which show the relationship between class *A* and *B* [6]. Their domain and range is shown in Table 2. For each min/max cardinality differential 0 and n on the relationship *R*, add corresponding min/max constraint to the object property just added, whose domain is corresponding class with that entity type.

3.4.2   Transformation of Binary Relationship with Attributes

Rule EER14: Transform each binary relationship *R* with attributes between two entities *A, B* into a class, add two inverse object properties: *AHasR, ROfA* shown in Table 3, add two inverse object properties *BHasR, RofB*. Attributes of relationship *R* are transformed into datatype property in class *R* [6]. Add function characteristic for their inverse object properties: *ROfA, ROfB*, min constraint set to one. Each

Table 2 Properties added when transformation of binary relationship without attributes

Identity	Domain	Range
ARB	A	B
BRA	B	A

Table 3 Properties added when transforming binary relationship with attributes

Identity	Domain	Range
AHasR	A	R
ROfA	R	A
BHasR	B	R
ROfB	R	B

Note Transformation of recursive relationship is similar to binary relationship. Using the name of the relationship to set identifier for object properties to distinguish two inverse properties to present new relationship of the class created itself

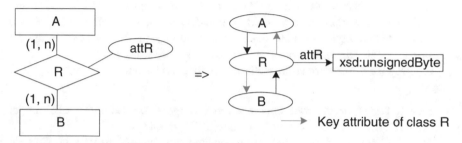

Fig. 4 Transformation of binary relationship with attributes

min/max cardinality differential 0 and n on the relationship *R* of entity *A* and *B*, add corresponding min/max constraint to the object properties *AhasR* and *ROfB*. If the binary relationship *R* is n-n, add two object properties *ROfA*, *ROfB* to set of key attributes of class *R* (Fig. 4).

3.4.3 The n-ary Relationship

Transformation of the n-ary relationship R among the entity types E_i is also executed as the binary relationship with attributes. Adding class *R* and pair of inverse object properties shows the relationship between class R and classes E_i together with the max/min min cardinality constraints. Also, the attributes of the relationship *R* is transformed into the datatype properties of class *R* and add the object properties which has just added with the domain is class *R* and the key set of class *R* (Fig. 5).

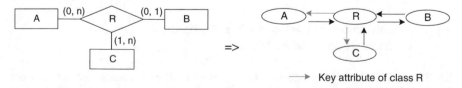

Fig. 5 Transformation of the n-ary relationship types. *Note* If relationship *R* has function constraint, remove the object property which has range is the corresponding with the entity appeared on the right of the function constraint out of the key set of class *R*. This is also an expansion in comparing to the previous researches

4   Transformation of TimeER Model into OWL Ontology

TimeER model is an extension of EER model by allowing quite fully support the temporal components in comparing to the other temporal ER model. Some temporal aspects that this model supported as the lifespan of an entity, the valid time of a fact, and the transaction time of an entity or a fact [2, 7].

This model have a convention that, temporal aspects of the entities in the database can be either the lifespan (LS), or the transaction time (TT), or both the lifespan and the transaction time (LT). The temporal aspects of the attributes of an entity can be either the valid time (VT), or the transaction time (TT), or both the valid time and the transaction time (BT). Besides, because a relationship type can be seen as an entity type or an attribute, consequentially, the designer can define the temporal aspects supported with this relationship type, if necessary.

We propose a transformation method TimeER model into OWL ontology in 3 steps:

- Step 1. Transform the components without temporal aspects on TimeER model (including the temporal entity type) into OWL ontology.
- Step 2. Create the OWL ontology to present the temporal aspects in TimeER model. This step will create some components on OWL for representing the data and temporal aspects' constraints on TimeER model (showed in Sect. 4.1).
- Step 3. Transform the components of a TimeER model into OWL ontology. The temporal aspects are: the temporal aspect of entity type, the temporal attributes, the temporal relationship and the temporal attributes of an entity type (showed in Sect. 4.2).

4.1   Initiate Ontology for Presenting Temporal Aspect

4.1.1   Initiate InstantDateTime Class

Create the class *InstantDateTime* are represented for a timeline. In this class, create the functional datatype property *hasDateTime* with min cardinality set to one, range is set to *xsd:dateTime* and it is the key attribute of *InstantDateTime* class.

4.1.2   Create Object Properties for Presenting the Temporal Constraints on TimeER Model

Six functional datatype properties are created with min cardinality set to one: *hasVTs*, *hasVTe*, *hasLSs*, *hasLSe*, *hasTTs*, *hasTTe* represent the relationship between class *owl:Thing* and class *InstantDateTime*, range set to the class *InstantDateTime* and domain set to class *owl:Thing*.

4.2 Transformation of Temporal Components on TimeER

4.2.1 Transformation of Temporal Entity Type

Rule TimeER1: Transformation of temporal aspect XX of entity E into class E_XX, adding two inverse object properties: *EHasXX* with domain set to class E and range set to class E_XX, *XXOfE* with domain set to class E_XX and range set to class E. In addition, specified functional for *XXOfE* and min cardinality set to one. Set of key attributes of class E_XX include *XXOfE* and some attributes represent temporal constraint depends on temporal aspect XX as in Table 4. Figure 6 illustrates Rule TimeER1.

4.2.2 Transformation of Temporal Attributes of an Entity Type

Rule TimeER2. For each temporal attribute A with temporal aspect XX of entity E, create a class A_XX, add two inverse object properties: *AHasXX* with domain set to class E and range set to class A_XX, *XXOfA* with domain set to class A_XX and range set to class E. Simultaneously, set functional characteristics for *XXOfA* and min cardinality set to one. Transform attribute A into a datatype property of class A_XX by applying those rules EER6, EER7 and EER8. Set of key attributes of class A_XX includes *XXOfA* and some attributes represent temporal constraint depends on temporal aspect XX as in Table 4. Figure 7 illustrates Rule TimeER2.

4.2.3 Transformation of Temporal Relationship

Rule TimeER3. For each attribute R between entities Ei with their temporal XX, create a class R. Corresponding to each entity E in the relationship R, create two inverse object properties to show the relationship between class R and E: *EhasR*

Temporal aspect	Key attribute
VT	hasVTs
LS	hasLSs
TT	hasTTs
LT	hasLSs, hasLSe, hasTTs
BT	hasVTs, hasVTe, hasTTs

Table 4 The key attribute corresponds with the temporal

Fig. 6 Transformation of temporal entity type

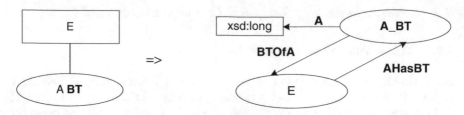

Fig. 7 Transformation of temporal attribute of entity type

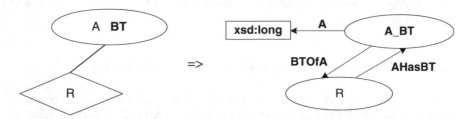

Fig. 8 Transformation of temporal attribute of relationship. *Note* If R is the n-ary relationship having functional constraint, set of key attributes will be removed the object property whose domain was set to class which corresponding to right entity in functional constraint

with domain set to class E and range set to class R, *RofE* with domain set to class R, and range set to class E, set functional characteristics for *XXOfA* and min cardinality set to one. In addition, for each min/max cardinality according to the time on entity type E different 0 and n, add min/max cardinality constraint corresponding to property *EHasR*. The non-temporal attribute of relationship R will convert to datatype property of class R similar to rule EER6, EER7 and EER8. If relationship R is binary or one-to-one recursive, key of class R include one of the two object properties has just added with domain set to class which corresponding to entity type in relationships type R and some properties which show temporal constraint, depending on temporal type XX showed in Table 4. Conversely, key of class R include object properties just added with domain set to class corresponding with entity which is in side n of relationship and some properties which show temporal constraint, depending on temporal type XX showed in Table 4. Figure 8 illustrates Rule TimeER3.

4.2.4 Transformation of Temporal Attribute of Relationship

Rule TimeER4: Transformation of attribute A has temporal aspect XX of relationship R into class A_XX, add two inverse object properties showing relationship of class R and A_XX, and min cardinality set to one: *AhasXX* with domain set to class R and range set to class A_XX, *XXOfA* with domain set to class A_XX and range set to class R, in addition *XXOfA* is function object property. Transform attribute A to datatype proprties of class A_XX by applying the rules EER6, EER7 and EER8, and key of class A_XX contains property *XXOfA* and some properties show temporal constraint depending on temporal type as Table 4.

5 Conclusion

In this paper we have proposed a method of converting the TimeER model to OWL ontology. This method is collected and improved from transformation methods ER and EER model into OWL ontology before, by adding key constraint for result class. And, this method has proposed class and properties for showing the temporal aspect corresponding on TimeER model when converting the components which relate to temporal aspect on TimeER model to OWL ontology.

The ontology result from the transformation is data conservation in corresponding database of TimeER model. The ontology result is OWL 2, which is considered as an extension of OWL DL, so that to ensure deciding ability. We successfully setup our rules for transforming TimeER model into OWL ontology. OWL results is compatible with the Protégé and HermiT 1.3.7 reasoner. This somehow proves the correctness of transformation method that we proposed in Sect. 4.

References

1. Motik, B., Patel-Schneider, P.F, Paria, B., Bock, C., Fokoue, A., Haase, P., Hoekstra, R., Horrocks, I., Ruttenberg, A., Sattler, U., et al.: OWL 2 Web Ontology Language: Structural Specification and Functional-Style Syn-tax. W3C Recommendation, 27, (2009)
2. Jensen, C.S.: Temporal Database Management, Dr.techn. Thesis, Aalborg University, http://www.cs.auc.dk/~csj/Thesis/
3. Fahad, M.: ER2OWL: generating OWL ontology from ER diagram. Intell. Inf. Process. IV, IFIP Int. Fed. Inf. Process. **288**, 28–37 (2008)
4. Myroshnichenko, M.S.I., Murphy, M.C.: Mapping ER schemas to OWL ontologies. In: IEEE International Conference on Semantic Computing, ICSC'09, pp. 324–329 (2009)
5. Chujai, P., Kerdprasop, N., Kerdprasop, K.: On transforming the ER model to ontology using protégé OWL tool. Int. J. Comput. Theory Eng. **6**(6), 484–489 (2014)
6. Upadhyaya, S.R., Kumar, P.S.: ERONTO: a tool for extracting ontologies from extended E/R diagrams. In: Proceedings of the 2005 ACM Symposium on Applied Computing, ACM, pp. 666–670 (2005)
7. Elmasri, N.: Fundamentals of Database Systems, 6th edn. Addison-Wesley Publishers, United States of America (2011)

Peer to Peer Social Network for Disaster Recovery

Duy Tai Nguyen, Pham Tran Vu and Quang Tran Minh

Abstract Keeping people connected even in a severe condition when main parts of communication infrastructures are destroyed by disasters is essential to loss mitigation and emergency relief. It is hard, however, to quickly recover communication infrastructures due to many difficulties on available resources such as time, equipment, man-power and so forth. This paper proposes a practical solution thereby victims in the disaster areas can easily connect with each other to share their safety status via the means of a social network, namely the **peer to peer social network for disaster recovery** (P2PSNDR). The P2PSNDR is designed so that it can feasibly run on top of a mobile multihop ad hoc network established on demand utilizing the beacon stuffing mechanism. This approach does not require additional hardware such as network interface cards (NICs). Instead, it leverages the available WiFi NIC on the mobile devices to listen the data embedded in the beacon frames sent by the neighbor nodes. As nodes can deal with the received messages by appropriately forwarding messages to the intended destination, multihop communication is established extending the communication coverage. The feasibility of the proposed network has been validated via simulations with various scenarios. The results reveal that the network can work properly with maximum 250 nodes which is large enough for common disaster recovery situations.

D.T. Nguyen (✉) · P.T. Vu · Q.T. Minh
Ho Chi Minh City University of Technology, Ho Chi Minh City, Vietnam
e-mail: duytai.cse@gmail.com

P.T. Vu
e-mail: ptvu@hcmut.edu.vn

Q.T. Minh
e-mail: quangtran@hcmut.edu.vn

© Springer International Publishing Switzerland 2016
D. Król et al. (eds.), *Recent Developments in Intelligent Information
and Database Systems*, Studies in Computational Intelligence 642,
DOI 10.1007/978-3-319-31277-4_11

1 Introduction

Disaster may occur in any circumstance (man-made or natural) causing loss of life while destroying infrastructures. In many situations, it is impossible to avoid disaster, specifically natural disasters. However we can diminish serious consequences caused by disasters by preparing better response and recovery plans. One of the most important plans is to quickly provide communication means for disaster victims to help them share their status or call for helps to nearby peers including rescue staffs. As mentioned, disaster may drastically destroy telecommunication infrastructure isolating victims in disaster area. This makes the situation more serious with more difficulties for emergency relief. Meanwhile, recovery of the network infrastructure takes a long time requiring a huge cost and man-power which are not always available at the disaster areas.

However, as users almost always carry a mobile device such as a laptop or a mobile phone, they could use these devices to connect to neighbors using the built-in WiFi interfaces. In turn, the connected neighbors may continue to extend the network topology by connecting to the further neighbors. Consequently, a mobile ad hoc network (MANET) is established allowing victims to share their safety information to further people. Furthermore, if a device in this connected network has the Internet connectivity, it can spread its Internet connectivity to the rest of network by acting as a Internet gateway (IG). Eventually, rescue team collects enough information in order to make correct decisions.

One of the difficulties in establishing the mobile multihop network mentioned above is that ordinary users could not manually configure networks as they are commonly non-technical users. They need an user-friendly application to communicate with other victims. This paper proposes a peer to peer social network for disaster recovery (P2PSNDR) solution to provide an easy means of network configuration and management to disaster victims. This approach leverages the ideas came from STARs [1] and PeerSON [2] while adding to specific constraints for disaster recovery applications. This paper also proposes basic theory of simple peer to peer social network (SP2PSN) and describes the extended beacon stuffing (EBS) model used for network establishment. This solution helps to overcome the bottleneck issues at the root node on the tree-based approaches such as DRANS [3], [4], NodeJoints [5] or even MA-Fi [6]. The feasibility of the proposed approach is analyzed using NS3.

The rest of the paper is organized as follows. Section 2 thoroughly analyzes notable related papers which deal with identical network scenarios. Section 3 describes the P2PSNDR and the EBS model. Section 4 verifies the feasibility of the proposed approach. Section 5 concludes our work and draws out the future directions.

2   Related Work

To the best of our knowledge, there are few works that mainly focus on social network for ad hoc systems/environments. STARS [1] allows users exchanging their information in the same star topology (i.e., network with a single hop is established, multihop communication is not supported). It also provides several properties of decentralized social network. The STARS has been implemented as a mobile application and experimented in the real world. Unlikely, the PeerSON [2] provides a solution for saving the data communication cost across many peers by using distributed hash table (DHT) as a look up service. It concentrates on privacy and security problems. Both of these approaches are conducted under an important assumption that underlying physical connectivity works smoothly. This constraint is broken in disaster scenarios.

Considering a severe condition in disaster environments where the main parts of communication infrastructures may have been heavily damaged, DRANS [3], [4], NodeJoints [5], and MA-Fi [6] attempted to establish physical connectivity between many nodes utilizing multihop communication technologies. They virtualized a single wireless network interface card (WNIC) to extend the star topology to tree-based topology. DRANs addressed de facto standard requirements for disaster recovery networks and proposed their own network model named wireless multihop access networks (WMANV) for WiFi based multihop ad hoc networks. It achieved a speed of 1.8 Mbps in several real life experiments. In contrast, NodeJoints focused on routing protocol and designed network architecture with tests on 10 laptops (10 hops for the best cases). However, similar to DRANS, it suffered from the topology changes. In Ma-Fi, router nodes (RONs) create the back-bone of ad hoc network while station nodes (STANs) connects to them. Ma-Fi's throughput is comparable with an infrastructure network.

As described none of the existing methods mentioned above can form a multihop MANET with minimal cost for network establishment and management for disaster recovery applications. In order to overcome this issue, we extend the beacon stuffing [7] idea to achieve MANET topology by leveraging the control beacon for carrying the necessary messages even in the phase of network establishment. Beacon stuffing was firstly introduced by Microsoft Research Lab for the original purpose of spreading advertisement messages such as coupons for a discount campaign, Wi-Fi advertisements, etc. In this work, readable data or information-carried messages are embedded to beacon frames, thus the nearby nodes can read the data without association while the network is being established. Obviously, with this design the a social network can run on top of the on demand multihop ad hoc networks established based on the EBS model. This paper combines social network and connectivity formulation into one unified system, the P2PSNDR. However, as discussed before establishing connectivity for multihop communications in severe environments as in disasters is challenging problem, this paper mainly focuses on resolving this issue utilizing EBS model.

3   Network Establishment for SP2PSN

3.1   Simple Peer to Peer Social Network (SP2PSN)

Firstly, it is believed that all nodes in disaster areas will provide trusted information. Therefore, identifying node is purely based on information node provided. There is no need to add a central server to validate information provided by participating nodes. The next inferences will be occurred in context of all nodes belonged to a MANET and no node had internet connectivity.

In SP2PSN, data is stored locally. When a node joins the SP2PSN, it will broadcast its profile, while other nodes conduct the process of profile identification. The profile includes: {*MAC address, full name, extra fields*}. *MAC address* is used to avoid profile duplication. Nevertheless, with strange *MAC addresses*, people do not know exactly who they are. Thus, the profile should also contain *phone number*, *full name* and some *extra fields* such as: *age, gender, job*, etc. This extension is provided to the user community as an option when they use the proposed system. The more information is provided in the profile the more probability node is identified by other nodes.

After the identification process occurred, every node searches in the database to determine relationship, which is *follower* or *stranger*. It is noticed that the relationship in SP2PSN is one-sided. Initially, A and B are *stranger*. When A recognizes B, it may send a following request to B to be a B's *follower*. It is quite different from popular social networks such as Facebook, Twitter, etc. Because physical connection of B is not always available and B's data is stored locally. Instead of A pulling messages away from B, B actively decides to send messages to A or not.

Messages are sent in three modes: *public, private* and *following*. In the *public* mode, messages belonged to a specific node are broadcasted to all other nodes in the SP2PSN. In the *private* mode, messages are unicasted to a particular node while in the *following* mode this node multicasts messages to a group of followers. However, it is impossible to build a physical connection between two nodes which are located in different networks. Instead of sending messages immediately, the sender caches data. After that, if the Internet connectivity is available, it will send the cached data to a specific global server (a specific server in the Internet). On the other hand, a receiver possibly pulls messages at any time later.

In order to extend the SP2PSN in the daily life, we add *phone number* to the profile. If user switches to another device, global server conducts following steps to identify user. The global server sends one-time key via telecommunication network, then user provides correct key to the server, then the global server accepts key and returns the cached data back. In addition, if the cached data is profile then user can continue using it to get other cached data.

As mentioned, SP2PSN is built on top of the mobile ad hoc network, the next section will present our approach on utilizing the EBS model to build an on demand MANET.

3.2 Extended Beacon Stuffing Model (EBS)

As aforementioned, the network is on-demand established. That means nodes will discover, negotiate to form the network. A node is free to choose its neighbors. A neighbor acts as a router that routes packets to the destination. If there is no node which has Internet connectivity, then the established network of a group of nodes is named isolated network (IN).

Figure 1 shows that there are three separate disaster areas: Area 1, Area 2 and Area 3. The dotted straight line is symbolic of wireless signal, the cloud is a black box which connects stations to the global server and be always alive. If any two devices directly connected, it is represented by 2 straight lines. In Area 3, the infrastructure is completely destroyed leading to form an IN, node {K, L, M, N, O} can not connect to the outside world. On the other hand, in Area 2, node G connects to a radio station, it unconsciously contributes the Internet connectivity to other nodes forming a zone network (ZN), it happened the same in Area 1. In order to link multiple ZNs together, a global server is added to route packets across multiple ZNs and store data of social network. Figure 1 shows all elements of possible networks in a disaster area. Every device located in the same ZN or across multiple ZNs can communicate with each other such as: E communicates with H by using path {C, A, radio station, global server, radio station, G, H}. Additionally,

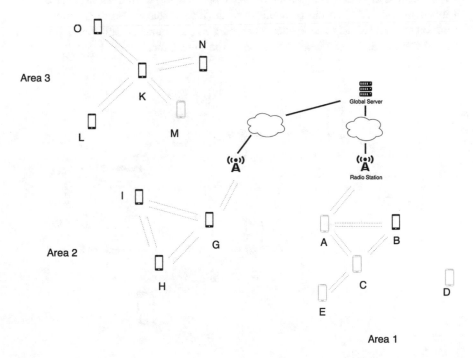

Fig. 1 All network elements

conversation in the IN is cached and is likely to sync with the global server when Internet connectivity is available such as:{*conversation between L and N is cached in L. After that, if L has internet connectivity, it is pushed to the global server and be available for synchronization process from N*}.

As the IN is the most severe situation in terms of the damage level on communication infrastructure when a disaster occurs, this paper focuses on solving the network establishment and manage for message communications in the IN. As mentioned before mechanism on beacon stuffing can be utilized to embed data messages while establishing the network. However, the original beacon stuffing uses one-hop routing protocol which is not directly applicable to build a multihop ad hoc network required in this work. We propose the means of extending the beacon stuffing model by adding AODV [8] algorithm, namely the EBS model, to route beacon frames through many nodes depicted in Fig. 2.

Figure 2 shows that all nodes were added the routing algorithm for message forwarding. When node A sends a beacon frame X, node D may receive and forwards it to C and C drops it based on C's routing table. Meanwhile {B, E} act the same as D. The EBS also naturally solves two difficult problems: *MANET connection establishment, dynamic IP allocation* [9]. When a node sends out a beacon frame, all in-range nodes can read without association, decryption key. In the other word, all possible connections are established or the MANET connection is constructed. In addition, beacon frame runs on layer 2 of the OSI model and MAC address is global unique. Therefore, EBS uses MAC address to route beacon frames instead of IP. Another advantage of the EBS is that it also runs concurrently with other MAC 802.11 mode such as ad hoc, AP, station. For example, a node can surf Facebook and route beacon frames at the same time.

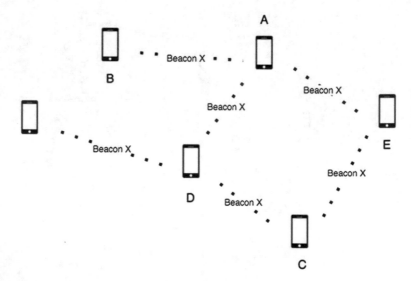

Fig. 2 Route beacon frame through multiple hops

3.3 Routing for Beacon Frame

There is a large number of works attempted to solve routing problem in MANET. AODV is a famous one among them. It provides several approaches in which we can customize in a specific scenario to be fit with the EBS. According to the EBS, AODV should be converted to work with beacon frame. Converting process must consider the beacon frame structure depicted in Fig. 3.

Figure 3 describes that a beacon frame includes one frame body reaching a maximum of 2312 bytes. The frame body contains a group of Information Elements (IEs) and an IE is formed by three parts: *Element ID* (1 byte), *Length* (1 byte), *Information* (variable length, maximum of 255 bytes).

As proposed in [7], there are three fields that data can be embed in: *SSID, BSSID, IE*. The *SSID* and *BSSID* allow embedding only 32 and 6 bytes respectively which are too small to be beneficial for data carrying. However, every beacon frame has frame body of maximum 2312 bytes, every IE contains 255 bytes in which 1 byte for IE ID, 1 byte for IE header, and the rest for IE body. The maximum number of IEs which belong to one beacon frame is 9. This means one beacon frame allows transporting up to $9 \times 253 = 2277$ bytes revealing that *IE* is the best component for data carrying.

If a node sends 1 beacon every 10 ms, during 1 s it sends about 100 beacons transporting $100 * 2277$ bytes $= 222.7$ kbytes. Therefore, maximum speed of EBS is about 223 kbytes/s. It could be a promising speed.

Packets in AODV belongs to one of the three basic types: *Routing Request (RREQ)*, *Routing Reply (RREP)* and *Data Packet (DP)*. All of them are filled one by one in the IE body of beacon frame. It could be across multiple IEs, multiple beacon frames if data is large. A beacon frame is treated as a fragment. RREQ, RREP are fit in a beacon frame. As DP is variable in length, DP could be broken into multiple fragments and be reassembled at the destination.

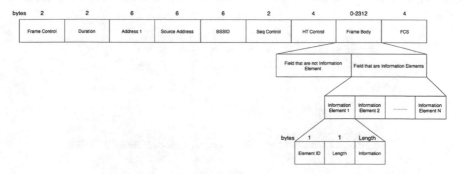

Fig. 3 Beacon frame structure

4 Evaluation

In this section, we evaluate the feasibility of P2PSNDR using basic chatting function via simulations on NS3. To conduct a suitable experiment which represents the feasibility of the proposed approach on real world application, we must design a suitable network scenarios. However, there are a large number of important factors which affect the result of simulation such as two-node distance, number of nodes, number of beacon frames, size of packet, etc. Therefore, we conduct a preliminary simulation to find suitable values for these factors. We start with a small number of nodes with a short range between the two nodes and then increase the number of nodes and the distance, respectively. The destination and source nodes (many nodes concurrently serve as source nodes) are randomly selected from all nodes in the network. Every beacon frame contains 1 kbyte data and each source node sends up to 500 beacons in a simulation. The result is described as in Fig. 4.

In Fig. 4, the X axis shows the number of nodes and the Y axis shows the average rate of missing of beacon frames. Two-node distance increases from 5, 10, 25, 50, up to 100 m. The result shows that when increasing number of nodes or two-node distance, missing percentage is intent to increase gradually, except in 100 m. Overall, the missing percentage is unstable because of randomly starting sending beacon frame and unexpected collision in the air. However, it seems to be stable when two-node distance is 10 or 5 m, where the missing rates are always less than 10 %.

After thoroughly analyzing these results, we design new appropriate network scenarios to evaluate the feasibility of P2PSNDR. Every node is randomly assigned to a particular position in a limited square area. As focusing on validating the feasibility of P2PSNDR using basic chatting function, packet size is set to 64 bytes which is similar to the length of daily chat messages. Only 30 % nodes are assigned as active nodes (source or destination) among which the destination is randomly selected while the rest are sources (it should be noted that the other 70 % of nodes

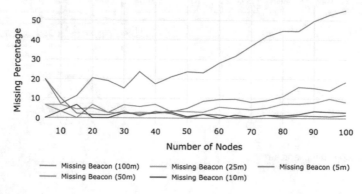

Fig. 4 Missing of beacon frames via multihop communication under the proposed EBS models

Table 1 The simulation result for five network scenarios

	Number of nodes	Sending packets	ARR(%)
Scenario 1	50	7500	98.44
Scenario 2	100	15,000	99.983
Scenario 3	150	22,500	96.35
Scenario 4	200	30,000	93.715
Scenario 5	250	37,500	96.357

serve as forwarding nodes). Every approximately 5 s, every active node serving as a source node sends a data beacon. This time interval could be similar to a period of time that people wait for chatting messages. One node will send 500 beacon frames in total. The size of the network will be increased by 50 nodes in every simulation. In total, we have conducted five network scenarios for this evaluation, namely networks with 50 nodes, 100 nodes, 150 nodes, 200 nodes, and 250 nodes.

The result of simulation is measured by **Average Receiving Rate** (ARR), which describes the percentage of packets from source nodes that reach correct destinations. Simulation result is showed in Table 1.

The sending packets in every simulation is a number of beacon frames which are sent by source nodes. Which means Table 1 does not include beacon frames forwarded by median nodes. Table 1 shows that all network scenarios accept an ARR > 93 %. When increasing the network size, beacon frame collision occurs more frequently, it decreases ARR gradually. In addition, every node starts sending beacons at different times in different simulations. Therefore, the results fluctuate. However, the results show that it is a little bit better than simulation result of AODV because every node in P2PSNDR does not move. It avoids the re-routing cost. Beacon interval is 5 s, which is a good number to avoid beacon frame collision in the air. The results are also reasonable for chatting function on P2PSNDR. It also proves that P2PSNDR is feasible in real world, specifically for disaster recovery.

5 Conclusion and Future Work

This work proposed an extension model for the beacon stuffing, the EBS, to efficiently establish an ad hoc network supporting for social network functions used for disaster recovery. It is completely possible to implement in mobile devices such as Android mobile phones by modifying Linux kernel. It could be written as built-in kernel module to parse, generate, and route beacon frames to exchange packets on the P2PSNDR. The simulation result shows that chatting function works well on the proposed networks. It is absolutely possible to extend this approach for more useful functions on P2PSNDR such as sharing images or files, conducting video calls, etc.

However, *beacon frame collision (BFC)*, *reliable connection (RC)* on the proposed EBS model are still challenging issues. In the future work, we should pay more focuses on resolving these issues. In addition, further evaluations should be conducted to confirm not only the feasibility but also the effectiveness, energy consumption, overhead, etc., of the proposed solution before implementing this approach on the real mobile devices.

Acknowledgment This research is funded in part by Ho Chi Minh City University of Technology under grant number TSDH-2015-KHMT-57 (FY. 2015–2016).

References

1. Long, T.Q., Pham, T.V.: STARS: Ad-hoc peer-to-peer online social network. In: 4th International Conference on Computational Collective Intelligence, pp. 385–394 (2012)
2. Buchegger, S., Schioberg, D., Vu, L., Datta, A.: PeerSoN: P2P social networking—early experiences and insights. In: Proceedings of the Second ACM Workshop on Social Network Systems Social Network Systems 2009, co-located with Eurosys 2009, pp. 46–52 (2009)
3. Quang, T.M., Kien, N., Yamada, S.: DRANs: resilient disaster recovery access networks. In: IEEE 37th Annual Conference on Computer Software and Applications (COMPSAC), pp. 754–759 (2013)
4. Quang, T.M., Kien, N., Cristian, B., Yamada, S.: On-the-fly establishment of multihop wireless access networks for disaster recovery. IEEE Commun. Mag. **52**(10), 60–66 (2014)
5. Sarshar, M.H., Poo, K.H., Abdurrazaq, I.A.: NodesJoints: a framework for tree-based MANET in IEEE 802.11 infrastructure mode. In: IEEE Symposium on Computers and Informatics (ISCI), pp. 190–195 (2013)
6. Hanno, W., Tobias, H., Robert, B., Klaus, W.: Establishing mobile ad-hoc networks in 802.11 infrastructure mode. In: Proceedings of the 6th ACM international workshop on Wireless Network Testbeds, Experimental Evaluation and Characterization, pp. 89–90 (2011)
7. Chandra, R., Padhye, J., Ravindranath, L., Wolman, A.: Beacon-stuffing: Wi-Fi without associations. In: Eighth IEEE Workshop on Mobile Computing Systems and Applications, HotMobile, pp. 53–57 (2007)
8. Perkins, C.E., Royer, E.M., Chakeres, I.D.: Ad hoc on-demand distance vector (AODV) routing. In: Second IEEE Workshop on Mobile Computing Systems and Applications, WMCSA '99, pp. 90–100 (1999)
9. Mohsin, M., Prakash, R.: IP address assignment in a mobile ad hoc network. Proc. MILCOM **2002**, 856–861 (2002)

Impact of Information System-Based Customer Relationship Management Strategy, Social Media Interaction on Customer Perception, Satisfaction, Loyalty, Corporate Image and Purchase Intention to XYZ Cellular Customers in Samarinda City

Wick Indiyarto and Harisno

Abstract This study aims to investigate the impact of information system-based Customer Relationship Management (CRM) strategy, social media interaction on customer point of view from XYZ cellular customers in Samarinda city, East Kalimantan, Indonesia. This study used Structural Equation Modeling-Partial Least Square (SEM-PLS) with two-step approach for second order confirmatory factor analysis from 246 customers as respondents. This paper found the direct and indirect impact of information system-based CRM strategy on variables related with customers such as perceived value, perceived service quality, customer satisfaction, customer loyalty, corporate image and purchase intention.

Keywords Customer relationship management · Information systems · Purchase intention · Structural equation modeling-partial least squares

1 Introduction

Despite many companies have implemented Customer Relationship Management (CRM) to maintain existing relationship with their customer, but they fail to develop a comprehensive strategy because they view CRM as a technological solution and ignore the business issues [1]. Moreover due to high cost of acquiring

W. Indiyarto (✉) · Harisno
Master of Information System Management, Binus Graduate Programs,
Bina Nusantara University, Jl. Kebon Jeruk Raya no. 27, Jakarta 11530, Indonesia
e-mail: wick.indiyarto@outlook.com; wickind@gmail.com

Harisno
e-mail: harisno@binus.edu

© Springer International Publishing Switzerland 2016
D. Król et al. (eds.), *Recent Developments in Intelligent Information and Database Systems*, Studies in Computational Intelligence 642,
DOI 10.1007/978-3-319-31277-4_12

new customers, companies are also giving efforts to maintain existing customers, especially the most profitable customers [1]. In order to do that, companies should have strategy or capability to be able the tendency of customer interest, the necessary information which is processed to identify trends to understand customer need, manage customer relationship and maintain customer loyalty from the implementation of Customer Relationship Management (CRM). According to Buttle [2], CRM is a business strategy integrates core internal processes, functions and external networks to create and deliver value to customers with the goal of achieving a profit [2]. Information system—based CRM provide continuous improvement in customer relations in hotel organisations [3]. From the information systems perspective, Business analytics (BA), business intelligence (BI) and especially on how to leverage big data from social media, network, mobile and web data are dynamic, fascinating and highly relevant field of research need to be discovered [4]. PT. XYZ aims the implementation of CRM to provide customer insights for better decision making, faster decisions, and has a high business performance, the company has been using big data and big data analytics [5]. Data analytics refer to the Business Intelligence and Analytics (BI&A) technologies with data mining techniques and statistical analysis [6, 7]. Business Intelligence (BI) is an information system, BI can be interpreted as a decision support, executive information systems, and management information systems management [8, 9]. This paper has contribution to investigate the effect of implementation information system-based CRM strategy which is measured from customer point of views such as perceived value and service quality, satisfaction, loyalty, corporate image and purchase intention in PT. XYZ cellular customers.

2   Research Background

The lack of knowledge about the impact of information system-based CRM in customer point of view in XYZ cellular customers in Samarinda city give rise to 4 research questions below:

1. Are there positive and significant influence between information system-based CRM on customer perceived value, perceived service quality, corporate image, customer satisfaction, and customer loyalty from XYZ cellular customers in Samarinda city?
2. Are there any positive and significant correlation between customer satisfaction and customer loyalty towards purchase intention from XYZ cellular customers in Samarinda city?
3. Is there any positive and significant influence between social media interaction on purchase intention from XYZ cellular customer in Samarinda city?
4. What factors that directly or indirectly affect customer satisfaction, customer loyalty, and purchase intention to XYZ cellular customer in Samarinda city?

2.1 Previous Study

Develop from relevant literature [10–12, 13]. Table 1 shows previous research framework used and developed in this study.

Information systems-based CRM strategy as exogenous variables in this study using Ngai et al. [11] classification framework for data mining techniques in CRM consist of 4 sub dimensions: customer retention, customer identification, customer development, customer attraction. Another exogenous variable was social media

Table 1 Previous research

Study	Research title	Research methodology	Result
Bayol et al. [10]	Use of PLS path modelling to estimate the European consumer satisfaction index (ECSI) model	Using Lohmöller's program LVPLS 1.8 to estimate ECSI	Presenting ECSI models with PLS approach used to estimate model parameters
Ngai et al. [11]	Application of data mining techniques in customer relationship management: a literature review and classification	Identified 24 journal and 87 papers from 2000 to 2006	A literature review and a comprehensive classification scheme for data mining techniques for CRM consisting of customer identification customer attraction, customer retention and customer development
Scheepers et al. [12]	The dependent variable in social media use	Using SEM-PLS method, for measurement model using confirmatory factor analysis and structural analysis models using SmartPLS	Findings of this study indicate that dependent variable of a sense of community reflects four sub construct that can identify the behavior of social media users is information seeking, hedonic activities, sustaining extending of strong ties and weak ties
Kurniati et al. [13]	The effect of customer relationship marketing (CRM) and service quality to corporate image, value, customer satisfaction, and customer loyalty	Using descriptive and GSCA (generalized structured component analysis)	CRM and quality of service can directly enhance the corporate image, value, customer satisfaction and loyalty, while the image of the company does not directly affect customer satisfaction and loyalty, but increase customer satisfaction and loyalty through value

interaction [12], while endogenous variables in this study were: corporate image [10], perceived value [14, 15] perceived service quality [16], customer satisfaction [10, 17, 18] customer loyalty [10, 19] and purchase intention [20, 21].

2.2 Research Hypothesis

The research hypothesis and model can be illustrated in Fig. 1. The four dimensions of CRM consist of:

i. Customer Retention: The purpose of customer retention is customer loyalty where customer satisfaction is a very important condition for retaining customers or make loyal customers. Customer retention is central concern of CRM [11].

ii. Customer Identification: In the identification phase, determined the target populations to determine which customers are most profitable [11], moreover customer identification provides information on the characteristics of profitable customers and analyze customer lost and how to get back to the customer [22].

iii. Customer Development: The goal of development is expand its customer consistently to the intensity of the transaction, transaction value and aims to enhance customer benefits [11].

iv. Customer Attraction: After identifying of potential customers segment, the company can directly approach the customer towards customers segment [11].

v. Hypothesis 1: information system based-CRM has positive relationship and significantly effect to customer perceived value.

vi. Hypothesis 2: information system based-CRM has positive relationship and significantly effect to customer perceived service quality.

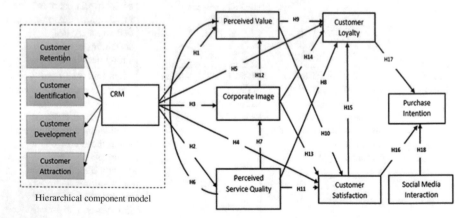

Fig. 1 Research hypothesis, path model with hierarchical component structure → : direct effect

vii. Hypothesis 3: information system based-CRM has positive relationship and significantly effect to corporate image.

viii. Hypothesis 4: information system based-CRM has positive relationship and significantly effect to customer satisfaction.

ix. Hypothesis 5: information system based-CRM has positive relationship and significantly effect to customer loyalty.

x. Hypothesis 6: Customer perceived service quality has positive relationship and significantly effect to customer perceived value.

xi. Hypothesis 7: Customer perceived service quality has positive relationship and significantly effect to corporate image.

xii. Hypothesis 8: Customer perceived service quality has positive relationship and significantly effect to customer loyalty.

xiii. Hypothesis 9: Customer perceived value has positive relationship and significantly effect to customer loyalty.

xiv. Hypothesis 10: Customer perceived value has positive relationship and significantly effect to customer satisfaction.

xv. Hypothesis 11: Customer perceived service quality has positive relationship and significantly effect to customer satisfaction.

xvi. Hypothesis 12: Corporate image has positive relationship and significantly effect to customer perceived value.

xvii. Hypothesis 13: Corporate image has positive relationship and significantly effect to customer satisfaction.

xviii. Hypothesis 14: Corporate image has positive relationship and significantly effect to customer loyalty.

xix. Hypothesis 15: Customer satisfaction has positive relationship and significantly effect to customer loyalty.

xx. Hypothesis 16: Customer satisfaction has positive relationship and significantly effect to purchase intention.

xxi. Hypothesis 17: Customer loyalty has positive relationship and significantly effect to purchase intention.

xxii. Hypothesis 18: Customer interaction social media has positive relationship and significantly effect to purchase intention.

3 Research Methodology

3.1 Population and Sample

This study used quantitative approach with purposive sampling technique because the respondent must meet the following requirements:

1. As a cellular customer of PT. XYZ.
2. Based in Samarinda city, East Kalimantan, Indonesia.

Sample size was determined using Ferdinand Formula as much as 5–10 times the number of parameters to be estimated [23]. In this study the questionnaires contain 44 observed indicators, so the minimum number of samples required is 220 respondents. The questionnaires were distributed to 250 respondents but only 246 were completed, so total samples in this study were 246 respondents and considered sufficient.

3.2 Respondents Profile

Overall, customer frequency by gender was composed of 48.37 % male and 51.63 % female. Respondent's education was dominated by the respondents who studying in high school 78.05 % followed by respondents who studying or finished bachelor degree and university 18.70 %. In terms of age majority of respondents were between 21 and 30 years old 55.28 %, followed by respondents were under 20 years old 40.24 %. In terms of respondent job, most of respondents was students 73.17 % followed by respondents who had job as worker 18.29 %. The majority of respondents declared that they got online promotion 3–5 times a month 36.18 %, followed by respondents got online promotion 6-8 times a month 15.85 % and they has become telecom operator customer 6–10 years 56.50 %, followed by respondents were less than 5 years 27.24 %.

3.3 Data Techniques Analysis

The Structural Equation Modeling—Partial Least Square (PLS-SEM) software was used to analyze the statistical data. Method for analyzing CRM strategy using second order confirmatory factor analysis with two stage approach using SmartPLS 3.0 Software. In the two-stage approach used the following steps [24, 25]:

1. Draw the construct and its indicators in the Low Order Constructs (LOC), then run PLS Algorithm to get scores of latent variables.
2. Input latent variable scores as indicators of higher-order constructs (HOC) and other constructs in the model and then be evaluated structural models.

Figure 2 in the step 1, repeated indicator approach is used to obtain the latent variable scores for the LOCs: customer retention, customer identification, customer development and customer attraction connected to CRM. After running PLS Algorithm, we obtained the latent variable scores result.

The evaluation result of first stage of CRM dimension measurement are as follows:

1. Convergent validity (loading factor): the result confirm all indicators has loading factor values 0.60 or higher, so all indicators were valid or already met the

Fig. 2 Step 1 of two stage approach on second order confirmatory factor analysis of this study

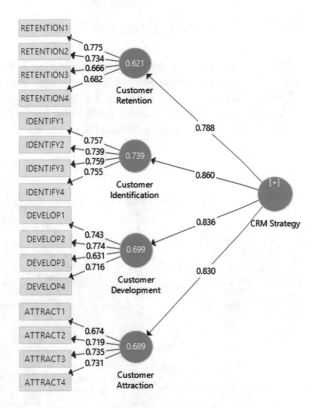

convergent validity. According to Hulland in exploratory studies loading factor value 0.4 are acceptable [26].

2. Discriminant validity (cross loading): The most conservative criterion recommended to evaluate discriminant validity is the Fornell and Larcker criterion. The method compares each construct's AVE value with the squared interconstruct correlation (a measure of shared variance) of that construct with all other constructs in the structural model [27]. The result confirm correlation value of each construct indicator is higher than construct correlation with other indicators, it proves that predicting latent constructs indicators on their block are better than the indicator in others block.

3. Reliability (composite reliability and Cronbach Alpha): an acceptable composite reliability is 0.70 or higher [24, 28] while an acceptable Cronbach Alpha is 0.60 or higher [29]. The result confirm all constructs expressed reliable.

In step 2, we took latent variable scores from the result of step 1, then input latent variable scores as indicators of higher-order construct (HOC): CRM and drawed also the other constructs (perceived value, perceived service quality, corporate image, customer satisfaction, customer loyalty, and purchase intention) in the model. Then result of step 2 of Two Stage Approach on Second Order CFA of this study:

3.4 Analysis of Measurement Model

The evaluation for measurement model from Fig. 3, to measure the validity of the outer model can be obtained from convergent validity which the value of loading factor is 0.70 or higher [28]. In the second stage evaluation, SERVQ2 and VALUE4 indicators have loading factor value below 0.70, so indicators: SERVQ2 and VALUE4 were removed in the study then the result become as Fig. 4.

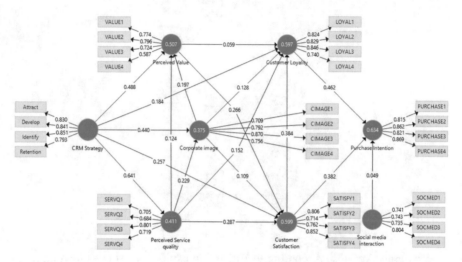

Fig. 3 Step 2 result of two stage approach on second order confirmatory factor analysis

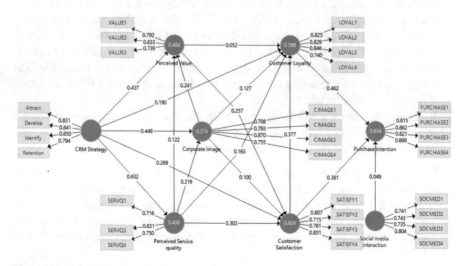

Fig. 4 Result of PLS Algoritm for measurement and structural model

After removing those 2 indicators, the loading Factor for all indicators have met the value of convergent validity (factor loading) >0.70. Then the discriminant validity were obtained from cross loading and $\sqrt{AVE}$ (square root of average variance extracted). AVE value is calculated as the mean of the squared loadings for all indicators associated with a construct [27]. The result of discriminant validity were valid because construct correlation to the indicator itself is higher than the correlation indicators to other constructs and $\sqrt{AVE}$ for each construct is greater than the correlation between the construct and other constructs [24, 28]. Then the result of the constructs were also reliable because AVE score were greater than 0.50 and composite reliability and Cronbach Alpha >0.70 [24, 28].

3.5   Analysis of Structural Model

This evaluation includes an examination of R-square and the significance of the estimated coefficients. Below were the result of R-square and path coefficient (Table 2).

Value of R-squares or adjusted R-square are ≤ 0.7, ≤ 0.50, and ≤ 0.25 that is strong, moderate and weak model [28]. The significant of the hypothesis in this study can be explained in Table 3.

To assess the path coefficients' significance, critical t-values for a two-tailed test are 1.65 (significance level = 10 %), 1.96 (significance level = 5 %), and 2.58 (significance level = 1 %) [28]. Table 3 shows the result for the proposed hypothesis: 16 hypothesis were significant and 2 hypothesis were not significant. Then we can answer some of the research questions in this study:

1. There were positive and significant influence between information system-based CRM on customer perceived value, perceived service quality, corporate image, customer satisfaction, and customer loyalty from XYZ cellular customers in Samarinda city.
2. There were positive and significant correlation between customer satisfaction and customer loyalty towards purchase intention from XYZ cellular customers in Samarinda city.

Table 2 R-square

Variables	R-square	Variance on the model
Corporate image	0.374	Moderate
Customer loyalty	0.598	Strong
Purchase intention	0.634	Strong
Customer satisfaction	0.604	Strong
Perceived service quality	0.400	Moderate
Perceived value	0.484	Strong

Table 3 Path coefficient result of this study

	Original sample (O)	Sample mean (M)	Standard error (STERR)	T statistics (|O/STERR|)	P values	result (P value <0.05 and <0.1)
CRM → corporate image	0.448	0.447	0.072	6.266	0.000	Significant
CRM → customer loyalty	0.190	0.190	0.066	2.885	0.002	Significant
CRM → customer satisfaction	0.269	0.267	0.066	4.078	0.000	Significant
CRM → perceived service quality	0.632	0.635	0.044	14.278	0.000	Significant
CRM → perceived value	0.437	0.434	0.065	6.707	0.000	Significant
Corporate image → customer loyalty	0.127	0.129	0.060	2.105	0.018	Significant
Corporate image → customer satisfaction	0.100	0.100	0.057	1.758	0.040	Significant
Corporate image → perceived value	0.241	0.245	0.062	3.917	0.000	Significant
Customer loyalty → purchase intention	0.462	0.459	0.060	7.721	0.000	Significant
Customer satisfaction → customer loyalty	0.377	0.376	0.069	5.472	0.000	Significant
Customer satisfaction → purchase intention	0.381	0.383	0.059	6.419	0.000	Significant
Perceived service quality → corporate image	0.219	0.219	0.081	2.717	0.003	Significant
Perceived service quality → customer loyalty	0.163	0.163	0.055	2.947	0.002	Significant
Perceived service quality → customer satisfaction	0.302	0.298	0.056	5.356	0.000	Significant
Perceived service quality → perceived value	0.122	0.122	0.075	1.628	0.052	significant
Social media interaction → purchase intention	0.049	0.056	0.044	1.110	0.134	Not significant
Perceived value → customer loyalty	0.052	0.050	0.064	0.815	0.208	Not significant
Perceived value → customer satisfaction	0.257	0.261	0.065	3.971	0.000	Significant

3. There was no positive and insignificant influence between social media interaction on purchase intention from XYZ cellular customer in Samarinda city.

Then result of indirect path values to analyze mediating effect that influence the
hypothesis can be seen in Table 4.

Table 4 Indirect path result of this study

Indirect path	Original sample (O)	Sample mean (M)	Standard error (STERR)	T statistics (|O/STERR|)	P values
CRM → corporate image	0.139	0.139	0.052	2.670	0.004
CRM → customer loyalty	0.471	0.471	0.056	8.380	0.000
CRM → purchase intention	0.567	0.566	0.039	14.717	0.000
CRM → customer satisfaction	0.418	0.420	0.050	8.407	0.000
CRM → perceived service quality					
CRM → perceived value	0.219	0.221	0.052	4.224	0.000
Corporate image → customer loyalty	0.073	0.073	0.027	2.711	0.003
Corporate image → purchase intention	0.154	0.156	0.044	3.541	0.000
Corporate image → customer satisfaction	0.062	0.064	0.024	2.527	0.006
Corporate image → perceived value					
Customer loyalty → purchase intention					
Customer satisfaction → customer loyalty					
Customer satisfaction → purchase intention	0.174	0.172	0.035	4.920	0.000
Perceived service quality → corporate image					
Perceived service quality → customer loyalty	0.176	0.176	0.036	4.882	0.000
Perceived service quality → purchase intention	0.297	0.296	0.043	6.840	0.000
Perceived service quality → customer satisfaction	0.067	0.069	0.029	2.339	0.010

(continued)

Table 4 (continued)

Indirect path	Original sample (O)	Sample mean (M)	Standard error (STERR)	T statistics (|O/STERR|)	P values
Perceived service quality → perceived value	0.053	0.054	0.026	2.068	0.019
Social media interaction → purchase intention					
Perceived value → customer loyalty	0.097	0.098	0.032	3.027	0.001
Perceived value → purchase intention	0.167	0.168	0.047	3.523	0.000
Perceived value → customer satisfaction					

4 Discussion

The result analysis of measurement and structural model can be discussed and explained as below:

1. In this study indicate that the purchase intention of the XYZ products in Samarinda city was influenced very strongly by loyalty (t-statistic 7.721) and customer satisfaction (t-statistic 6.419), which was in line with research Chinomona and Dubihlela [21], where customer loyalty encourage or influence customer purchase intention. While customer interaction in social media did not affect purchase intentions (t-statistic 1.110) which maybe the customers accessed social media to interact with their friends or to find information for private purpose only and not related to mobile products and services provided by PT. XYZ in Samarinda city
2. In this study indicate that customer loyalty was influenced by customer satisfaction (t-statistic 5.472), CRM strategies (t-statistic 2.885), customer perceived service quality (t-statistic 2.947), and corporate image (t-statistic 2.105). While customer perceived value did not affect customer loyalty directly, it was not in accordance with the research from Kurniati et al. [13] because customer perceived value influence indirectly on customer loyalty after the customer get satisfaction in advance on products and services provided by PT. XYZ in Samarinda city.
3. In this study indicate that customer satisfaction was influenced by the customer service quality (t-statistic 5.356), customer perceived value (t-statistic 3.971), corporate image (t-statistic 1.758), and CRM strategy (t-statistic 4.078), which was in line with research from Kurniati et al. [13]. Customer retention, which was part of the CRM aims to maximize the value of customer lifetime value that can create customer satisfaction [13].

4. In this study indicate that corporate image, service quality and customer perceived value was influenced directly by the CRM strategy, which was in line with research Kurniati et al. [13] where CRM is how to increase customer value by using a marketing tool such as communication, trust, commitment and empathy that aims to make customers feel safe, to achieve the company's profitability and improve the corporate image [13]. In this study the strategy of Customer Relationship Management which was formed from 4 sub dimensions of identification, attraction, development and retention of the most powerful influence in shaping the corporate image, customer service quality, perceived value, satisfaction and customer loyalty in sequence were customer identification, attraction, development, and retention.

5 Conclusions

Based on the discussion, then we can conclude that:

1. There were positive and significant influence between information system-based CRM on customer perceived value, perceived service quality, corporate image, customer satisfaction, and customer loyalty from XYZ cellular customers in Samarinda city.
2. There were positive and significant correlation between customer satisfaction and customer loyalty towards purchase intention from XYZ cellular customers in Samarinda city.
3. There was no positive and insignificant influence between social media interaction on purchase intention from XYZ cellular customer in Samarinda city.
4. Customer perceived value, perceived service quality, and corporate image have mediating effect between information system-based CRM to customer satisfaction, customer loyalty, and purchase intention to XYZ cellular customer in Samarinda city.

This study has limitation number of latent variables measured and can be explored from different point of view, it is suggested for the future studies are expected to use different variables such as the addition another exogenous and endogenous to enrich the results of analysis of other factors that can affect corporate image, customer perception, satisfaction, loyalty and purchase intention.

Acknowledgment The Authors gratefully acknowledge to Prof. Dr. Harjanto Prabowo and Prof. Dr. Gerardus Polla for their kind opportunities and support so that this paper can be finished.

References

1. Almquist, E., Bovet, D., Heaton, C.J.: What Have We Learned so Far? Making CRM Make Money—Technology Alone Won't Create Value. Collaborative Customer Relationship Management Taking CRM to the Next Level. Springer, Berlin (2004)
2. Buttle, F.: Customer Relationship Management; Concept and Technologies. Elsevier Ltd, USA (2009)
3. Ku, E.C.S.: The impact of customer relationship management through implementation of information systems. Total Qual Manag Bus Excellence **21**(11), 1085–1102, (2010). doi:10.1080/14783360903250514
4. Marjanovic, O., Ariyachandra, T., Dinter, B.: Introduction to HICCS-47 business analytics, business intelligence and big data minitrack. In: 47th Hawaii International Conference on System Science, p. 3727 (2014)
5. XYZ: Annual Report. PT. XYZ, Jakarta (2014)
6. Chaudhuri, S., Dayal, U., Narasayya, V.: An overview of business intelligence technology. Commun. ACM **54**(8), 88–98 (2011)
7. Chen, H., Chiang, R.H.L., Storey, V.C.: business intelligence and analytics: from big data to big impact. MIS Q. **36**(4), 1-XX (2012)
8. Negash, S.: Business intelligence. Commun. Assoc. Inf. Syst. **13**, 177–195 (2004)
9. Thomsen, E.: BI's promised land. Intell. Enterp. **6**(4), 21–25 (2003)
10. Bayol, M.P., Foye, A.D.l., Tellier, C., Tenenhaus, M.: Use of PLS path modelling to estimate the European consumer satisfaction index (ECSI) model. Statistica Applicata **12**, 361–375 (2000)
11. Ngai, E.W., Xiu, L., Chau, D.C.: Application of data mining techniques in customer relationship management: a literature review and classification. Expert Syst. Appl. **36**, 2952–2602 (2009)
12. Scheepers, H., Stockdale, R., Scheepers, R., Nurdin, N.: The dependent variable in social media use. J. Comput. Inf. Syst. Winter **54**(2), 25–34 ()
13. Kurniati, R.R., Suharyono, H.D., Arifin, Z.: The effect of customer relationship marketing (CRM) and service quality to corporate image, value, customer satisfaction, and customer loyalty. Eur. J. Bus. Manag. 107–119 (2015)
14. Pura, M.: Linking perceived value and loyalty in location-based mobile services. Managing Service Qual. **15**(6), 509 (2005)
15. Zeithaml, V.A.: Consumer perception of price, quality, and value: a means-end model and synthesis of evidence. J. Mark. **52**(3), 2–22 (1988)
16. Parasuraman, A., Zeithaml, V.A., Berry, L.L.: SERVQUAL: a multiple-item scale for measuring consumer perceptions of service quality. J. Retailing 12–40 (1988)
17. Fornell, C., Johnson, M.D., Anderson, E.W., Cha, J., Bryant, B.E.: The American customer satisfaction index: nature, purpose, and findings. J. Mark. **60**(4), 7 (1996)
18. Malik, S.U.: Customer satisfaction, perceived service quality and mediating role of perceived value. Int. J. Mark. Stud. **4**(1), 68–76 (2012)
19. Wang, C.-Y., Wu, L.-W.: Customer loyalty and the role of relationship length. Managing Serv. Qual. **22**(1), 58–74 (2012) doi:10.1108/09604521211198119
20. Rahman, M.S., Haque, M.M., Khan, A.H.: A conceptual study on consumers' purchase intention of broadband services: service quality and experience economy perspective. Int. J. Bus. Manag. 115–129 (2012)
21. Chinomona, R., Dubihlela, D.: Does customer satisfaction lead to customer trust, loyalty and repurchase intention of local store brands? The case of Gauteng Province of South Africa. Mediterr. J. Social Sci. 23–32 (2014)
22. Kracklauer, H., Mills, A., Seifert, D.: Collaborative Customer Relationship Management Taking CRM to the Next Level. Springer, Heidelberg (2004)

23. Supriaddin, N., Palilati, A., Bua, A., Patwayati, J.H.: The effect of complaint handling toward customers satisfaction, trust and loyalty to bank rakyat Indonesia (Bri) southeast Sulawesi. Int. J. Eng. Sci. (IJES), **4**(6) 01–10 (2015)
24. Jan-Michael, B., Klein, K., Wetzels, M.: Hierarchical latent variable models in PLS-SEM: guidelines for using reflective-formative type models. Long Range Plann. **45**, 359–394 (2012)
25. Ringle, C.M., Sarstedt, M., Straub, D.W: A critical look at the use PLS-SEM in MIS Quaterly. MIS Q. **36**(1), iii–xiv (2012)
26. Hair, J.F., Sarstedt, M., Ringle, C.M., Mena, J.A.: An assessment of the use of partial least squares structural equation modeling in marketing research. Acad. Mark. Sci. **40**, 414–433 (2012)
27. Sarstedt, M., Ringle, C.M., Smith, D., Reams, R., Hair, J.F., Jr.: Partial least squares structural equation modeling (PLS-SEM): a useful tool for family business researchers. J. Family Bus. Strategy **5**, 105–115 (2014)
28. Hair, J.F., Ringle, C.M., Sarstedt, M.: PLS-SEM: indeed a silver bullet. J. Mark. Theory Pract. **19**(2), 139–150 (2011)
29. Ismail, M.B., Yusof, Z.M.: The impact of individual factors on knowledge sharing, quality. J. Organ. Know. Manag. **2010**(327569), 13 (2010)

Lexical Matching-Based Approach for Multilingual Movie Recommendation Systems

Xuan Hau Pham, Jason J. Jung and Ngoc Thanh Nguyen

Abstract Recommendation systems (RecSys) have been developed for personalized users interaction process to deal with overload information. Movie Content-based recommendation approaches try to measure similarity between movie or users based on relevant information. Nowadays the amount of information on the web exists in several languages. The items description on the RecSys may be not only native languages but also multilingualism. Besides, users interact to the system come from many countries in different languages. However, most of these recommendation systems lack mechanisms to support users overcoming the language problem. Thus, in this paper, we propose a lexical matching-based approach to deal with multilingualism in our process and show efficient experiment for multilingual recommendation system in movie domain.

Keywords Recommendation systems · Multilingual movie · User profile

X.H. Pham
Department of Technology, Quangbinh University, Dong Hoi, Vietnam
e-mail: pxhauqbu@gmail.com

J.J. Jung (✉)
Knowledge Engineering Laboratory, Department of Computer Engineering,
Chung-Ang University, Seoul, Korea
e-mail: j2jung@gmail.com

N.T. Nguyen
Division of Knowledge and System Engineering for ICT, Faculty of Information Technology,
Ton Duc Thang University, Ho Chi Minh City, Vietnam
e-mail: nguyenngocthanh@tdt.edu.vn; Ngoc-Thanh.Nguyen@pwr.edu.pl

N.T. Nguyen
Faculty of Computer Science and Management, Wroclaw University of Technology,
Wroclaw, Poland

© Springer International Publishing Switzerland 2016   149
D. Król et al. (eds.), *Recent Developments in Intelligent Information
and Database Systems*, Studies in Computational Intelligence 642,
DOI 10.1007/978-3-319-31277-4_13

1 Introduction

With the rapid growth of the Web and the overload information that we have to deal with, the systems have to be responded customer needs without taking them a lot of time and many declarations. Users have to face an evident difficulty in selecting interested items. Recommendation systems (RecSys) is the best choice. RecSys is a automatically mechanism to provide to users interested items (e.g., web pages, documents, movies, books, music and so on) from a large information repository based on analyzing user profile. These systems can generate a set of items "potentially" to show to users. These items are shown to help users finalize their decisions (e.g., products to buy, music to listen to, movies to watch, and news to read). Thus, it is important for the RecSys to collect as a lot of user feedback and user interactions as possible and to compute similarities among them. These contents not only relevant users but also items information.

Nowadays, most of the e-commerce systems can be accessed that not be restricted by language barriers. As we known, in the real world customers come from many countries and they can interact to the system in different languages including native language. They also have many foreign friends. Therefore, how the systems can generate profiles and measure the similarity among them? Besides, there are many of the target items contents have been internationalized, such contents can be localized. We consider the following example:

and these movies are expressed as follows:

Taking Tables 1 and 2 into account, we assume that u_1 and u_3 are friends, u_1 is French and u_3 is Korean. We know that i_1 and i_5 are the same one, i_1 is original edition and i_5 is Korean languages edition. If the system cannot identify i_1 and i_5, it has a mistake for recommending. For this case, how can user profiling and recommend interested movies to these users? In order to deal with this, we will consider multilingual contents in the recommendation process. In this paper, we propose a lexical matching-based approach for multilingual movie recommendation systems. The recommendation performance consists of two main stages: user profiling and recommending with multilingual content of movies.

The outline of paper is organized as follows. In Sect. 2, we represent the previous studies on content-based recommendation systems and user profiling.

Table 1 User-item multilingual information

User ID (Language)	Movie ID (Title)	Language	Rating
u_1 (French-fr)	i_1 (Avatar)	English	4
	i_2 (Taxi (1998))	French	4
u_2 (English-en)	i_3 (Taxi (1998))	English	5
	i_4 (A beautiful mind)	English	2
u_3 (Korean-ko)	i_5 (아바타)	Korea	3
	i_6 (Titanic)	English	5

Table 2 Multilingual movie information

Movie ID	Genre	Director	Actor	Country
i_1	Action, Adventure	J. Cameron	S. Worthington Z. Saldana	USA
i_2	Comédie	G. Pirès	S. Naceri F. Diefenthal	Français
i_3	Comedy	G. Pirès	S. Naceri F. Diefenthal	French
i_4	Drama	R. Howard	R. Crowe Ed Harris	USA
i_5	모험, 행동	제임스 캐머런	샘 워딩턴 조 샐다나	미국
i_6	Drama	J. Cameron	L. DiCaprio Kate Winslet	USA

In Sect. 3, we describe our proposal and present formal modeling with mathematical definitions and properties. Section 4 presents some experimental results and our discussion. Finally, Sect. 5 draws our conclusion of this work.

2 Related Work

Content-based recommendation systems is a suitable approach for personalizing items based on analyzing user profile [1, 2]. In order to recommend, content-based method extracts the content of items to predict relevant items based on the users profile. Thus, in this approach, the systems collect user feedback and user interactions for user profiling. There are many user profiling approaches have been proposed.

In [3], a content-based document representation based on requested pages to build a user's interest model. The system builds the user model as a semantic network. It has been used a multilingual database to allows navigation multilingual agents. During the matching phase, the system obtains the senses representation of a document and the user interest model, and the recommendation results as an prediction of the relevant document. WebShopper+ [4] is a system to help user find and compare products using different languages. It designed with a multilingual ontology to deal with the language barriers. A semantic search mechanism to help user to find interested products based on concept similarity. The semantic similarity of concepts and lexical based on ontology have been presented in [5–7].

Multilingual information retrieval in the extracting keywords of document has been proposed by [8]. The authors developed a new algorithm to extract multiple languages and effective keywords with high degree, uniquely identify a document. The approach has been implemented by using a keyword search algorithm on a Japanese-English bilingual corpus and the Reuter's corpus. In [9], the authors have been proposed an approach to integrate content of document across languages for presentation of retrieved. The form of dynamic multilingual documents generated to respond to user interest. In [10], they have been proposed user profiling based on ontology and the RDF/OWL EuroWordNet. This approach recommends relevant information as a personalize retrieval systems according to user preferences with multilingual data of the user. Flickling [11], a multilingual search interface have been presented to collect a large search log of multilingual image searches. It has two ways for searching, mono and multilingualism, that allow to represent Flickr images annotated in different languages. The interface is able to automatically translate it into several languages based on user preference. The integrated multilingual entities dataset for recommendation system has been proposed in [12].

3   Multilingual Recommendation Systems

In this approach, we assume that the multilingual values of movies have been extracted. It means that multilingual movie attributes and values are described as Tables 1 and 2.

3.1   Multilingual Recommendation Process

We denote as follows:

- U is a set of users
- I is a set of movies
- A is a set of movie attributes
- V is a set of values
- L is a set of languages
- R is a set of rating

In order to be easy understand our approach and also perform the matching up among languages. The following table will show the list of languages codes and describe them (Table 3).

Table 3 The languages codes

Language name	Native language	Code
English	English	en
French	français	fr
German	Deutsch	de
Korean	한국어	ko
Italian	Italiano	it

Definition 1 (*Item representation*) Each certain movie, denoted $i \in I$ can be expressed as follows:

$$i = \{(a, v, l) | \forall a \in A, v \in V, l \in L\}$$

In the above mentioned example as Tables 1 and 2, "Avatar" English movie is directed by "J.Cameron", movie genre is "Action" and "Adventure", movie starting is "S. Worthington" and "Z. Saldana", and recommended language is English, denoted i_1,

$i_1 = \{(id,$ "tt049954", $o),(language,$ "English", en), (title, "Avatar", en), (director, "J.Cameron", en), (starting, "S. Worthington", en), (starting, "Z. Saldana", en), (genre, "Action", en), (genre, "Adventure", en) $\}$

The "아바타" Korean edition is the same one, "Avatar" movie. It is described as follows:

$i_5 = \{(id,$ "tt0499549",$o),(language,$ "미국", $ko), (title,$"아바타", $ko), (director,$"제임스 캐머런", $ko), (starting,$"샘 워딩턴", $ko), (starting,$"샘 워딩턴", $ko), (genre,$"모험", $ko), (genre,$"행동", $ko)\}$

With a certain movie, we define a multilingual movie collection as follows:

Definition 2 (*Multilingual item*) Given movie $i \in I$, $L_1 \subset L$ is a set of movie languages in different editions, D is a set of languages codes. The multilingual movie collection can be defined as follows:

$$C(i) = \{(d, l', l) | (d, l) = (d, l') : l \neq l', d \in D, <l', l> \in L_1\}$$

For example, The "Avatar" movie has seven editions in English, French, Korea and so on. It can be described in a multilingual movie collection:

$C("Avatar") = \{(en,$ "English", "English"), (fr, French, "Français"), (ko, Korean, "한국어")\}

Definition 3 (*Transaction*) Given user $u \in U$, u rated a certain movie, denoted $i \in I$, each user-movie interaction can expressed as follows:

$$T(u) = \{(i, l, r) | \exists i \in I, l \in L, r \in R, r \in [1..5]\}$$

For example, we can see user-movie matrix in Table 1. The table has three users and six movies. User u_1 watched two movie i_1, i_2 with English and French languages of editions, respectively. We have:

$$T(u_1) = \{(i_1, English, 4), (i_2, Vietnamese, 2)\}$$

Similarly,

$$T(u_2) = \{(i_3, English, 5), (i_4, English, 2)\}$$

$$T(u_3) = \{(i_5, Korean, 3), (i_6, English, 5)\}$$

The set of selected movies of certain user without ratings and languages is denoted by $I(u)$ and a set of users who selected the same movie is denoted by $U(i)$.

3.2 User Profiling

In our approach, we try to extract user profile based multilingual movie description and user-movie transaction. Each movie is described by a tuple (attribute, attribute value, language) as Def. 1. The systems obtain user preference based on the measurement the dominant attribute values of item. This means that the dominant values will transform from movie information into user interest by analyzing user transactions. In order to user profiling, we have to extract item contents as values and the similarity between them.

Definition 4 (*User model*) A model of user as a triple is described as follows:

$$\mathcal{M} = \langle T, C \rangle$$

where T is a set of user transactions, C is a set of multilingual movies.

After collecting user data, the systems have to extract which interested items that will be recommended to user. In order to do this, it try to analyze and find out user preference. It means that we have to measure to identify which value is dominant. The list of movies candidate will be identified based on the matching between them and user preference. In order to find out the dominant value, we have to compute the weight of each value with respect to correlative attribute.

Definition 5 (*Weighted value*) The weight of certain value in the user profile of multilingual data is defined as follows:

$$w(I(u), v_a, l) = \frac{card(v_a, l)}{card(v_a, u)} \tag{1}$$

where $card(v_a, l)$ is number of v_a values in languages l and $card(v_a, u)$ is number of v_a values of user u.

The weight of value on a set of selected movies of certain user u, obtained by:

$$w(v_a, I(u)) = \sum_{l \in L} (w(I(u), v_a, l)) \tag{2}$$

Definition 6 (*User preference*) Each given user u, u preference can be expressed as follows:

$$f(u) = \{(a, v, l)|w(v_a, I(u)) \geq \tau : a \in A, v_a \in V, l \in L\}$$

where τ is threshold and $\tau \in [0..1]$

3.3 Recommendation Process

There are four steps for recommendation process:

- Collecting user-movie transaction
- Extracting user preference with multilingual collected data.
- Matching and finding potential items.
- Representing multilingual recommended movies.

In content-based recommendation process approach, the main task is to find out which items are closed to user preference based on their information. It means to compute the similarity between a set of item values and user preference. There are many measures have been proposed to compute the similarity of two vectors. The cosine similarity is the most widely used among them. The similarity between two set of values can obtained by cosine normalization [2, 13, 14].

Definition 7 (*Value similarity*) Give two values, v_1 and v_2, the value similarity is computed by:

$$sim(v_1, v_2) = \begin{cases} 1 & \text{if } M_{l \in L}(v_1, l) = M_{l \in L}(v_2, l) \\ 0 & \text{otherwise} \end{cases}$$

where $M_{l \in L}(v_1, l)$ is function to get a set of the same value in the set of languages L.

Based on the computed similarity, we extend it to propose the definitions for content-based multilingual recommendation systems. In collaborative filtering approach, item similarity usually measures based on user rating data. Otherwise, in content-based approach, item similarity take into account the value similarity in different item attributes.

User preference contains a set of dominant values that the system extracts from user profile. Thus, instead of measure similarity on users' profiles we measure on users' preferences.

The recommendation process will extract a set of items recommendation based on user preference and measuring the user similarity and movie similarity. User similarity extracts the set of user neighbors who have relevant dominant value to current user on each interested value and take in into account all of the values of the user preference. Movie similarity finds out a set of movies which become potential movies to recommend to user. In the next section, we present the datasets that we implemented for our proposal.

Table 4 The multilingual movies

Attribute	#Attribute value	#Languages
Title	497	20
Actor	2469	88
Director	599	66

Table 5 Structure of profile

User ID	Native language	Multilingualism
u_1	German	German, English
u_2	Vietnamese	Vietnamese
u_3	Vietnamese	Vietnamese, French, English

4 Experimental Results and Discussion

In this paper, in order to implement our proposal, we have to collect and integrate datasets from multi-sources:

1. IMDB[1] dataset contains movies and related information about movies.
2. DBpedia[2] contains multilingual movie information.
3. collected users' profiles describe native language and multilingualism.

The multilingual related information of movies have been described in Table 4.

In the Table 4, we can see that the number of movie title languages is more different than movie actor and movie director. Based on each attribute value on each movie, we try to extract a set of multilingual values on DBpedia repository. However, the quantity of languages on each entity is not the same depending on the original source, Wikipedia.[3]

In order to perform our proposal, we try to construct the structure of profile in the Table 5.

We implement 50 users on our dataset. We compare between content-based approach (CB) and multilingual content-based approach (MCB). We separate user profile into two types based on their languages for evaluating our proposal as follows:

1. Monolingual user profiling for content-based approach (e.i., these user can only watch movies with native language)
2. Multilingual user profiling for multilingual content-based approach.

We assume that English is language of system. We can predict three scenarios for these cases on our dataset:

---

[1]www.imdb.com.

[2]www.dbpedia.org.

[3]www.wikipedia.org.

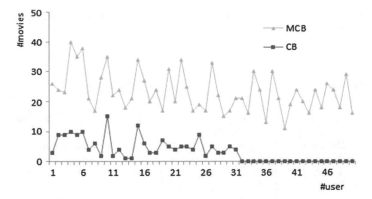

Fig. 1 The comparison between CB and MCB

1. User cannot get any movies for non English users.
2. System can show a list of recommendation movies to user for English users
3. The list of multilingual movies will be recommended to user.

For monolingual user profiles case, we extract a list of movies in English movies repository. There are 31 users who can get candidate movies. However, there are 19 users who cannot get any related movies. The simple reason is that their profile do not match any English movies. It means that they cannot watch English movies (e.g., Spanish, French and so on). For multilingual user profiles, system can recommend movies to 19 users from multilingual movies dataset. The result is shown in the Fig. 1

5 Conclusion

In this paper, there is a new multilingual content-based recommendation systems approach in movie domain have been proposed. We extract user's preference in difference languages based user feedback and user interactions. The similarity between items or users based on movie attributes, attribute values and a set of languages have been defined to find out a set of potential movies that can recommend to user. A multilingual dataset have been generated to perform our approach from several sources. In this approach, we not only take into account users preference but also integrate multilingualism into them. The experimental results show that our approach more efficient and flexible. In future work, we will show evaluation of recommendation accuracy.

Acknowledgment This research is funded by Vietnam National Foundation for Science and Technology Development (NAFOSTED) under grant number 102.01-2013.12. This work was also supported by the National Research Foundation of Korea (NRF) grant funded by the Korea government (MSIP) (NRF-2014R1A2A2A05007154).

References

1. Pazzani, M.J., Billsus, D.: Content-based recommendation systems. In: The Adaptive Web. Springer, Berlin, pp. 325–341, (2007)
2. Lops, P., De Gemmis, M., Semeraro, G.: Content-based recommender systems: state of the art and trends. In: Recommender Systems Handbook. Springer, Berlin,pp. 73–105, (2011)
3. Magnini, B., Strapparava, C.: Improving user modelling with content-based techniques. In: User Modeling 2001. Springer, Berlin, pp. 74–83, (2001)
4. Huang, S.L., Tsai, Y.H.: Designing a cross-language comparison-shopping agent. Decis. Support Syst. **50**(2), 428–438 (2011)
5. Sriharee, G.: An ontology-based approach to auto-tagging articles. Vietnam J. Comput. Sci. **2**(2), 85–94 (2014)
6. Duong, T.H., Nguyen, N.T., Jo, G.S.: A method for integration of wordnet-based ontologies using distance measures. In: Knowledge-Based Intelligent Information and Engineering Systems, Springer, Berlin, pp. 210–219, (2008)
7. Pham, X.H., Nguyen, T.T., Jung, J.J., Nguyen, N.T.: (a, v)-spear: a new method for expert based recommendation systems. Cybern. Syst. **45**(2), 165–179 (2014)
8. Bracewell, D.B., Ren, F., Kuriowa, S.: Multilingual single document keyword extraction for information retrieval. In: Natural Language Processing and Knowledge Engineering, 2005. IEEE NLP-KE'05. Proceedings of 2005 IEEE International Conference on, IEEE, pp. 517–522, (2005)
9. Jones, G.J., Wade, V.: Integrated content presentation for multilingual and multimedia information access. New Dir. Multilingual Inf. Access **40**, 31–39 (2006)
10. De Luca, E.W., Plumbaum, T., Kunegis, J., Albayrak, S.: Multilingual ontology-based user profile enrichment. In: MSW, Citeseer, pp. 41–42, (2010)
11. Peinado, V., Artiles, J., Gonzalo, J., Barker, E., López-Ostenero, F.: Flickling: a multilingual search interface for flickr. In: Working Notes for the CLEF 2008 Workshop, Aarhus, Denmark, Citeseer (2008)
12. Pham, X.H., Jung, J.J.: Recommendation system based on multilingual entity matching on linked open data. J. Intell. Fuzzy Syst. **27**(2), (2014)
13. Ekstrand, M.D., Riedl, J.T., Konstan, J.A.: Collaborative filtering recommender systems. Found. Trends Hum.-Comput. Interact. **4**(2), 81–173 (2011)
14. Sidorov, G., Gelbukh, A., Gómez-Adorno, H., Pinto, D.: Soft similarity and soft cosine measure: similarity of features in vector space model. Computación y Sistemas **18**(3), 491–504 (2014)

Part III
Web Services, Cloud Computing, Security and Intelligent Internet Systems

Analyzing and Conceptualizing Monitoring and Analytics as a Service for Grain Warehouses

Tien-Dung Cao and Hong-Linh Truong

Abstract In developing countries, like Vietnam, it is vital to ensure high quality of grain with traceable origins, because grain is of paramount importance for business and food security purposes. However, it is very challenging to build information systems to track, monitor, analyze and manage grain warehouses in scattered flood of Mekong delta—Vietnam, given under developed physical and informational infrastructures. In this paper, we analyze requirements and conceptualize monitoring and analytics as a service for grain warehouses, with a focus on Vietnam's environment. We analyze different stakeholders and their needs for monitoring and analytics features for grain warehouses. Leveraging the cloud computing model, big data analytics and Internet of Things (IoT), we design a conceptual framework to monitor various important information for grain warehouses and present different data analytics services that should be provided.

Keywords Grain warehouse monitoring · Monitoring and analytics as a service · IoT · Cloud computing

1 Introduction

Recently, advances in cloud computing, big data management, business intelligence, and Internet of Things (IoT) have been introduced to various domains, including transportation, logistics, and smart agriculture [1–3]. Furthermore, substantial reduction of investment costs in smart devices and software services in the cloud has enabled the further integration between physical worlds with software services in the cloud. Apart from such emerging technologies, shared economy

T.-D. Cao (✉)
School of Engineering, Tan Tao University, Đức Hòa, Vietnam
e-mail: dung.cao@ttu.edu.vn

H.-L. Truong
Distributed Systems Group, TU Wien, Vienna, Austria
e-mail: truong@dsg.tuwien.ac.at

© Springer International Publishing Switzerland 2016
D. Król et al. (eds.), *Recent Developments in Intelligent Information and Database Systems*, Studies in Computational Intelligence 642,
DOI 10.1007/978-3-319-31277-4_14

models have shown that assets will be shared and managed with joint interesting from different partners [4]. In this trend, we have also observed that novel technological and economic models are needed to pave the way for smarter, but cheaper, logistics and management, enabling shared economy models to provide a lot of benefits for farmers, small and medium enterprises (SMEs), and government in agriculture in developing countries, e.g., Vietnam.

We are interested in investigating how such novel techniques and models could be utilized for grain warehouse management in the context of Vietnam, in particular, in the Mekong delta. In such a context, we face several problems in managing grains: (i) poor storage infrastructures with high air humidity; (ii) severe floods in the rain season; (iii) farmers are lack of knowledge in post harvest; (iv) and poor role of modern scientific and technological methods and tools in post grain harvest. Such problems, on the one hand, cause loss of grains due to poor quality management, and reduce readiness in logistics and grain trading. On the other hand, these problems prevent stakeholders in producing and selling grains to follow standard models, such as VietGAP/GlobalGAP,[1] to provide high quality of grains with clear sources of origins in order to be able to gain benefits in a highly competitive environment, being governed by international treaties, like the future Trans-Pacific Partnership (TTP).[2]

In this paper we analyze requirements for building a grain warehouse information network by levering advanced concepts on everything-as-a-service [5], cloud computing and IoT. First, we contribute to an analysis of requirements for grain warehouse information under the service model by examining various stakeholders and their needs. Second, we propose a monitoring and analytics as a service framework, in which the SMEs/farmers can rent a warehouse to store their grain and use this framework to monitor the quality of grain as well as to receive early warning about the risk. In this framework, we enable the gathering of various types of monitoring data of grains, stakeholders, grain input/output flows and grain knowledge. This enables us to provide analytics and management features for grain warehouses under cloud services in data center.

The rest of this paper is organized as follows: Sect. 2 presents the motivating scenario. Section 3 analyzes stakeholders of monitoring and analytics as a service. Section 4 describes our conceptual framework. We discuss related work in Sect. 5. Section 6 concludes the paper and outlines our future work.

2   Motivating Scenario

Despite being one of the leading countries in rice producer and exporter, the rice value in Vietnam is low and, therefore, the life of farmer is still challenging. Among some reasons (e.g., type of grain, there is no brand) leading to the low value, an

---

[1]http://vietgap.gov.vn/, http://www.globalgap.org/.
[2]https://en.wikipedia.org/wiki/Trans-Pacific_Partnership.

important reason is that the process of harvest and storage is poor. According to Jayas [6], the post-harvest losses of grains is approximately 20–50 % in case of poorly managed storage systems comparing to the case of well managed facilities with aeration and drying capabilities. In Vietnam, the government currently allows state enterprises or private enterprises who have their warehouse with the capacity to store at least 5000 tons of grain to export the rice [7]. However, in fact, most enterprises, who have license to export the rice, do not store the grain at their warehouse.[3] When they get an order from customers, they will collect rice or grain from smaller enterprises or directly from farmers, packaging and exporting the requested grain/rice. It means that the grain after harvest is stored by farmers or small enterprises in different warehouses. The main problem is that such a storage infrastructure, i.e., warehouse, is not adequate for keeping quality of grain. Moreover, currently it is not easy to identify, if not impossible, the origin (i.e., where the rice is planted) due to the lack of various information sources and analytics because the rice is mixed from many different sources. This is very difficult to build the rice brand for farmers. Apart from these technical problems, current business models in grain warehouses are also not flexible. While there is no standard warehouse in which the SMEs/farmers use to store their grain, building their own standard warehouse is out of their capacity because they need not only money but also the knowledge to operate it and to share information to others. The shared economy models [4] is a good solution for this situation, in which the government or large enterprises can build a network of grain warehouse reached grain storage standard. The SMEs/farmers then use it to store their grain and pay as they use. However, to realize this, online collaboration and sharing, two major characteristics of the sharing economy must be enabled. It means that the grain warehouse network must support advanced monitoring and analytics features to allow various stakeholders involved in the grain warehouses to interact with their warehouse providers, their renting warehouse, domain experts, as well as retrieve their grain history to the potential rice consumer. All the above-mentioned problems call for the development of novel grain information network in Vietnam with the following features:

- allow providers to govern their grain warehouses in a seamless and smart manners (e.g., changing the temperature, the humidity of air, or the light inside warehouse).
- allow consumers to monitor the quality of grain by through various sensors and analytics in a near-realtime manner.
- enable early warning for consumers about the quality of grain if the properties such as temperature, humidity, weather, are changed or the warehouse is flooded.
- manage the grain warehouse contract and the grain history.
- provide the grain history, such as origin, storage period, weather/land/water of place during the rice growing, to potential rice customers.

---

[3]As private discussion with Prof. Tong-Xuan Vo.

In order to support the above-mentioned features, in the next sections, we will present our detailed requirement analysis and propose a novel monitoring and analytics as a service framework.

3 Stakeholder Analysis

We determine in this section relationships between stakeholder and the activities of grain warehouse information framework. Of course, two main stakeholders are warehouse provider and user. Domain experts play an important role in this framework since they can advise to warehouse provider during the operation. The grain users concern to this framework since they can query the grain history. Because the warehouse network is built from many providers by providing their resources, so a macro decision, e.g., distributing grain, on whole network is impossible for a single provider. However the government agencies can do it. In summary, there are five groups of stakeholder including: provider (large enterprises and government agencies), user (SMEs and farmers), scientist (domain expert), government agencies and grain user. The activities of framework are grouped into five services: sensor operation, grain information, grain knowledge, contract management and data analytics (Fig. 1). Figure 2 details relationship between stakeholders, activities, data sources and services, in which:

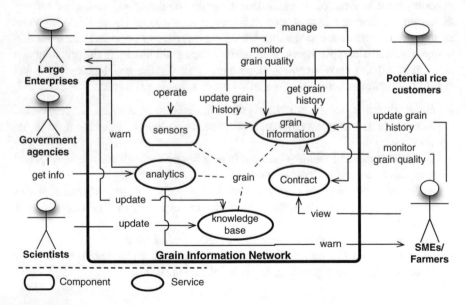

Fig. 1 The stakeholders and their interaction in the framework

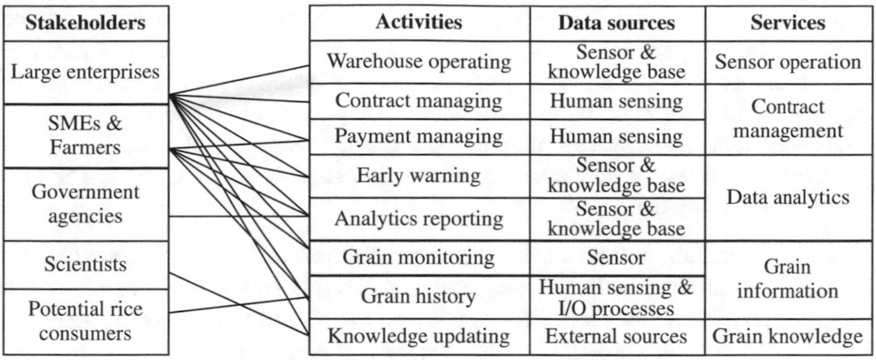

Stakeholders		Activities	Data sources	Services
Large enterprises		Warehouse operating	Sensor & knowledge base	Sensor operation
SMEs & Farmers		Contract managing	Human sensing	Contract management
		Payment managing	Human sensing	
Government agencies		Early warning	Sensor & knowledge base	Data analytics
		Analytics reporting	Sensor & knowledge base	
Scientists		Grain monitoring	Sensor	Grain information
Potential rice consumers		Grain history	Human sensing & I/O processes	
		Knowledge updating	External sources	Grain knowledge

Fig. 2 Relationship between stakeholders, activities and data sources

- *Large Enterprises*: large enterprises are the providers who provide the warehouse for rent. It means that they must: (i) define and manage warehouse renting contracts; (ii) operate and monitor their warehouses to control the quality of grain; (iii) update knowledge base, e.g., grain seed and its properties, weather, from different sources. They also plays the role of consumer to use their warehouse to store their grain;
- *SMEs and Farmers*: they need monitoring and analytics features for grain information to monitor their grain and update their grain history.
- *Government agencies*: the government agencies do not only play the role of a grain warehouse provider, but also use various types of information (e.g., the quality and quantity of grain, and history of grain) to plan the use of the stored grain, e.g., for the food security program, or to allocate the rice export quota. In urgent cases (e.g., flooding), the government agencies can also make macro decisions on distributing grains.
- *Scientists*: scientists play the role of domain expert who provides their knowledge that helps the providers to operate the warehouse and the framework to produce early warning about quality of grain.
- *Potential rice customers*: they need information of grain history to decide their purchases of rice.

4 IoT Cloud Based Platform for the Grain Warehouse

4.1 Monitoring Data

To support different analytics for the grain warehouse, several types of monitoring data must be captured. We consider two types of monitoring data that need to be provisioned:

- *grain warehouse monitoring data*: it is the data about grain status. Examples would be temperature, air humidity, moisture sensors, air quality sensors for both indoor and outdoor of the grain warehouses. Such IoT-based data is needed for continuously monitoring grain warehouse and grain conditions.
- *external data services*: several important external services will be utilized. In our framework, they are the weather information service, and the grain seeds and rice knowledge bank.

All of these data will be collected and provisioned as a service in the cloud to enable near-realtime monitoring and analytics. To this end, we leverage IoT technology to provide the first type of data whereas the second type of data is integrated from data-as-a-service cloud, such as weather and grain seeds.

4.2   Warehouse Information System

Another type of information is the traditional warehouse information which stores information about stakeholders, volumes of grains, and number of bags of grains and containers of grains. This kind of data is provided through a grain input/output management system within warehouses as well as through management activities for stakeholders and their roles.

4.3   Grain Knowledge and Expert Inputs

One important information is the domain knowledge about grains. Such knowledge have been collected over the time for grains produced in the region. This type of knowledge differs from the global grain knowledge bank from external services, as it incorporates specific knowledge obtained from the region. In addition, we also allow human expert as well as people involved in the management of grain warehouse and stakeholders to provide inputs into the knowledge base. In our system, the knowledge therefore will be managed in a separate service where knowledge comes from people during warehouse operations as well as from experts through internet.

4.4   Monitoring and Analytics Services

To enable the monitoring of grain quality and operations for early warning, we design a near-realtime monitoring service. This service mainly utilizes the IoT data with streaming data processing capabilities to enable the detection of problems for grains and grain warehouses. Monitoring data can be informed to stakeholders through external services integration. Furthermore, monitoring data is used to make decisions in controlling the IoT system within the grain warehouse. For example, the control is executed to assure the following conditions inside the warehouse based on domain knowledge.

Several analytics features will be needed to support different use cases. An analytics can leverage big data processing techniques to deal with various types of data, mentioned above. Based on that, different customized analytics can be built, such as for grain quality analytics and logistics and recommendation.

4.5 Pricing and Contract Models

Another aspect is about the pricing and contract models.[4] Here we leverage concepts of data contracts in data marketplaces [8] where we allow the owner defined various pricing schemas for warehouses. Different from typical pricing models for warehouses where the customer pay only for the space they take from the warehouse and from typical cloud services. The pricing models combine two distinguisable aspects:

- pricing for using a warehouse: the pricing models will be based on contemporary grain warehouse contracts enriched with different guaranteeing and liability conditions. In general, different from existing grain warehouse contracts, given new monitoring and analytics features, the owner of grain warehouses could offer better liability conditions for their customers and, in parallel, could charge higher costs for grain storage.
- pricing for using additional services for grains monitoring and analytics: this is a new type of pricing models for grain warehouse customers, although it is not really new for customers in contemporary IT services in the cloud. The key difference for the customers is that now they also can opt in pay-per-use features to utilize advanced services for their grains.

4.6 Customized Services

Based on the core services about monitoring, analytics, and contract management, a set of different customized services is designed for different stakeholders:

- Warehouse warning service: it can be used to provide warning information related to the quality and quantity of grains.
- Grain quality service: this provides regular quality analytics for grain to support the logistics, purchase and recommendation for grains.

---

[4]Detailed pricing and contract models are out of the scope of this paper.

- Logistics planning and recommendation: this service helps the grain owner to make a plan on uses of grain based on information of analytics, storage history and/or contract termination. It can also recommend the warehouse providers how to govern the warehouse processes.
- Warehouse usage scheduling: based on contract management, this service helps the warehouse providers to make schedule to optimize the uses of warehouse.
- Grain history service: it provides the grain history to grain/rice consumers.

4.7 IoT Cloud Architecture for Grain Warehouses

Figure 3 presents our conceptual framework. The architecture is followed the concept of IoT cloud systems [9] in which we have several IoT elements to provide monitoring data of grains and input/output flows and monitoring data and analytics are provisioned as cloud services in data centers. At the warehouse side, we have three different subsystems: IoT with sensors, actuators, and gateways for monitoring grains and warehouse; the Grain Input/Output Management System for managing grain flowing in/out the warehouse; and Human Sensing for getting knowledge about grain during the warehouse operations (e.g., inspection, daily management). Our IoT system in grain warehouses is designed based on common sensors and lightweighted gateways, such as Raspberry PI. Mobile networks are used to push the monitoring data to the cloud.

At the cloud, we have three different layers: data-as-a-service, core services, and customized services. In the data-as-a-service layer, we have different services

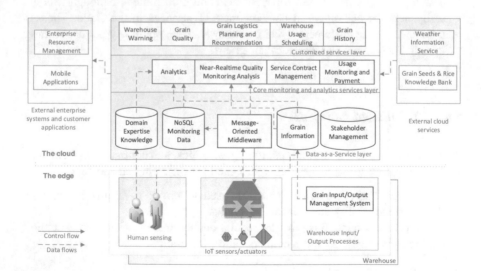

Fig. 3 IoT cloud based platform for grain warehouses

managing different types of data and knowledge, such as Domain Expertise Knowledge, Monitoring Data, Grain Information Service, Stakeholder Management Service. Core services include Analytics, Near-Realtime Quality Monitoring, Service Contract Management, and Usage Monitoring and Payment. Customized services include Warehouse Warning Service, Grain Quality, Grain Logistics Planning and Recommendation, and Warehouse Usage Scheduling. Furthermore, external data services, such as Weather Information Service and Grain Seeds and Rice Knowledge Bank, are integrated for core services to support advanced monitoring and analytics. Customized services can also be integrated with external enterprise systems of customers.

In the cloud, we utilize NoSQL database (MongoDB/Cassandra) for near-realtime monitoring data, grain information and grain knowledge, while relational databases are used for stakeholder management. Near-realtime monitoring service and analytics service will be based on stream data processing and big data analytics based on Spark and Hadoop. All of our services are designed as Web services with REST interfaces.

5 Related Work

Integrated Data for Smart Agriculture and Logistics: Applying Internet of Things (IoT) technology for agriculture has been emerging in recent years.[5] A review of the technical and scientific state of the art of wireless sensor technologies and standards for wireless communications in the agriculture food sector was presented in [10]. Wang et al. [11] introduced several examples of wireless sensors and sensor networks applied in agriculture and food production for environmental monitoring and precision agriculture. Sugahara [12] developed a traceability system for agricultural products, based on the innovative technology, RFID, and mobile phones. While Ko et al [13] presented a monitoring system for agricultural products' yields and distribution based on wireless sensor network (WSN). Hwang et al. [14] used both these technologies, i.e., RFID and WSN, to develop a food traceability system. Several industrial frameworks have also integrated IoT data for smart agriculture, such as Thingworx,[6] Senseye[7] and Carriots.[8] Although most of these works focus on food traceability, the data is not integrated with a knowledge base or external data courses to make an early warning about quality of food.

---

[5]https://www.foreignaffairs.com/articles/united-states/2015-04-20/precision-agriculture-revolution, https://www.microsoft.com/enterprise/industry/caglayan-arkan-blog/articles/how-iot-enables-smart-agriculture.aspx#fbid=XzhcfZ2Thhn.

[6]http://www.thingworx.com/markets/smart-agriculture.

[7]http://www.senseye.io/internet-of-things-sensors-for-smart-farming-part-1/.

[8]https://www.carriots.com/use-cases/agriculture.

Cloud service models for Warehouse Management: Borstell et al. [15] presented a system to provide real-time information and visual assistance to workers or operators involved in warehouse operations (e.g. storage, retrieval, rearrangement and picking), while Rohrig and Spieker [16] introduced a technique to monitor the manual transportation processes of goods in a warehouse. The Westeel Company[9] provided a grain storage monitoring system that can help us maintain optimum temperature and moisture. It is easy to recognize that these works only focus on warehouse management/monitoring, they do not focus on analytics problems. They also do not consider to share their warehouse as a service, i.e., missing the business functions such as pricing and contract models.

Analytics for Grain warehouse: Capturing and analyzing domain knowledge about agricultural processes, soil, climatic condition and farmers experiences, Dutta et al. [17] develop an architecture framework for knowledge recommendation using computational intelligence and semantic web technology. From our view, this analytics should be integrated with warehouse management system to provide this work and warehouse as the services, in which the individual farmers and SMEs can use to monitor and analyze their grain quality.

6 Conclusions and Future Work

It is expected that combining IoT and various sources of data will enable efficient monitoring and analytics for grain warehouses, thus improving the current poor situations in managing grain warehouses in Vietnam. Furthermore, leveraging the cloud service model will simplify and reduce the operation cost of grain warehouses management systems. In this paper, we introduce a conceptual framework that enables the integration of various data, including grain quality monitoring, grain expert knowledge, human inputs, etc. to enable near-realtime monitoring and analytics for grain warehouses.

Although we have focused on grain management in the Mekong delta, we believe that our conceptual frameworks could be applied/customized for similar areas. Currently, we are elaborating our design and working on a prototype to realize our framework.

Acknowledgement We thank Prof Tong-Xuan Vo and the ITA-Rice company (http://tantaorice. com) for fruitful discussion and useful knowledge on grain. We thank Quang-Hieu Vu and Tran-Vu Pham for their fruitful discussion on IoT and data analytics.

---

[9]http://www.westeel.com/easycheck-grain-storage-monitoring-system.

References

1. Institute, M.G.: The internet of things: mapping the value beyond the hype. Technical report, MCKinsey & Company, June 2015
2. Atzori, L., Iera, A., Morabito, G.: The internet of things: a survey. Comput. Netw. **54**, 2787–2805 (2010)
3. Chen, S.L., Chen, Y.Y., Hsu, C.: A new approach to integrate internet-of-things and software-as-a-service model for logistic systems: a case study. Sensors **14**, 6144–6164 (2014)
4. Hamari, J., Sjöklint, M., Ukkonen, A.: The sharing economy: why people participate in collaborative consumption. J. Assoc. Inf. Sci. Technol. (2015)
5. Schaffer, H.: X as a service, cloud computing, and the need for good judgment. IT Prof. **11**(5), 4–5 (2009)
6. Jayas, D.S.: Storing grains for food security and sustainability. Agric. Res. **1**(1), 21–24 (2012)
7. Ministry of Industry and Trade of Vietnam: Decree No. 109/2010/N-CP. http://www.moit.gov.vn/Images/Upload/ND%20109-2010-CP.pdf
8. Truong, H.L., Comerio, M., Paoli, F.D., Gangadharan, G.R., Dustdar, S.: Data contracts for cloud-based data marketplaces. Int. J. Comput. Sci. Eng. **7**(4), 280–295 (2012)
9. Truong, H.L., Dustdar, S.: Principles for engineering IoT cloud systems. IEEE Cloud Comput. **2**(2), 68–76 (2015)
10. Ruiz-Garcia, L., Lunadei, L., Barreiro, P., Robla, I.: A review of wireless sensor technologies and applications in agriculture and food industry: state of the art and current trends. Sensors **9**, 4728–4750 (2009)
11. Wang, N., Zhang, N., Wang, M.: Wireless sensors in agriculture and food industry—recent development and future perspective. Comput. Electron. Agric. **50**(1), 1–14 (2006)
12. Sugahara, K.: Traceability system for agricultural products based on RFID and mobile technology. Comput. Comput. Technol. Agric. II, 2293–2301 (2009)
13. Ko, D., Kwak, Y., Song, S.: Real time traceability and monitoring system for agricultural products based on wireless sensor network. Int. J. Distrib. Sens. Netw. (2014)
14. Hwang, Y.M., Moon, J., Yoo, S.: Developing a RFID-based food traceability system in Korea ginseng industry: focused on the business process reengineering. Int. J. Control Autom. **8**(4), 397–406 (2015)
15. Borstell, H., Kluth, J., Jaeschke, M., Plate, C., Gebert, B.: Pallet monitoring system based on a heterogeneous sensor network for transparent warehouse processes. In: Sensor Data Fusion: Trends, Solutions, Applications, pp. 1–6 (2014)
16. Röhrig, C., Spieker, S.: Tracking of transport vehicles for warehouse management using a wireless sensor network. In: IEEE/RSJ International Conference on Intelligent Robots and Systems (2008)
17. Dutta, R., Li, C., Smith, D., Das, A., Aryal, J.: Big data architecture for environmental analytics. In: Environmental software systems. Infrastructures, Services And Applications. IFIP Advances in Information and Communication Technology, pp. 578–588 (2015)

Review of Current Web Service Retrieval Methods

Adam Czyszczoń and Aleksander Zgrzywa

Abstract This paper provides a review on current Web Service Retrieval methods. Firstly, it includes description of this research area and introduces its systematics against many various research on general concept of service discovery. Secondly, it provides an overview of current research on Web Services Retrieval. Analysis of Web Service Retrieval methods refers to current study that use Information Retrieval techniques that addresses the topics of Web crawling and indexing. In addition, analyzed study also includes a number of specific issues without which the retrieval of Web Services is not possible. It includes: indexing terms extraction, tokenization, construction of test collection, classification and clustering of services, and extraction of relevant information.

Keywords Web service retrieval · Review · Web service · Indexing · Web crawling · Service discovery

1 Introduction

Recently, a number of various studies related to service discovery and service searching have appeared. In these studies, one can observe a strong distinction between different areas of computer science. In order to arrange this studies we provide in this paper systematics against many research on finding Web Services. Additionally, we provide a description of Web Service Retrieval (WSR) research area which despite the fact is widely used, is still not sufficiently described.

A. Czyszczoń (✉) · A. Zgrzywa
Faculty of Computer Science and Management, Wrocław University of Technology, Wybrzeże Wyspiańskiego 27, 50-370 Wrocław, Poland
e-mail: adam.czyszczon@pwr.edu.pl
URL: http://www.wiz.pwr.edu.pl

A. Zgrzywa
e-mail: aleksander.zgrzywa@pwr.edu.pl

© Springer International Publishing Switzerland 2016
D. Król et al. (eds.), *Recent Developments in Intelligent Information and Database Systems*, Studies in Computational Intelligence 642,
DOI 10.1007/978-3-319-31277-4_15

Presented in this paper analysis of Web Service Retrieval methods refers to current research that use Information Retrieval (IR) techniques to address key issues such as collecting information about the services through Web crawlers and indexing. In addition to the basic issues, analyzed study also includes a number of specific problems such as indexing terms extraction, tokenization, construction of test collection, classification and clustering of services, and extraction of relevant information, without which the Web Service Retrieval is not possible.

In the literature one can distinguish many different approaches and techniques of WSR. However, current methods, although in many cases are presented extensively, they describe the topic of WSR only on selected aspects and do not treat the problem as a whole. First of all, these studies do not take into account Web Service Retrieval methods of both SOAP and RESTful services—the research is carried out separately for each service class. Therefore, the analysis service retrieval methods is divided into two areas: analysis of Web Service Retrieval methods of SOAP Web Services and analysis of Web Service Retrieval methods of RESTful Web Services.

2   Web Service Retrieval and Service Discovery

Web Service Retrieval is a research area which concerns methods that use Information Retrieval models for full-text and ranked retrieval and take into account the problems of indexing and Web crawling. Web Service Retrieval is also a branch of the broader concept, namely, Web Service Discovery. It is an overall process of finding existing services that meet the criteria of given request, based on functional and non-functional semantic descriptions [1].

There is a wide variety of service discovery approaches, which vary depending on the supported service description format and chosen data representation model. They can be divided into two main groups: approaches based on the descriptions in the WSDL (Web Services Description Language) and approaches based on ontological descriptions. In the first group one can distinguish the following approaches: textual, structural and semantic. Semantic search approaches were based on the extension of interface descriptions using third-party tools such as lexical databases, concept lattice or data mining methods. Structural approaches were based on the methods of finding similarities in service interfaces, based on techniques from the field of software engineering. They utilized service structure representation in the form of a tree or a graph. Text approaches initially allowed only searching by keywords that were assigned by service suppliers. Afterwards, one proposed an extension to text search solutions that allowed to match strings to keywords and in result search for words with different spelling.

The last group includes approaches based on IR models that enable full text search with ranking, and take into account the Web crawlers and indexing. Therefore, this group belongs to Web Service Retrieval approaches. Overview of approaches to service discovery is shown on Fig. 1, where WSR is located in the lower left corner of the figure.

Fig. 1 Directions of research in the field of Web Service Discovery [2]

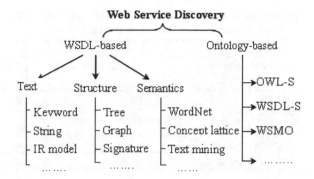

3 SOAP Web Service Retrieval Methods

Review of methods about SOAP Web Service Retrieval distributed on the Web is presented in Table 1. On the basis of those methods many tools were developed that were needed to conduct experiments on effectiveness. Those are: *Woogle* [3], *VitaLab* [4], *merobase* [5], *WSCE* [6], *WSRS* [7], *CFWSFinder* [8], *seekda* [9].

Research on SOAP WSR can be divided into three areas: research on finding similarities between services, research in web crawlers that are focused on finding services on the Web and research taking into account text search of Web Services, among which one can distinguish modeling services in the Vector Space (VSM), Latent Semantic Indexing (LSI), and various other methods such as Artificial Neural Networks (ANN), which extend the retrieval capabilities. For example using SOM (Self-Organising Map) [7] or KBSOM (Kernel Batch Self-Organising Map) [8].

4 RESTful Web Service Retrieval Methods

With the increasing popularity of RESTful Web Services more and more research papers appeared that focused on their architecture and the rules for their implementation [28–30]. What is more, many of the proposals was also created that concerned methods for creating machine-readable descriptions of RESTful services, such as: WADL (Web Application Description Language) [31], hRESTS (HTML for RESTful Services) [32], ReLL (Resource Linking Language) [33], WRDL (Web Resource Description Language) [34], NSDL (Norm's Service Description Language) [35], SMEX-D (Simple Message Exchange Descriptor) [36], RSDL (RESTul Service Description Language) [37] or method for semi-automatic creation of semantic models based on sample data [38].

However, none of these proposals have been widely adopted. Detection of RESTful Web Services on the Internet is much more difficult than detecting WSDL files of SOAP services. These services are described by the API documentation in the form of HTML documents, as well as regular pages, differing only by the fact

Table 1 Review of SOAP Web Service Retrieval Methods. The names of the tools developed on the basis of these methods are shown in bold

Usługi klasy SOAP

Method authors	Similarity matching	Web crawlers	Text search	
			VSM	LSI
Woogle Dong et al. [3]	✓	✓	–	–
Wu and Wu [10]	✓	–	–	–
Zhuang et al. [11]	✓	–	–	–
Fan and Kambhampati [12]	–	✓	–	–
VitaLab Platzer et al. [4, 13]	–	✓	✓	–
Lu and Yu [14]	✓	–	–	–
Li et al. [15]	–	✓	–	–
Lausen and Haselwanter [16]	–	✓	–	–
Song et al. [17]	–	✓	–	–
Merobase Atkinson et al. [5]	–	✓	✓	–
Peng [18]	–	–	✓	–
Ma et al. [19]	–	–	–	✓
WSCE Al-Masri et al. [6, 20]	–	✓	–	–
Wu et al. [2, 21–23]	✓	✓	✓	✓
Hao et al. [24, 25]	–	–	✓	–
WSRS Chan et al. [7]	–	✓	✓	✓
Seekda Scicluna et al. [9]	–	✓	–	–
Wu [26]	–	✓	✓	–
CFWSFinder Chen et al. [8]	–	–	✓	✓
Li et al. [27]	–	–	✓	✓

that they provide functions that can be triggered by a shared URL. These documents are very heterogeneous structure and very diverse level of detail of descriptions. Crawlers looking for such services on the Web must be able not only to identify these documentations but also must be able to extract relevant information from heterogeneous HTML documentation.

Review of methods about RESTful Web Services Retrieval is presented in Table 2. Based on the presented methods one developed many tools necessary to conduct experiments: *RestDescribe* [34], *APIHut* [39], *DEIMOS* [40], *RESTler* [33] or *iServe* [41].

Table 2 Review of RESTful Web Service Retrieval Methods. The names of the tools developed on the basis of these methods are shown in bold

RESTful web services			
Method authors	Text search	Identification	Information extraction
RestDescribe Steiner [34]	–	–	✓
APIHut Gomadam et al. [39]	✓	–	–
DEIMOS Ambite et al. [40]	–	✓	✓
Steinmetz et al. [42, 43]	–	✓	–
RESTler Alarcón and Wilde [33]	–	✓	–
iServe Pedrinaci et al. [41]	–	✓	–
Karma Taheriyan et al. [38]	–	✓	✓
Ly et al. [44]	–	–	✓
feaLDA Lin et al. [45]	–	✓	–

5 Conclusions

Similarity matching methods that are based on software engineering techniques give the possibility to compose services and allow to find services with similar operations. These methods are beyond the scope of WSR and belong to the field of Web Service Discovery. Algorithms that are used there suffer from high complexity and are therefore inefficient. In addition, because of the wide variety of parameter names, these methods are not very effective because operation matching is not always possible. The use of more efficient methods that assess the lexical similarity was however associated with low effectiveness because the terms used in the descriptions and names of service parameters were too diverse. In order to increase the effectiveness and efficiency of Similarity Matching, in many studies techniques from the field of IR were employed. Authors proposed i.a. methods using operation corpus that allowed for faster calculation of the operation similarities. The proposed algorithms have proved to be complex and their efficiency was still low (about 50 %). This show gradual shift from similarity matching methods to IR methods, which can be observed on Table 1.

The analysis of Web Service Retrieval methods that use Information Retrieval techniques involved:

- for SOAP Web Services—research in the field of web crawlers and indexing enabling text search with ranking that uses indexes in VSM and LSI models,

– for RESTful Web Services—research in the field of text search with ranking, identification methods of RESTful Web Services API documentation on the Web and methods of information extraction from these documentations.

The analysis of current research on Web Service Retrieval identified the following problems and concepts:

1. there are two main sources, which currently can be used to search for Web services: public repositories of Web services and services distributed on the Internet.
2. Current studies conducted over the Web crawlers allow only to traverse specific repositories and do not include issues related to the possibility of finding services distributed on the Internet. To collect information about the distributed services, crawlers must have the following functional characteristics: including website changes, scalability, load minimization.
3. Present studies conducted over the WSR of SOAP and RESTful services include a wide variety of problems, but none of them takes into account distributed indexing. These methods are therefore limited to data processing for indexing as a single process. As a result, developed methods are not suitable for large-scale search and efficiency of the indexing process is greatly restricted.
4. Ongoing studies conducted over the WSR which include indexing, apply only to SOAP Web Services and use the classic VSM model. In addition, studies were conducted that proposed a distributed VSM model, methods that included weights modifications based on WSDL document structure and methods of conceptual indexing. Research conducted over service indexing with VSM model often overlooked used indexing techniques such as variant of TF-IDF scheme or selected elements of the services structure that were indexed.
5. Current studies conducted over the text search of RESTful Web Services do not include indexing and ranking. On the other hand, analysed studies presented the alternative method of faceted classification with novel ranking algorithm based on the ProgrammableWeb directory. It has been shown that the effectiveness of this method is higher than search using ProgrammableWeb or Google.
6. Test collections used in research on Web Service indexing using VSM model was incomplete because it did not contain a set of queries with statistics necessary to assess the effectiveness such as the number of relevant documents for each query. Furthermore, collections were described in terms of service repositories which made them impossible to re-use in experiments. For this reason, analysis of the effectiveness of WSR methods was often overlooked or was presented too generally. As a result, the interpretation of the results and their comparison to the results of other methods very difficult or even impossible. Despite these problems, studies on the effectiveness of service retrieval provided a test collection *Assam*, which consists of 294 WSDL documents divided into 26 categories.
7. Studies conducted on the service indexing using the LSI model shown that although the index using this model is based on the VSM model, certain patterns of TF-IDF scheme allowed to achieve greater retrieval efficiency only

for the LSI index and vice versa. Furthermore, analyzed studies have demonstrated that the LSI model is generally less effective than VSM, however, it is its complement in cases of low effectiveness or lack of search results using VSM model. Thanks to the LSI method, it is possible to find higher order relations also for queries. Indexing all of the WSDL elements may allow for higher search effectiveness than indexing only of selected items.

8. In most research on WSR methods the method of indexing terms extraction was ignored making it impossible to recreate the experiments. This is due to the fact that term extraction method is crucial to the index structure and search effectiveness. Among the term extraction methods one can distinguish basic methods, that use classical techniques and tokenization methods taking into account the problem camelCase that are necessary for indexing Web Services. Comparative review of camelCase tokenization methods has shown that the MDL and MMA methods are the most effective, while maintaining high performance.

9. RESTful Web Services are described by the API documentation in the form of HTML documents. These documents have very heterogeneous structure and very diverse level of detail of the descriptions. Web crawler looking for RESTful services must be able to identify their API documentation and extract relevant information. Therefore, the current research on text search for RESTful Web Services mainly concern methods for identifying and extracting information from the API documentation.

10. Identification of RESTful Web Services relies on determining whether a HTML document is an API documentation of Web Service. Current research on identification is divided according to the methods of classification: NB and SVM, LDA, pLDA and feaLDA. Experiments conducted on a large collection that contained 1547 websites showed that the feaLDA method is most effective.

11. Information extraction of RESTful Web Services relies on the analysis of heterogeneous structure of API documentations in the form of HTML pages, in order to extract relevant information. Previous studies on the extraction of information have been divided into methods that concern the analysis of service URIs and methods that concern segmentation of API documentation. Current methods of URI analysis proved to be little effective. For the investigated segmentation methods one can distinguish two approaches: Tag-based Segmentation and Template-based Segmentation. However, due to the large heterogeneity of API documentations they are not very effective. Higher efficiency was showed by an algorithm that was a combination of NLP techniques and mentioned segmentation approaches.

References

1. Klusch, M.: Service discovery. In: Alhajj, R., Rokne, J. (eds.) Encyclopedia of Social Network Analysis and Mining, pp. 1707–1717. Springer, New York (2014)
2. Wu, C., Chang, E.: Searching services "on the web": A public web services discovery approach. In: Third International IEEE Conference on Signal-Image Technologies and Internet-Based System, SITIS 2007, pp. 321–328, 16–18 Dec 2007. Shanghai, China (2007)
3. Dong, X., Halevy, A., Madhavan, J., Nemes, E., Zhang, J.: Similarity search for web services. In: Proceedings of the Thirtieth international Conference on Very Large Data Bases, VLDB '04, VLDB Endowment, vol. 30, pp. 372–383 (2004)
4. Aiello, M., Platzer, C., Rosenberg, F., Tran, H., Vasko, M., Dustdar, S.: Web service indexing for efficient retrieval and composition. In: CEC/EEE, p. 63 (2006)
5. Atkinson, C., Bostan, P., Hummel, O., Stoll, D.: A practical approach to web service discovery and retrieval. In: ICWS, IEEE Computer Society, pp. 241–248 (2007)
6. Al-Masri, E., Mahmoud, Q.H.: Investigating web services on the world wide web. In: Proceedings of the 17th International Conference on World Wide Web. WWW '08, pp. 795–804. ACM, New York (2008)
7. Chan, N., Gaaloul, W., Tata, S.: A web service recommender system using vector space model and latent semantic indexing. In: IEEE International Conference on Advanced Information Networking and Applications (AINA), (March 2011), pp. 602–609 (2011)
8. Chen, L., Yang, G., Zhu, W., Zhang, Y., Yang, Z.: Clustering facilitated web services discovery model based on supervised term weighting and adaptive metric learning. Int. J. Web Eng. Technol. 8(1), 58–80 (2013)
9. Scicluna, J., Blank, C., Steinmetz, N., Simperl, E.: Crowd sourcing web service annotations. In: AAAI Spring Symposium: Intelligent Web Services Meet Social Computing. Volume SS-12-04 of AAAI Technical Report., AAAI (2012)
10. Wu, J., Wu, Z.: Similarity-based web service matchmaking. In: Proceedings of the 2005 IEEE International Conference on Services Computing, SCC '05, vol. 01, pp. 287–294. IEEE Computer Society, Washington, DC (2005)
11. Zhuang, Z., Mitra, P., Jaiswal, A.: Corpus-based web services matchmaking. In: Proceedings of the Workshop on Exploring Planning and Scheduling for Web Services, Grid and Autonomic Computing, pp. 46–52 (2005)
12. Fan, J., Kambhampati, S.: A snapshot of public web services. SIGMOD Rec 34(1), 24–32 (2005)
13. Platzer, C., Dustdar, S.: A vector space search engine for web services. In: Proceedings of the 3rd European IEEE Conference on Web Services (ECOWS€™05), IEEE Computer Society Press, pp. 14–16 (2005)
14. Lu, J., Yu, Y.: Web service search: Who, when, what, and how. In: Weske, M., Hacid, M.S., Godart, C. (eds.) WISE Workshops, pp. 284–295. Springer, Berlin (2007). (Volume 4832 of Lecture Notes in Computer Science)
15. Li, Y., Liu, Y., Zhang, L., Li, G., Xie, B., Sun, J.: An exploratory study of web services on the internet. In: IEEE International Conference on Web Services, ICWS 2007, pp. 380–387 (2007)
16. Lausen, H., Haselwanter, T.: Finding web services. In: Proceedings of the 1st European Semantic Technology Conference (ESTC) (2007)
17. Song, H., Cheng, D., Messer, A., Kalasapur, S.: Web service discovery using general-purpose search engines. In: IEEE International Conference on Web services, ICWS 2007, pp. 265–271 (2007)
18. Peng, D.: Automatic conceptual indexing of web services and its application to service retrieval. In: Jin, H., Rana, O.F., Pan, Y., Prasanna, V.K. (eds.) ICA3PP. Volume 4494 of Lecture Notes in Computer Science. pp. 290–301, Springer, Berlin (2007)
19. Ma, J., Cao, J., Zhang, Y.: A probabilistic semantic approach for discovering web services. In: Proceedings of the 16th International Conference on World Wide Web. WWW '07. pp. 1221–1222, ACM, New York (2007)

20. Al-Masri, E., Mahmoud, Q.H.: Wsce: A crawler engine for large-scale discovery of web services. In: IEEE International Conference on Web Services, ICWS 2007, pp. 1104–1111 (2007)

21. Wu, C., Chang, E., Aitken, A.: An empirical approach for semantic web services discovery. In: Australian Software Engineering Conference, IEEE Computer Society, pp. 412–421 (2008)

22. Wu, C., Potdar, V., Chang, E.: Latent semantic analysis—the dynamics of semantics web services discovery. In: Advances in Web Semantics I: Ontologies, Web Services and Applied Semantic Web. IEEE Computer Society, pp. 346–373 (2009)

23. Wu, C., Dillon, T.S., Chang, E.: Intelligent matching for public internet web services towards semi-automatic internet services mashup. In: IEEE International Conference on Web Services, ICWS 2009, IEEE, pp. 759–766 (2009)

24. Hao, Y., Cao, J., Zhang, Y.: Efficient IR-Style search over web services. In: Advanced Information Systems Engineering, 21st International Conference, CAiSE 2009, pp. 305–318 (2009)

25. Hao, Y., Zhang, Y., Cao, J.: Web services discovery and rank: An information retrieval approach. Future Gener. Comp. Syst. **26**(8), 1053–1062 (2010)

26. Wu, C.: Wsdl term tokenization methods for IR-Style web services discovery. Sci. Comput. Program. **77**(3), 355–374 (2012)

27. Li, C., Zhang, R., Huai, J., Guo, X., Sun, H.: A probabilistic approach for web service discovery. In: IEEE International Conference on Services Computing (SCC), pp. 49–56 (2013)

28. Rodriguez, A.: Restful web services: The basics. IBM developerWorks (2008)

29. Pautasso, C., Zimmermann, O., Leymann, F.: Restful web services vs. "big"' web services: Making the right architectural decision. In: Proceedings of the 17th International Conference on World Wide Web. WWW '08, pp. 805–814. ACM, New York (2008)

30. Pautasso, C.: On composing RESTful services. In: Leymann, F., Shan, T., van den Heuvel, W. J., Zimmermann, O. (eds.) Software Service Engineering. Number 09021 in Dagstuhl Seminar Proceedings, Dagstuhl, Germany, Schloss Dagstuhl - Leibniz-Zentrum fuer Informatik, Germany (2009)

31. Marc, H.: Web application description language, member submission subm-wadl-20090831, World Wide Web Consortium. http://www.w3.org/Submission/wadl/ (2009). Access: February 2014

32. Kopecký, J., Gomadam, K., Vitvar, T.: hRESTS: An HTML microformat for describing RESTful web services. In: Proceedings of the 2008 IEEE/WIC/ACM International Conference on Web Intelligence and Intelligent Agent Technology, WI-IAT '08, vol. 01, pp. 619–625. IEEE Computer Society, Washington, DC (2008)

33. Alarcón, R., Wilde, E.: RESTler: Crawling RESTful services. In: Rappa, M., Jones, P., Freire, J., Chakrabarti, S. (eds.) WWW, pp. 1051–1052. ACM, New York (2010)

34. Steiner, T.: Automatic multi language program library generation for REST APIs. PhD thesis, Karlsruhe Institute of Technology (2007)

35. Walsh, N.: Witw: Nsdl—norm's service description language. Norman.Walsh.name **8**(40) (2005). Access: February 2015

36. Bray, T.: Smex-d. http://www.tbray.org/ongoing/When/200x/2005/05/03/SMEX-D (2005). Access: February 2015

37. Jonathan, R., Cavicchio, R., Sinnema, R., Wilde, E.: Restful service description language (rsdl): Describing restful services without tight coupling. In: Proceedings of Balisage: The Markup Conference 2013, Balisage Series on Markup Technologies, vol. 10 (2013)

38. Taheriyan, M., Knoblock, C.A., Szekely, P., Ambite, J.L.: Semi-automatically modeling Web APIs to create linked APIs. In: Proceedings of the First Linked APIs workshop at the Ninth Extended Semantic Web Conference (2012)

39. Gomadam, K., Ranabahu, A., Nagarajan, M., Sheth, A., Verma, K.: A faceted classification based approach to search and rank web apis. In: IEEE International Conference on Web Services, ICWS '08, pp. 177–184 (2008)

40. Ambite, J.L., Darbha, S., Goel, A., Knoblock, C.A., Lerman, K., Parundekar, R., Russ, T.: Automatically constructing semantic web services from online sources. In: Proceedings of the 8th International Semantic Web Conference, ISWC '09. Springer, Berlin, pp. 17–32 (2009)
41. Pedrinaci, C., Liu, D., Lin, C., Domingue, J.: Harnessing the crowds for automating the identification of web apis. In: AAAI Spring Symposium: Intelligent Web Services Meet Social Computing. Volume SS-12-04 of AAAI Technical Report, AAAI (2012)
42. Steinmetz, N., Lausen, H., Brunner, M.: Web service search on large scale. In: Baresi, L., Chi, C.H., Suzuki, J. (eds.) Service-Oriented Computing. Lecture Notes in Computer Science, vol. 5900, pp. 437–444. Springer, Berlin (2009)
43. Steinmetz, N., Lausen, H., Brunner, M., Martinez, I., Simov, A.: D5.1.3 second crawler prototype. Service Oriented Architectures for All, project number: 215219, 7-th Framework Programme (2008)
44. Ly, P.A., Pedrinaci, C., Domingue, J.: Automated information extraction from web apis documentation. In: The 13th International Conference on Web Information System Engineering (WISE 2012), pp. 497–511 (2012)
45. Lin, C., He, Y., Pedrinaci, C., Domingue, J.: Feature lda: A supervised topic model for automatic detection of web api documentations from the web. In: Cudré-Mauroux, P., Heflin, J., Sirin, E., Tudorache, T., Euzenat, J., Hauswirth, M., Parreira, J., Hendler, J., Schreiber, G., Bernstein, A., Blomqvist, E. (eds.) The Semantic Web—ISWC 2012. Volume 7649 of Lecture Notes in Computer Science, pp. 328–343. Springer, Berlin (2012)

Intelligent Secure Communications Between Systems

Azahara Camacho, Pablo C. Cañizares and Mercedes G. Merayo

Abstract Almost every day a new intelligent system is developed and connected to the Internet using a wireless network. However, there is a great number of hazards that turn them into vulnerable systems. Hackers, social engineering, software and hardware errors are the main reasons for this phenomenon. In the case of intelligent systems, the most important aspect that can be affected, in terms of security, is the communication. In this paper we present a tool, implementing an effective technique, to avoid these security problems. The main goal of the tool is to determine if the communication among different intelligent systems is unprotected and prevent unauthorized access from external actors. In order to do it, the tool allows us to check online that the observed behaviour of the systems, in a specific scenario, fulfils some properties.

1 Introduction

Due to the increase of intelligent systems in the market (digital televisions, smart meters, point-of-sale terminals, etc.), the volume and the nature of communications has increased too and there is a need to protect the information traffic from

Research partially supported by the Spanish MEC projects ESTuDIo and DArDOS (TIN2012-36812-C02-01 and TIN2015-65845-C3-1-R) and the Comunidad de Madrid project SICOMORo-CM (S2013/ICE-3006).

A. Camacho (✉) · P.C. Cañizares · M.G. Merayo
Departamento de Sistemas Informáticos y Computación, Universidad Complutense de Madrid, Madrid, Spain
e-mail: mariaazc@ucm.es

P.C. Cañizares
e-mail: pablocc@ucm.es

M.G. Merayo
e-mail: mgmerayo@fdi.ucm.es

© Springer International Publishing Switzerland 2016
D. Król et al. (eds.), *Recent Developments in Intelligent Information and Database Systems*, Studies in Computational Intelligence 642,
DOI 10.1007/978-3-319-31277-4_16

interception and exploitation. We have to ensure that the systems are robust and that no undesirable actions can violate their security. The application of testing techniques with a formal basis [1, 2] help us to increase the confidence in the correctness of a system. Although most of these techniques involve the interaction with the system under test, there exist some approaches, based on the observation of the behaviour of the system, that are more adequate for achieving our goal: *passive testing* techniques. Formal passive testing is already a well established line of research and extensions of the original frameworks [3–6]. We propose the use of this technique to keep the safety and privacy of the information and present a tool based on a formal approach to perform passive testing with asynchronous communications [7]. Basically, the methodology consists in the observation of the behaviour of the system by a monitor that checks whether the observed actions satisfy certain required properties. These properties are relatively simple and allow us to perform the checking process in real-time. The application of passive testing in real-time has a significant benefit: if an error is detected, then it can be immediately notified to the operators of the system for the application of the appropriate measures.

In this paper we present the tool SCOIS (Secure COmmunication for Intelligent Systems) that automates this passive testing approach. The tool focuses on the control of the communication among different systems, with the goal of preserving information privacy. The main functionality of SCOIS is the control and detection of unauthorized accesses to the information involved in communications. In order to evaluate the validity of the proposal, we have applied it to a typical scenario in which a user try and saves a photo taken with a smartphone.

The rest of the paper is structured as follows. In Sect. 2 we introduce the hazards for the security of our communications and which ones are more dangerous. Section 3 explains the scenario for the application of our tool and the vulnerabilities that can suffer the user of the scenario. In Sect. 4 we present the main features of SCOIS. Finally, in Sect. 5 we present our conclusions and provide some lines for future work.

2 Hazards Information Security

Since the popularization of Internet we are immersed in an *information society* in which new information is continuously added, used and manipulated. The main means for distribution are communication technologies. Therefore, we have seen an increase in the use of intelligent systems that allow us to access the information. Online information has turned into an essential and powerful resource for a high percentage of the society and, during the last years, it has led to an exponential increase in the number of threats and hazards to which these systems are exposed. In this section we review the main hazards and analyze which ones are the most dangerous for intelligent systems.

2.1 Software

The development of software should be a process well documented and methodical, similar to the one that architects use to build a house [8]. It is essential to define all the requirements, check that they are fulfilled by the system, design the initial appearance and enhance it with additional improvements. One of the most important aspects that must be taken into account is the control system security. As is the case of a house, users need to feel that the software system that they use is secure, they need to trust not only its functionalities but also its security level and they need to know that the information privacy will be preserved.

Smartphones are the devices which store more information about our lives and controlling its protection is extremely important to keep our privacy. Since 2014, the number of users of these devices worldwide reached almost 5 billions [9]. Therefore, the relevance of software security in mobile devices, specifically in the functionalities related to communications, is unquestionable. Currently, the operative system (OS) most widely used in mobile devices is Android [10]. Due to the support of open source communities, it is easier and faster to solve software problems. However, this opening-up increases its exposition to malware compromising information privacy. Our tool helps us to overcome this vulnerability. The online checkout of the behaviour of the system will allow us to control if a specific property is being fulfilled. In the case that an anomalous action is detected, the problem can be fixed and the necessary measures to prevent security failures and malware actions can be taken.

2.2 Social Engineering

Although there exist many technical hacks, one of them does not require a specific knowledge for being applied: *social engineering*. It refers to the psychological manipulation of people with the aim of abusing their trust for gathering information or system access [11]. This kind of scam happens due to a very simple reason: people assume that everybody is honest. This assumption is not valid in professional environments because the most insignificant question could reveal confidential information that only a limited group of people should know [12]. According to this definition, we can identify different methods to apply this technique:

- *Direct requests*: this is the simplest and most effective method because the receiver of the request tends to not have time to check the veracity of the transmitter.
- *Trust and emotion*: it consists in the manipulation of people gratitude. The victim has a problem and the social engineer tries to help, but instead of that, the trickster gathers information with *technical* questions whose aim is to obtain personal information.

- *Impersonation and research*: this option is applied if the previous method is ineffective. To form a trusting relationship, social engineers get knowledge of potential victims and research their background in order to talk the same professional language.
- *Public clues*: it bases on obtaining system access data or user information from emails or shared documents that can be accessed by everybody.

The detection of these unauthorized accesses is a difficult task and protecting the information should be a priority. For this reason, we advocate for the use of a tool that controls the source of the requests in case of a remote access.

2.3 Hackers

All systems have vulnerabilities of different types and some of them are even published in the Internet. This information is used by hackers for planning attacks to computer systems and networks. Hackers are experts in breaking security and obtaining information for their own benefit. In the case of communication among systems, it is easy to be hacked due to the fact that there are many applications to carry out these attacks [13]. Communication attacks against privacy can be classified as:

- *Monitor and eavesdropping*: it is based on the capture of the network configuration. It is the most common attack and it is difficult to detect if the hacker does not play an active role in the transmission of information.
- *Traffic analysis*: it is based on the analysis of the communication content even if it is encrypted. The analysis of patterns and server responses allows hackers to figure out the sent messages.
- *Camouflage adversaries*: in this case an unauthorized system is inserted in a node network. This addition allows the user of this system to access all the resources of the network and capture the communication packets.

One of the main problems with the control of the communication among systems is the codification of the messages from the different available protocols. There exists software that allow us to classify them by protocol [14]. But despite this feature, it is difficult to control if the order of the messages is the correct or if someone is applying one of the previous attacks.

Our proposal can be applied also in this field. As in the case of software, our tool can be use to check online properties related to communication protocols. In this way, it allows us to detect unauthorized requests or unapproved reception of packets.

3   Case Study

In this section we present a non-trivial case study to illustrate some vulnerabilities that can be detected by our tool. We assume a scenario with two actors: a user and a camera application (CA) which is installed in an Android smartphone. Figure 1 illustrates this scenario. Next, we present the sequence of actions corresponding to the interaction between the actors.

1. The user requests the opening of the CA.
2. The CA turns on the camera and makes the application available.
3. The user takes a picture.
4. The CA asks whether the user wants to save the resource in the memory of the phone or delete it.
5. The user decides to save it.
6. The CA requests the name the user wants to use for identifying the image.
7. The user writes the name.
8. The CA saves the image in the memory and turns on the camera again, making it available for a new picture.
9. The user closes the application.

According to the list of hazards that we enumerated in the previous section, the considered case study can be classified as a software problem. When a new application is installed in an Android OS device, it is not possible to be completely sure that it is trustworthy. The fact that Android is open source entails a high risk of infection by malware. In our use case, we assume that the CA installation requires permissions for accessing Internet and the user accepts the conditions. This implies the activation of the option that allows users to share images by making them public. However, this is an automatic action and the user does not control and does not have any evidence of its activation. Therefore, there is a *malware* in our mobile and the images that are captured will be shared unless we detect the sequence of actions performed in the mobile when we use the CA. Next we present the sequence of actions that corresponds to the new scenario:

1. The user requests the opening of the CA.
2. The CA turns on the camera and makes the application available.
3. The user takes a picture.
4. The CA asks whether the user wants to save the resource in the memory of the phone or delete it.

Fig. 1 Scenario between a user and his smartphone

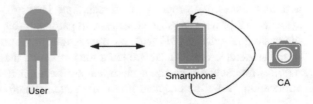

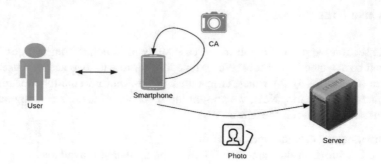

Fig. 2 Scenario derived from the action of the malware

5. The user decides to save it.
6. The CA requests the name the user wants to use for identifying the image.
7. The user writes the name.
8. **The CA requests to an external server to save a resource**.
9. **The external server accepts the request**.
10. **The CA sends the image to the server**.
11. The CA saves the image in the memory and turns on the camera again, making it available for a new picture.
12. The user closes the application.

In the evolution of this scenario we find a new actor: the external server, in charge of collecting all the images captured by the CA. This server could be used in undesirable ways: espionage, spam, pornography or in web sites with sensitive content. In Fig. 2 we present a schematic representation of the new scenario. In order to avoid this kind of malware, in the next section we present the main features of the tool that we propose to detect unauthorised actions.

4 SCOIS: Secure Communication for Intelligent Systems Tool

SCOIS automates a formal approach to perform passive testing with asynchronous communications [7]. Basically, its behaviour consists in observing the interactions between a system and the users to try and detect unexpected accesses. Essentially, in passive testing we have a property and we check that the observed sequence of actions satisfies that property. This technique is very appropriate to be applied in systems that present a restricted access, in particular due to security issues, because the system is running 24/7 and our interaction might produce undesirable changes in the associated data. These are situations in which the active testing schema, that requires interaction with the system, cannot be applied. In the case of the CA, the application cannot be checked in an idle state. Therefore, only the analysis of the flow of messages during its operation, without interacting or influencing it, will be effective.

SCOIS is a graphical user interface (GUI) that help us to keep the safety in scenarios like the previously mentioned. Its main components are:

- *Properties management*: it allows users to define properties to check whether an unauthorized actor, that might produce security problems, makes an appearance in a specific scenario.
- *Testing management*: this component allows users to determine *online* whether a property, previously defined, is satisfied during the communication among actors. In our scenario, the analysis of the messages exchange would help us to detect an undercover server.

4.1 Properties Management

The properties manager provides the most important functionality of SCOIS. The appropriate definition of properties guarantees the correct analysis of the communications. In Fig. 3 we present a capture of the GUI in which we can see the tab devoted to the definition of properties that will be checked.

This tab gives access to the interface used to indicate all the elements required to define a property: a descriptive name, the system with which the property is associated, the sequence of messages that should be exchanged between the actors and the set of authorised actions that can be observed after it. The definition of a property related to our case study is depicted In Fig. 3 and it includes the following information:

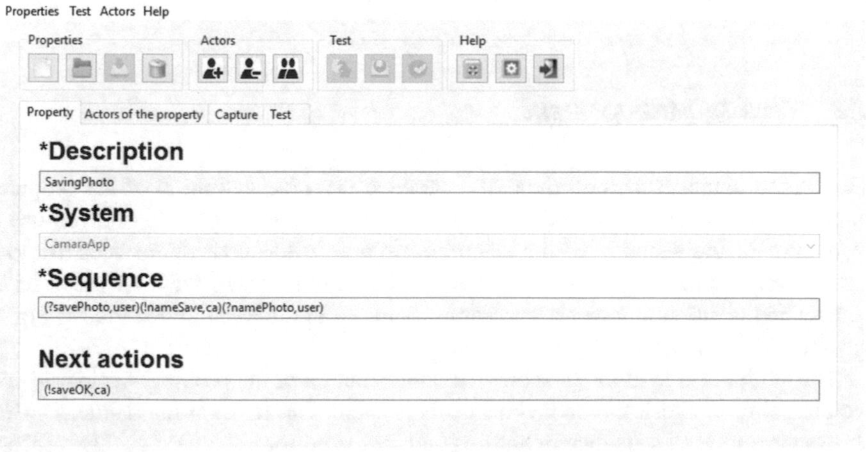

Fig. 3 Definition of a property

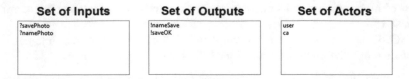

Fig. 4 Actors and actions in the case study property

- *Description*: **SavingPhoto**.
- *System*: **CamaraApp**.
- *Sequence*: (**?savePhoto**, **user**)(**!nameSave**, **ca**)(**?namePhoto**, **user**), which represents the interaction between the user and the CA. The user requires saving a picture, the CA asks for a name for the file to store it and the user inputs it.
- *Next actions*: (**!saveOK**, **ca**). This set establishes that the only allowed action after the observation of the previous sequence corresponds to the confirmation by the CA of the successful storage of the image.

As usual, in the previous enumeration of actions we have that inputs are preceded by ? and outputs are preceded by !. The specified property can be updated or deleted and new properties associated with the same system can be defined. The actors involved in a property are automatically obtained from its definition. The analysis of both, the sequence and the next actions set, will group the different actors and the input and output actions associated with both of them. Figure 4 presents the actions corresponding to our case study. Input actions are associated with the actions of the user (*?savePhoto* and *?namePhoto*), while output actions correspond to the actions of the CA (*!nameSave* and *!saveOK*). The set of actors only includes those that have access to the system. However, it can happen that an authorised actor does not appear in the property. In this case, if we want to consider it as a valid actor, in the case that it participates in the communication, we can manually add it to the set of actors.

4.2   Testing Management

The testing manager uses all the data generated from the definition of the property to check whether the communication among the actors is correct. In order to use it to determine the security of the communications in the considered scenario, we have connected the tool with the network of the mobile phone. This will allow us to control and check if any external actor participates in the communication. In Fig. 5 we show the testing manager interface.

The left hand side of the tab shows the information of the property against which we are going to check the behaviour of the system. On the right hand side, SCOIS shows the analysed sequence of actions and, at the bottom, the verdict. The process

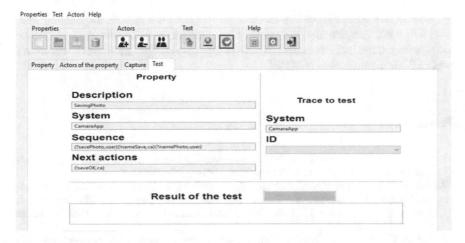

Fig. 5 Capture of the tab test of the tool

can be carried out either online or offline. In the second case, SCOIS stores the sequence of observed actions, so that it can be analyzed later against different properties associated to the system.

The property associated with our scenario can be used to detect unexpected actions performed by unauthorised actors during the communication between the user of the mobile and the CA. In addition, it can also be used to determine the correct behaviour of the system according to the defined property. In the case that the CA requests to an external server to save a resource, in our case an image, and a server accepts it, SCOIS will detect that an unauthorised actor is participating in the communication. This will allow us to immediately notify a possible security failure. In this way, the necessary measures to prevent malware actions could be taken.

5 Conclusions and Future Work

In this paper we have presented SCOIS, a tool for preventing the consequences of a malware. SCOIS checks the actors involved in the communication and the behaviour of the system, according to previously defined properties. The technique implemented in SCOIS is based on a passive testing approach where communications are asynchronous. This makes possible to check online the operation of a system. If an attack is detected, then the user can apply the appropriate measures in order to avoid the capture of private information.

As future work we plan to extend the features of the tool in order to deal with other hazards such as eavesdropping and access from forbidden systems. Currently, SCOIS is an independent software which is connected to the device to test.

We would like to integrate it with other software as the camera application introduced in this paper. Finally, we would like to extend our framework, and then our tool, with additional characteristic allowing to specify distributed, timed and probabilistic behaviours [15–20].

References

1. Cavalli, A.R., Higashino, T., Núñez, M.: A survey on formal active and passive testing with applications to the cloud. Ann. Telecommun. **70**(3–4), 85–93 (2015)
2. Hierons, R.M., Bogdanov, K., Bowen, J.P., Cleaveland, R., Derrick, J., Dick, J., Gheorghe, M., Harman, M., Kapoor, K., Krause, P., Luettgen, G., Simons, A.J.H., Vilkomir, S., Woodward, M.R., Zedan, H.: Using formal specifications to support testing. ACM Comput. Surv. **41**(2), 9 (2009)
3. Andrés, C., Merayo, M.G., Núñez, M.: Formal passive testing of timed systems: theory and tools. Softw. Test. Verification Reliab. **22**(6), 365–405 (2012)
4. Bayse, E., Cavalli, A., Núñez, M., Zadi, F.: A passive testing approach based on invariants: application to the WAP. Comput. Netw. **48**(2), 247–266 (2005)
5. Cavalli, A., Gervy, C., Prokopenko, S.: New approaches for passive testing using an extended finite state machine specification. Inf. Softw. Technol. **45**(12), 837–852 (2003)
6. Lee, D., Netravali, A.N., Sabnani, K.K., Sugla, B., John, A.: Passive testing and applications to network management. In: 5th IEEE International Conference on Network Protocols, ICNP'97, pp. 113–122. IEEE Computer Society (1997)
7. Hierons, R.M., Merayo, M.G., Núñez, M.: Passive testing with asynchronous communications. In: IFIP 33rd International Conference on Formal Techniques for Distributed Systems, FMOODS/FORTE'13, LNCS 7892, pp. 99–113. Springer, Berlin (2013)
8. Lamport, L.: Who builds a house without drawing blueprints? Commun. ACM **58**(4), 38–41 (2015)
9. Statista.: Number of mobile phone users worldwide from 2012 to 2018 (2015). http://www.statista.com/statistics/274774/forecast-of-mobile-phone-users-worldwide/
10. Statista.: Market share worldwide smartphone shipments by operating system from 2014 to 2019 (2015). http://www.statista.com/statistics/272307/market-share-forecast-for-smartphone-operating-systems/
11. Thompson, S.T.C.: Helping the hacker? Library information, security, and social engineering. Inf. Technol. Libr. **25**(4), 222–225 (2013)
12. Stajano, F., Wilson, P.: Understanding scam victims: seven principles for systems security. Commun. ACM **54**(3), 70–75 (2011)
13. Padmavathi, G., Shanmugapriya, D.: A survey of attacks, security mechanisms and challenges in wireless sensor networks. Int. J. Comput. Sci. Inf. Secur. (2009)
14. Harris, B., Hunt, R.: TCP/IP security threats and attack methods. Comput. Commun. **22**(10), 885–897 (1999)
15. Gregorio, C., Núñez, M.: Denotational semantics for probabilistic refusal testing. In: Workshop on Probabilistic Methods in Verification, PROBMIV'98, ENTCS 22. Elsevier (1999)
16. Hierons, R.M., Merayo, M.G.: Mutation testing from probabilistic and stochastic finite state machines. J. Syst. Softw. **82**(11), 1804–1818 (2009)
17. Hierons, R.M., Merayo, M.G., Núñez, M.: Testing from a stochastic timed system with a fault model. J. Logic Algebraic Program. **78**(2), 98–115 (2009)

18. Hierons, R.M., Merayo, M.G., Núñez, M.: Implementation relations and test generation for systems with distributed interfaces. Distrib. Comput. **25**(1), 35–62 (2012)
19. Hierons, R.M., Merayo, M.G., Núñez, M.: Timed implementation relations for the distributed test architecture. Distrib. Comput. **27**(3), 181–201 (2014)
20. López, N., Núñez, M., Rodrguez, I.: Specification, testing and implementation relations for symbolic-probabilistic systems. Theor. Comput. Sci. **353**(1–3), 228–248 (2006)

TooKie: A New Way to Secure Sessions

Péter Vörös and Attila Kiss

Abstract Nowadays cloud systems are widely spread, those are parts of our life even if we do not recognize it. These services store our pictures, documents, we post our fitness data there, we do our shopping, and post our thoughts to different clouds, or we just let apps to track our position and send it to some servers. Those services which are not mainly browser-based often offer a way to check our data from browsers through HTTP or HTTPS. Here comes a security vulnerability which we will study in this paper. Because HTTP is a stateless protocol there has to be something what keeps persistence, for this purpose they invented session cookies. We will be working with a special type of attack called session hijacking, which targets these cookies. If the attacker gets those cookies he can act like an authenticated user. We will show multiple ways to steal and to protect these cookies. In this paper we show our measurements, how vulnerable different sites are to this type of attack. We will study the existing methods how to protect our servers against this threat. Finally we introduce our embedded authenticator TooKie (which is a mix-word of token and cookie), which is a token based application. We will show that it gives a great protection against session hijacking, yet it's simple to set up and use.

Keywords Web authentication · One-Time-Token · Session cookie · Session hijacking · Cross site scripting · Network eavesdropping

1 Introduction

As a service provider, securing the users and their data must be the first priority task to do, but in practice websites are often left vulnerable to the most common attack types. The majority of the websites use the encrypted channel (HTTPS) for only the

P. Vörös (✉) · A. Kiss
Eötvös Loránd University, Budapest, Hungary
e-mail: vopraai@inf.elte.hu

A. Kiss
e-mail: kiss@inf.elte.hu

© Springer International Publishing Switzerland 2016   195
D. Król et al. (eds.), *Recent Developments in Intelligent Information
and Database Systems*, Studies in Computational Intelligence 642,
DOI 10.1007/978-3-319-31277-4_17

login page, which is understandable on one hand, because securing everything can be quite exhausting, but the leakage of security leaves a great attacking area especially if session cookies are not protected enough.

HTTP is a stateless protocol, which means users can only be recognized from a specific user-ID which is stored as a cookie, this specific cookie which identifies the user session is called session cookie. Session hijacking is an attack type where the attacker somehow gets the victim's the session cookie, and with that cookie the attacker can act like the victim, and do whatever the victim was qualified to do.

Firstly in this paper we will show our experimental results abut how different companies protect their session cookies, and how vulnerable are they against different type of session hijacking attacks (e.g.: Network eavesdropping, and XSS). After the evaluation of the results we introduce our authenticator called TooKie, which is an embedded solution for securing session cookies without the need of secured connection for the whole conversation.

In this paper we assume that HTTPS is safe and secure, man in the middle is not possible, so no 3rd person can get any data from between the server and the client. This assumption is necessary, because without it, raw user names and passwords could easily be stolen, which leads to a far worse problem than what we're working with.

The paper is organized as follows. First in Sect. 2 we start with some related work, then in Sect. 3 we give a detailed overall picture about how websites authenticate users, what are cookies, and how session hijacking works. In Sect. 4 we show what cookies do different sites (Gmail, Facebook, etc.) use to authenticate their users. We will also run a tests about what can an attacker do if he somehow gets our cookies. So we measure how serious issue is session hijacking in big websites. Later in Sect. 5 we present our token based authenticator TooKie, and show through examples how communication between client and server works, and how can the user authenticate himself through HTTP, while he doesn't send anything vulnerable through the network. At Sect. 5 we summarize the paper.

2   Related Work

Several surveys [1–5] have studied web authentications' security. Interestingly Cross Site Scripting (XSS), and Cross Site Request Forgery (CSRF), and network eavesdropping are still in the most common attack types against websites, like they were a very serious threat in the past decades. Those methods can be used in a very wide range of attacks, we now focus on their impact to the session hijacking attack. Session hijacking attacks show back to the creation of HTTP, and stays an existing hazard nowadays as well. However through the years several different methods are created to make sessions more secure, none of these are widely deployed.

SessionShield [6] acts as a proxy between the browser and the network. SessionShield strips the HTTP headers from incoming HTTP packets and stores them in a database, and puts them back to outgoing packets. This means on client side browsers don't store any cookies, therefore they cannot be stolen by client side scripts. Bugliesi et al. [7] created a Chrome extension caller CookieExt. After each response if a session cookie is found then, if it was an HTTPS response, the identified session cookies marked as isSecure and httpOnly. If it was not HTTPS but an HTTP response the session cookies are removed from the headers. This method keeps the cookies secure on client side. They work fine, but both SessionShield and CookieExt require installation or special attention from users, which we believe is the biggest reason why they are not widely used.

Not just client side solutions were implemented, there are plenty of new protocol ideas to prevent session hijacking [8]. Bortz et al. [9] introduce origin-cookies, which are a cookie-like functionality that make web applications able to secure their cookies. Dacosta et al. [10, 11] show a new protocols which add security tokens to the HTTP headers, this token based authentication is very similar to our method, but instead of writing a new protocol we implemented an embedded java application, which runs in the background on the user side, without the need to install anything, or the need of extending the browsers.

3 Background

To understand session hijacking attacks, first of all we summarize what are cookies, and how does web authentication work.

3.1 Cookies

Webpages have no memories [12]. If a user goes from page to page it will be treated by the website as a completely new visitor. For keeping persistence HTTP uses cookies. Cookies are small pieces of data which is sent by the server, stored in the user's browser. Cookies are sent back to the server with every request the user makes. Session cookies are those specific HTTP cookies, which stores an ID which with the website is able to keep track of users' movement from one page to another.

3.2 Web Authentication

Logging in into websites is always a similar procedure [13]. The user types in the username and password, and send it to the server. (Most of the servers use HTTPS for securing login information, so we can assume that our data is sent in a safe channel here.) The server calculates a specific hash from the received data, and checks if this hash matches the hash which is stored in the users database. After the server validated the login request, it generates a unique session ID for the client, and sends it back with the response. This session ID will be stored as a session cookie on client side. Cookies are always sent to the server with every request, so from the session cookie the server can easily recognize clients.

3.3 Session Hijacking

It's a commonly used technique to use cookies as the only identifier for user sessions, which means if the attacker owns the victim's session cookie, the website grants the attacker to act as the victim. Session hijacking [14] is the process, when an attacker steals the session cookie from the victim, and gain unauthorized access to information or services in a computer system, and do what the original owner was qualified to do. There are several ways to steal cookies, in the followings we introduce the most common methods.

Network Eavesdropping If network traffic is not encrypted (e.g.: HTTP) traffic can be read by any participant in the network where the data goes through, not just by the sender and receiver [15]. Attackers can observe the communication of other users by being a man in the middle on the network. To be a man in the middle can be quite easy, or very hard, it depends on the circumstances.

For example put a WiFi router without a password into a crowd area, some people will start to use it. Everything which travels through this router can be logged, and if the connection is not encrypted, usernames and passwords can be stolen easily. HTTPS traffic can also be logged, and observed in this way but since it's encrypted it cannot be understood.

XSS—Cross-site Scripting Another common attack type is Cross-site scripting (XSS) [16]. This occurs when the attacker posts program code (mostly JavaScript) to a website which does not escape input fields right. By posting malicious code, the attacker can cause the victim's web browser to send the victim's cookies to a website which is controlled by the attacker.

Listing 1.1 Example XSS code, which will redirect us to attacker.com, and sends our cookies to the attacker, where he can store it

```
<script>
window.location='attacker.
com/stole?cookie='+document.cookie;
</script>
```

It is mostly used in forums or where users are allowed to post content, which others see. The posted malicious script is interpreted as an executable code, so if anyone loads the post, their browser will automatically run the script, and do whatever the attacker wanted us to do.

Cross-site Request Forgery with URL Redirection A URL redirection is used to navigate the browser from one URL to another. In some websites there is a page, which does the URL redirection to the page given as a parameter, which is harmless by default.

Listing 1.2 Example URL redirection. This website brings the users to: http://
www.example.com/home.php

```
http://www.example.com/login.php?redirect='http://www.
example.com/home.php?page=1'
```

Unless the attacker crafts a hyperlink like in Listing 1.3, where he gives his
webpage as a target of the redirection. In this case users will be navigated to the
attacker's page even if the URL they clicked started with the domain name of a
trusted company. Paying more attention is simply not enough in some cases,
because URL-s are usually shortened in browsers, as there is no room for long texts
at the bottom of a browser. There are also many url shorteners, which make it really
difficult to notice the malicious redirect (as it is usually not as trivial as in our
example).

Listing 1.3 Example crafted cookie stealing URL, using the harmless website's
URL redirection

```
http://www.example.com/login.php?redirect='http://www.
attacker.com/stole?cookie='+document.cookie;
```

3.4 Defending Cookies

Cookies can be protected by 2 flags, isSecure and httpOnly [17]. If the service
provider encrypts the communication between the user's computer and the server
by employing Transport Layer Security (HTTPS protocol), the server can specify
the isSecure flag while creating a cookie. This will cause the clients' browser to
send the cookie over only an encrypted channel, such as an SSL connection.

Smaller companies don't use HTTPS all the time, they mostly use it only for the
login page. If they do so, they cannot make their session cookies secure, because to
keep users logged in, they have to send session cookies back all the time in an
unencrypted channel.

The other flag httpOnly is less likely to be used. This flag "tells the browser" to
keep the cookie inaccessible from client side scripts. One problem with that flag is
that it does not protect against network eavesdropping, the bigger problem is it
doesn't used in most of the websites for some reason.

4   Measurements

In our measurements we assume that the attacker stole the session cookies by any
method, the actual method is not important now. To measure site vulnerability we
introduce 2 test scenarios. In the first one we simulate the following: we log into the
website with a browser (Firefox in our case), then copy all cookies to a different

browser (Opera), then we check if we can act like the authenticated user. If the website allows us to do that it fails this test.

The second test is a similar but a bit more realistic than the first one. Since our data is mostly stolen over the web, we simulate what happens if our cookies are getting used with a second computer with different IP, MAC address, hostname. This test is more specific then the first one, so those sites who passes the first test automatically passes this one as well.

4.1 Test Targets

We aimed to check multiple different sites, with different purpose and different amount of users. Unfortunately we didn't have the permission to test banks, or governmental sites. But we believe they would not fail our tests since the information stored in those sites are much more sensitive, and must be protected with several more authentication steps.

Google (GMail) Our first test target is Google, especially their Gmail service. After logging into their system, GMail creates 29 different cookies under 5 different domains. Most of them are under accounts.google.com, myaccount.google.com and .google.com, but as google owns youtube, the login also generates the cookies for that site, and (because of I'm Hungarian) some cookies under .google.hu also appears.

We dropped the unnecessary cookies, and kept only those which were required to hijack a session. 3 cookies under accounts.google.com and 6 under .google.com so totally 9 cookies needed to do that (Table 1).

With these cookies, the identity of any user can be stolen. We have to consider that getting those could be really difficult as Google uses HTTPS all the time, and there is no cookie which is not protected either with httpOnly or isSecure flag.

Facebook The next target is the biggest social website Facebook. As expected Facebook does not create cookies under different domains like Google did. It only

Table 1 List of necessary cookies to act as a logged in user in GMail

Domain	httpOnly	isSecure	Name
accounts.google.com	false	true	ACCOUNT_CHOSER
accounts.google.com	false	true	LSID
accounts.google.com	false	true	GALX
.google.com	true	false	NID
.google.com	true	false	SID
.google.com	true	false	HSID
.google.com	true	true	SSID
.google.com	true	false	APISID
.google.com	true	true	SAPISID

uses .facebook.com to store the 9 different cookies. 3 of them were enough to compromise any account.

Notice that there is no cookie which uses httpOnly. Therefore with javascript it is easy to access and steal everything necessary to act like the victim. To show that there is a small code which is enough to get the required data to control the account. Due to facebook is a post oriented site, XSS can always be a big threat (Table 2).

Listing 1.4 Pasting this code to the URL bar will write out every cookie which are not having the httpOnly flag `javascript:alert(document.cookie);`

MyFittnessPal After seeing some of the biggest websites we targeted some of the smaller companies which still contain some vulnerable user data. Let's begin with one of the biggest fitness site MyFitnessPal with it's 75 million users. While in Google and Facebook everybody knows what are the risks of a compromised account, on a fitness page it may need some explanation.

These pages track several things, MyFitnessPal aims mostly the nutrition data and the workouts. Nutrition data can be used to give specific ads, or monitor your daily routine, when do you eat breakfast, so how long do you usually sleep. Workout data can be used well for example guessing when is nobody at home.

MyFitnessPal uses 15 cookies per a logged in account, 6 under www. myfitnesspal.com and the rest 9 at .myfitnesspal domain. Unfortunately, or fortunately as the point of view, there is only one cookie which is required to authenticate ourselves (Table 3).

That cookie _session_id is not even protected with any of the security flags, so getting these can be quite easy.

RunKeeper Another example of fitness apps with browser support, this site tracks the running, biking, etc. activities, and even provides a map showing the route of the exercise. The attacker will not just know whether the victim is home or not, they can easily meet the target while the activity.

From the 14 cookies again there is only one which is necessary to keep anyone logged in, therefore getting this will do the job for the attacker. As we seen before the cookie is not secured with any flag (Table 4).

Amazon One of the most known webshops is amazon, there are 11 cookies appearing after a successful login attempt, 10 with .amazon.com, and 1 with www. amazon.com domain. However only 2 of them are enough to fail our test.

Table 2 List of necessary cookies to act as a logged in user in Facebook

Domain	httpOnly	isSecure	Name
.facebook.com	false	true	c_user
.facebook.com	false	true	xs
.facebook.com	false	true	s

Table 3 List of necessary cookies to act as a logged in user in MyFitnessPal

Domain	httpOnly	isSecure	Name
www.myfitnesspal. com	false	false	_session_id

Table 4 List of necessary cookies to act as a logged in user in RunKeeper

Domain	httpOnly	isSecure	Name
.runkeeper.com	false	false	checker

None of the two is protected anyhow, but we have to say that it's not likely to have an XSS vulnerability in a webshop, and by using HTTPS all the time packet sniffing is not likely to happen (Table 5).

Aliexpress Another big online store with 289 million monthly active users is Aliexpress. 4 cookies required to authenticate ourselves from the 26 cookies under 6 different domains which were generated after logging in (Table 6).

A big difference between Amazon and Aliexpress is while none of them secures their cookies, Amazon uses HTTPS by default to every single page, but Aliexpress does not use encryption after login.

4.2 Results

Every protection is as effective as the least effective part of it. This rule is true with cookies as well, so first of all we summarize what flags do the most secure cookie has in the different sites (Table 7).

As it was expected the results shows us that the bigger the site the better the protection. The most secure page in our test is Google. While it uses cookies with both isSecure and httpOnly flag, it's really difficult to steal information about the sessions, not to mention that all communication goes over HTTPS.

Facebook is protecting cookies quite well, they don't let us to choose between secure and insecure channels, they redirect to HTTPS even if HTTP is explicitly written before the URL. In every forum like site where posting is enabled just like with Facebook, there is always a big risk of XSS attacks, due to the shortage of httpOnly flag it can lead to serious issues. We also classify Amazon as semi-safe

Table 5 List of necessary cookies to act as a logged in user in Amazon

Domain	httpOnly	isSecure	Name
.amazon.com	false	false	ubid-main
.amazon.com	false	false	session-id

Table 6 List of necessary cookies to act as a logged in user in Aliexpress

Domain	httpOnly	isSecure	Name
.aliexpress.com	false	false	xman_us_t
.aliexpress.com	false	false	xman_t
.aliexpress.com	false	false	xman_f
.aliexpress.com	false	false	xman_us_f

Table 7 This table show what flags do the most secure cookie has in different sites

Target	httpOnly	isSecure
Gmail	Yes	Yes
Facebook	No	Yes
Amazon	No	No
Aliexpress	No	No
MyFitnessPal	No	No
RunKeeper	No	No

Table 8 Test results of different sites

Target	Test 1	Test 2	Easy to get cookies
Gmail	Fail	Fail	No
Facebook	Fail	Fail	No
MyFitnessPal	Fail	Fail	Yes
RunKeeper	Fail	Fail	Yes
Amazon	Fail	Fail	No
Aliexpress	Fail	Fail	Yes

because while they don't protect their cookies with any flag and it can explicitly be used with pure HTTP (while the default communication is using SSL).

The fitness sites (MyFitnessPal and RunKeeper) and Aliexpress both use HTTP by default, HTTPS is used for only the login form. With the insecure cookies, they can easily be targeted by a prepared attacker. Both with packet sniffing and with JavaScript injection, the identity of a logged in user can be stolen.

Our every test target failed both the tests, yet it would be interesting to check how banks, or governmental sites perform (Table 8).

5 TooKie

The biggest problem with the sites we tested above is that they do not use any client side stored data, which is not sent through the network with every request. We created a token based authenticator which extends the casual sessionid based identification, with a special token which is generated by the client for every request. This authenticator is called TooKie (from the words Token and Cookie), and it's implemented as an embedding java application (Fig. 1).

A high level behavior can be summarized as follows:

1. Client logs in
2. Server generates a session cookie, a secret, and a generator-string (both are long random strings)
3. Client stores the session cookie, and put the secret into TooKie (where it cannot be queried from), and stores the generator-string as a cookie.

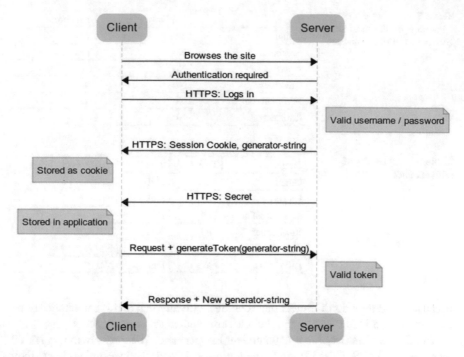

Fig. 1 Sequence diagram about how TooKie works

4. The client authenticates itself with it's session cookie, and a token generated with TooKie from the generator-string.
5. Server calculates the token itself, if it matches the client's token, the client is authenticated.
6. Server generates a new generator-string, and send it back to the Client with the response.

5.1 TooKie in Details

TooKie must be embedded in the main frame of a website, and the original content of the site shall be moved to a new frame, and some JS scripts shall be included, and configured. It's compatible with most of the sites, and does not require too much work to make it work. On client side, no extra attention is required, it's not even visible for the average users.

After setting those things up, the server is ready to go. Users can browse unsecured pages without the need of logging in, but secured pages require password based authentication. After a successful attempt, the user gets 3 data in the response. A session-id, and a generator-string, which are stored as cookies without any flags, and a secret which is given to the application as soon as the user receives it. This secret is only sent to the client at the login phase, which can only be made through HTTPS. As we assumed before HTTPS is safe, because if it's not, raw passwords can be stolen which leads to a lot more serious issue. The secret cannot be accessed after the login, so stealing it is not an option.

A logged in user can browse everything just like before without TooKie, but after each page load the application appends the One-Time-Token (OTT) parameter to the end of each URL under the domain of the server. The OTT is sent back to the server with every request, as an HTTP GET parameter. The attacker can only get the Session cookie and the OTT-s with eavesdropping. But due to these tokens cannot be reused, without knowing the secret, the attacker cannot generate OTT-s, therefore cannot act like an authenticated user.

If a server gets a request it identifies if there is any need for permissions to serve the request. If so, it checks if all 3 parameters are given with the request: session-id, generator-string, and the OTT. If anything is missing it redirects to the login page, because there is something wrong. If everything is given it identifies the user with it's session-id. Checks if the user sent back generator-string matches the one which is stored in the server. Then looks up the user's secret, and with the generator-string and secret it also computes the OTT. If it matches the one the user attached to the request, the user is authenticated itself for this request and the request can be served. Among with the reply the server changes the generator-string cookie if the client to a new one, making it impossible to use the old one again. If any check fails above, the user is redirected to the login page.

Performance is a key thing when computing OTT-s. TooKie uses one hash computation per request at server side, and one after each response at client side, which results in an immeasurably low overhead on both sides.

5.2   Under Attack

TooKie keeps the secret secure from the users as well, due to security reasons users don't have the right to query their secret. The application gives the access to the users to 2 functions:

- `void setSecret(secret)`
- `String generateOTT(generator-string)`

We believe that because of this our solution gives a great protection against session hijacking.

XSS—Cross Site Scripting TooKie makes XSS attacks inefficient. The attacker cannot trick or force the users to send their secret to anywhere. Due to this secret after send to the user via HTTPS, is immediately stored in the TooKie, which does not let users to query their secret even if they wanted to. Cross-site request forgery is made impossible because of the same reason.

Network Eavesdropping The traffic which goes over HTTP can be eavesdropped by attackers, but it does not contain anything vulnerable, which they can use up later. Among with the request there is a session-id, a generator-string, and the OTT. If the attacker steals both he cannot use them up later, because the OTT expires after the first use, and from these data new OTT-s cannot be generated.

6 Conclusion

In this paper we gave a short introduction about web authentication, cookies, and the most common attack types targeting them. We ran tests on well known sites, we measured if we can act like a logged in user if we somehow get all of their cookies. We also tested what happens on each site if the "stolen" cookies are used with a different computer, with different properties (IP, MAC, hostname, etc.).

We selected and kept only the sufficient cookies, which were enough to hijack a session. We classified those what security flags (isSecure, and httpOnly) are set on them. Also we measured what type of connection does each site use by default, HTTP or HTTPS, and if they use HTTPS is there a way to explicitly change the security back to pure HTTP. The results weren't too satisfying, Google seems to have a hard protection, but smaller sites like Aliexpress and RunKeeper can be targeted easily by a well prepared attacker.

After the measurements, we introduced our application TooKie, which we believe gives a strong protection against session hijacking. With our method no vulnerable data goes through the network after the login phase. We made our application secure, so even client side cannot access their vulnerable data, in practice the app does not let users to query their secrets. Under the development it was a big goal to keep it simple and easy to set up on server side, and to be able to run without any installation or special attention on client side.

Acknowledgment Authors thank Ericsson Ltd. for support via the ELTE CNL collaboration.

References

1. Fu, K., Sit, E., Smith, K., Feamster, N.: The dos and don'ts of client authentication on the web. In: USENIX Security Symposium, pp. 251–268 (2001)
2. Visaggio, C.: Session management vulnerabilities in today's web. IEEE Secur. Priv. **5**, 48–56 (2010)
3. Wedman, S., Tetmeyer, A., Saiedian, H.: An analytical study of web application session management mechanisms and http session hijacking attacks. Inf. Secur. J. Glob. Perspect. **22** (2), 55–67 (2013)
4. Khan, W.: Web session security: Formal verification, client-side enforcement and experimental analysis (2015)
5. Calzavara, S., Tolomei, G., Bugliesi, M., Orlando, S.: Quite a mess in my cookie jar! WWW 2014 (2014)
6. Nikiforakis, N., Meert, W., Younan, Y., Johns, M., Joosen, W.: Sessionshield: Lightweight protection against session hijacking. In: Engineering Secure Software and Systems, pp. 87–100. Springer, Berlin (2011)

7. Bugliesi, M., Calzavara, S., Focardi, R., Khan, W.: Automatic and robust client-side protection for cookie-based sessions. In: Engineering Secure Software and Systems, pp. 161–178. Springer, Berlin (2014)
8. Liu, A.X., Kovacs, J.M., Huang, C.-T., Gouda, M.G.: A secure cookie protocol. In: Proceedings of the 14th International Conference on Computer Communications and Networks, ICCCN 2005, pp. 333–338. IEEE (2005)
9. Bortz, A., Barth, A, Czeskis, A.: Origin cookies: Session integrity for web applications. In: Web 2.0 Security and Privacy (W2SP) (2011)
10. Dacosta, I., Chakradeo, S., Ahamad, M.: One-time cookies: Preventing session hijacking attacks with disposable credentials (2011)
11. Dacosta, I., Chakradeo, S., Ahamad, M., Traynor, P.: One-time cookies: Preventing session hijacking attacks with stateless authentication tokens. ACM Trans. Internet Technol. (TOIT) 12(1), 1 (2012)
12. Park, J.S., Sandhu, R.: Secure cookies on the web. IEEE Internet Comput. 4, 36–44 (2000)
13. Ma, Y.-N., Qian, H.-Y., Sun, Y.-M.: Research on cookie's application in web authentication. MINIMICRO SYSTEMS-SHENYANG- 25(2), 207–210 (2004)
14. Kolšek, M.: Session fixation vulnerability in web-based applications. Acros Secur., 7 (2002)
15. owasp.org. (2012) Network eavesdropping (Online). Available: https://www.owasp.org/index.php/Network_Eavesdropping
16. Di Lucca, G.A., Fasolino, A.R., Mastoianni, M., Tramontana, P.: Identifying cross site scripting vulnerabilities in web applications. In: 26th Annual International Telecommunications Energy Conference, INTELEC 2004, pp. 71–80 (2004)
17. Barth, A.: Http state management mechanism (2011)

Surface of Articular Cartilage Extraction Using Fuzzy C-means Segmentation

Jan Kubicek, Iveta Bryjova, Marek Penhaker, Michal Kodaj and Martin Augustynek

Abstract The article deals with complex segmentation approach which is focused on detection of pathological changes on the articular cartilage surface. There is a significant problem, in the clinical practice with recognition of individual structures of cartilage surface. Cartilages are normally investigated either by ultrasound or MRI. Both methods give output images in shade level spectrum. This fact is severe problem for detection especially small changes which indicate surface deterioration. This pathological stage often leads to further development of disease which cause to total loss of articular cartilage. The proposed software approach partially solves mentioned problem. The core of segmentation is based on fuzzy C-means algorithm which is very sensitive even in the noisy environment and for tiny structures as well. Furthermore, color transformation is implemented. This key benefit of proposed software allows transformation physiological cartilage to color spectrum, other tissues are suppressed. In the output, we give mathematical model of articular cartilage with detection of pathological discontinuities.

Keywords Articular cartilage · Fuzzy C-means · Image segmentation · Color transformation · Chondromalacia

J. Kubicek (✉) · I. Bryjova · M. Penhaker · M. Augustynek
FEECS, VSB—Technical University of Ostrava, K450 17, Listopadu 15,
708 33 Ostrava, Poruba, Czech Republic
e-mail: jan.kubicek@vsb.cz

I. Bryjova
e-mail: iveta.bryjova@vsb.cz

M. Penhaker
e-mail: marek.penhaker@vsb.cz

M. Augustynek
e-mail: martin.augustynek@vsb.cz

M. Kodaj
Podlesí Hospital, a. s. Konská 453, 739 61 Třinec, Czech Republic
e-mail: michal.kodaj@gmail.cz

© Springer International Publishing Switzerland 2016 209
D. Król et al. (eds.), *Recent Developments in Intelligent Information
and Database Systems*, Studies in Computational Intelligence 642,
DOI 10.1007/978-3-319-31277-4_18

1   Introduction to Articular Cartilage

Cartilage is a specialized type of fibrous tissue. It is composed of different sub-stances, each of them responsible for its overall integrity, deformability, hardness and the ability to repair itself. Cartilage is created from mesenchymal cells at the ends of the epiphyses of bones during embryonic development in human foetuses.

Histologically, it is classified into three basic types: elastic, fibrous and hyaline. The contact surfaces of joints are covered with hyaline cartilage. Cartilage is organized into a layered structure, which is functionally and structurally divided into four layers. The surface layer is responsible for its smoothness and is resistant to friction. It makes up about 10–20 % of the total depth of the cartilage. It can compress about 25 times more than the middle layer.

Adult cartilage is composed of 75 % water and 25 % solid compounds. The metabolism of cartilage is mainly anaerobic. Due to cartilage's shortage of its own vascular supply and nerve fibres, it has a very low ability to repair itself or remove metabolites. Despite its low metabolic turnover, the replacement and the continuous exchange of cells occurs in cartilage [1–3].

2   Examination of Articular Cartilage by MRI

Magnetic resonance imaging (MRI) is the most common investigative method. Thanks to its high distinctive ability and spatial resolution, it is the optimal non-invasive method for viewing the soft tissues of joints and cartilage. Chondral separations manifest as vertical defects in cartilage that extend deeply to the sub-chondral bone and are sharply outlined against the surrounding cartilage. The best results in chondral pathology imaging are achieved via a proton density weighted sequence with fat suppression, a gradient spin-echo sequence, SPGR sequences, MR arthrography and 3D fast spin echo imaging. During readings and evaluations by radiologists or orthopaedists, small chondral lesions may remain undiagnosed. Post-processing methods based on color coding can significantly contribute to more accurate diagnostic conclusions.

2.1   Fat Suppression Techniques

To improve the dynamic range of signal intensity from an image, fat suppression sequence is implemented in the routine protocol for cartilage imaging pulse sequences. There is better detection of minor signal intensity modifications. A T1-weighted inversion recovery sequence, termed as short tau inversion recovery (STIR), is another technique by which the effect of fat suppression is obtained. Combination of fat suppression together with three-dimensional (3D) spoiled gradient echo sequence only depicts the articular cartilage as a bright structure. The sequence with the best possible potential to assess the articular cartilage accurately is spoiled gradient echo sequence (SPGR).

2.2 Volumetric Sequences

Optimal spatial resolution of images can be obtained by acquiring thinner slices (0.2–0.4 mm thickness) producing isotropic voxels with equal dimensions in the three planes. The advantage of volumetric image acquisition is that it facilitates multiplanar reformation without loss of the spatial resolution, allowing a better quantitative measurement of cartilage abnormalities and reducing partial volume artifacts.

2.3 Gradient Echo Sequences

Gradient echo sequences allow volumetric image acquisition with reduced imaging times and improved spatial resolution. Technically, a refocusing pulse separates two or more gradient echoes, eventually combining the echoes to generate an image. Gradient-recalled echo (GRE) sequences suffer from a potential fallacy, as they are unusually sensitive and susceptible to intra voxel dephasing, especially in patients with previous surgical intervention or some hardware placement (Figs. 1 and 2).

Fig. 1 A 3D GRE, sagittal plane image of knee joint revealing articular cartilage

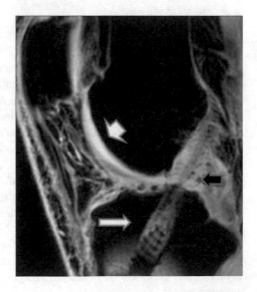

Fig. 2 Sagittal 3D SPGR
(TR/TE 14.1/5 ms)

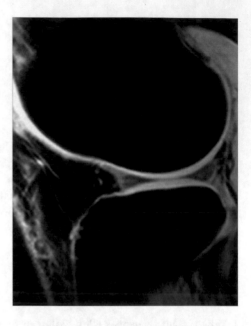

Excellent details of articular cartilage (short arrow). The additional advantage of this sequence is in imaging of joints with prior interventions or any grafts or screws fixations (long arrow). The tiny blooming intensities (black arrow) around the BPB graft, likely postoperative bone fragments.

2.4 SPGR Sequences

Spoiled gradient echo sequences such as SPGR or fast low angle shot (FLASH) provide T1-weighted images with better contrast between cartilage (hyperintense) and intra-articular fluid (hypointense). Standard SE MRI has a lower accuracy as compared to the fat-suppressed SPGR. The sensitivity of the sequence approaches 93 %, with articular cartilage appearing bright and remainder of the structures appearing relatively dark. 3D SPGR is recommended by the International Cartilage Repair Society (ICRS) as a standard sequence in imaging for evaluation of the articular cartilage lesions, especially in post cartilage repair status. The images are primarily T1-weighted sequences obtained at a low flip angle. The disadvantages are longer acquisition times and inadequacy to depict surface defects as well as the other joint structures. Finally, 3D gradient echo methods are less useful for the diagnosis of ligament or meniscal tears than SE techniques. Despite these limitations, 3D SPGR imaging is considered the standard for morphologic imaging of cartilage.

At flip angle 60° of knee provides excellent contrast between cartilage and subchondral bone along signal-intensity cartilaginous surfaces in tibiofemoral compartments.

2.5 MR Arthrography

Post cartilage repair, MR arthrography (MRA) is recommended in evaluation of cartilage integrity. The instilled contrast outlines cartilage defects and their contours. A small amount of 2 mmol/l gadolinium contrast mixed with saline is injected within the joint space to increase its sensitivity to detect cartilage abnormalities. MRA has been shown to be more sensitive than the standard 3D T2-weighted gradient echo sequences for the detection of cartilage lesion (Fig. 3).

3 The Proposed Method for Surface of Articular Cartilage Segmentation

The main task of software implementation is extraction of pathological lesions of articular cartilage. The main problem is, that pathological lesions are represented be very week contract and therefore they are with difficulty recognizable by human's eyes. There are four stages of chondromalacia lesions. From the perspective image processing is the most challenging chondromalacia of first degree. In this particular case it is very complicated to detected and recognize pathological lesion from physiological cartilage.

The overall structure of segmentation process is illustrated on Fig. 4. Data of articular cartilage are acquired from MRI. MRI usually generates longer data sequence. Firstly it is needed perform selection of data where pathological lesions are manifested. After taking region of interest (RoI), pixel's interpolation is applied. RoI extraction is needed for increasing area of articular cartilage. Important problem is that articular cartilage takes small part of native image and therefore individual

(a)   **(b)**

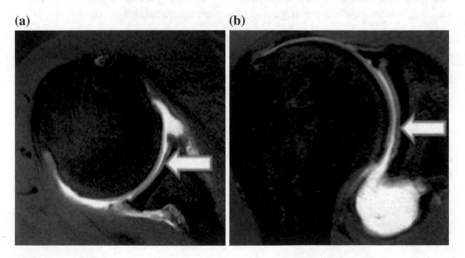

Fig. 3 A T1W FS Axial (**a**) and coronal (**b**) image of MR Arthogram [3–5, 9]

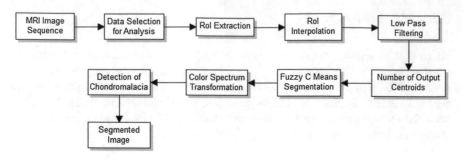

Fig. 4 The block diagram of articular cartilage segmentation

structures are badly observable. For this particular task, linear interpolation of tenth order is used. Interpolation technique significantly influences contrast of extracted image and mainly individual structures. If we used lower level of interpolation, we would obtain badly recognizable image. On the other hand grater level of interpolation leads to higher computational time. Interpolation of tenth order is appeared as good compromise for sufficient image contrast. The low-pass filtering is used for suppression of higher frequencies which in some cases impair recognition of articular cartilage structure. Native data contain adjacent structures such nodes and bones which should be suppressed by filtration process. For this purpose Gaussian filter has been used. After taking preprocessing steps it is approached to segmentation process. Firstly, number of centroids is selected. This step specifies number of output classes. We need to specify first class for physiological cartilage, second for chondromalacia area, other classes represent adjacent structures. Namely, they are soft tissues and bones. In the concluding step filtration of adjacent tissues in segmentation model is performed. Segmentation process separates individual tissues to classes which are coded to color spectrum. By filtration process we keep only classes witch represent articular cartilage and chondromalacia interruptions.

3.1 C-means Penalized Segmentation

The major benefit of used segmentation approach is incorporating the neighborhood information into C means centroid segmentation during classification process. Segmentation approach uses penalized objective function by regularization term in order to incorporate spatial context. The objective function of the used algorithm is defined by following equation:

$$J = \sum_{k=1}^{n} \sum_{i=1}^{c} (u_{ik})^q d^2(x_k, v_i) + \gamma \sum_{k=1}^{n} \sum_{j=1}^{n} \sum_{i=1}^{c} (u_{ik})^q (1 - u_{ij})^q w_{kj} \tag{1}$$

where w_{kj} is defined as:

$$w_{kj} = \begin{cases} 1, & \text{if } x_j \text{ and } x_k \text{ are neighbors and } j \neq k \\ 0, & \text{otherwise} \end{cases}$$

The parameter γ (≥ 0) controls effect of the penalty term. The major benefit of the fuzzy C means algorithm is that penalty term should be minimized in order to satisfy the principle of fuzzy C means (FCM) algorithm. The mentioned penalty term is minimized when the membership value for a particular class is large and the membership values for the same class at neighboring pixels is also large.

The objective function J can be minimized similar to standard FCM algorithm. Derivation of iterative algorithm for minimizing is performed by evaluating the centroids and membership function that satisfy a zero gradient condition. The Lagrange multiplier is used for solving constrained optimization from Eq. (1):

$$\varphi_q = \sum_{k=1}^{n} \sum_{i=1}^{c} (u_{ik})^q d^2(x_k, v_i) + \gamma \sum_{k=1}^{n} \sum_{j=1}^{n} \sum_{i=1}^{c} (u_{ik})^q (1 - u_{ij})^q w_{kj} + \lambda (1 - \sum_{i=1}^{c} u_{ik}) \tag{2}$$

After taking first order derivation according to u_{ik} we obtain following expression:

$$\left[\frac{d\varphi_k}{du_{ik}} = q(u_{ik})^{q-1} d^2(x_k, v_i) + \gamma q(u_{ik})^{q-1} \sum_{j=1}^{n} (1 - u_{ij})^q w_{kj} - \lambda \right]_{u_{ik} = u_{ik}^*} = 0 \tag{3}$$

The parameter u_{ik}^* is represented by following equation:

$$u_{ik}^* = \left(\frac{q(d^2(x_k, v_l) + \gamma \sum_{j=1}^{n} (1 - u_{ij})^q w_{kj})}{\lambda} \right)^{\frac{-1}{q-1}} \tag{4}$$

Since $\sum_{l=1}^{c} w_{lk} = 1$, this constraint equation is expressed by form:

$$\sum_{l=1}^{c} \left(\frac{q(d^2(x_k, v_l) + \gamma \sum_{j=1}^{n} (1 - u_{ij})^q w_{kj}))}{\lambda} \right)^{\frac{-1}{q-1}} = 1 \tag{5}$$

The parameter λ from Eq. (5) is given by following expression:

$$\lambda = \frac{q}{\left(\sum_{l=1}^{c} \left(\frac{1}{d^2(x_k, v_l) + \gamma \sum_{j=1}^{n} (1 - u_{ij})^q w_{kj}} \right)^{\frac{1}{q-1}} \right)^{q-1}} \tag{6}$$

By incorporating zero gradient condition, membership estimator is rewritten to following term:

$$u_{ik}^* = \frac{1}{\sum_{l=1}^{c} \left(\frac{d^2(x_k, v_i) + \gamma \sum_{j=1}^{n} (1-u_{ij})^q w_{kj}}{d^2(x_k, v_l) + \gamma \sum_{j=1}^{n} (1-u_{ij})^q w_{kj}} \right)^{\frac{1}{q-1}}} \tag{7}$$

Similarly, by taking Eq. (2) with respect to parameter v_i and setting the result to zero, we will obtain expression:

$$v_i^* = \frac{\sum_{k=1}^{n} (u_{ik})^q x_k}{\sum_{k=1}^{n} (u_{ik})^q} \tag{8}$$

Expression is identical to that of FCM because the penalty function from (1) is not depended on v_i. Generally, penalized FCM algorithm is possible summarize to four essential stages:

Stage 1   Setting the centroids v_i, parameter of fuzzification q and the value of c.
Stage 2   The calculation of membership values from Eq. (7).
Stage 3   The calculation of cluster centroids.
Stage 4   Return back to stage 2 and repeat it until convergence is reached.

As soon as algorithm converges, defuzzification process is performed in order to convert the fuzzy partition matrix to crisp partition. The described method works on the principle of assigning of object k to the class C with the highest membership [6–8]:

$$C_k = arg_i\{\max(u_{ik})\}, \quad i = 1, 2, \ldots, c. \tag{9}$$

4   Data Analysis and Segmentation Results

As it is mentioned above, chondromalacia is often manifested by very week contrast in images which are generated by MRI. There are four stages of those pathological lesions. Chondromalacia is classified by level of diseases in scale 1–4. Higher stage corresponds with advanced seriousness. For ensuring prevention, that chondromalacia won't be spread, early diagnosis is crucial. From this reason, we are mainly focused on detection chondromalacia of first stage. Furthermore, as regards mentioned stages, pathological lesions in advanced stages are clearly observable by human eyes and using segmentation methods appeared as useless. Articular cartilages are usually examined by MRI sequences either with fat suppression or proton dense sequence. For purposes of our analysis 30 patient's records have been used. Effectiveness of segmentation algorithm is shown on three cases bellow. Mentioned cases represent chondromalacia first degree where pathological changes are badly observable. The main intention was focused on clear separation of physiological cartilage and pathological changes. Segmentation approach allows assignment individual tissue's structures to output classes, so pathological changes should be conclusively recognizable (Figs. 5 and 6).

Articular cartilage is normally represented by white color. Pathological interruptions are often represented by week contrast changes (a). Segmentation process differentiates individual tissue's structures into color spectrum where individual tissue's structures are clearly observable. By suppressing adjacent structures, cartilage is manifested by red color and white gap presents chondromalacia's lesion (c). In the last step, contour of detected cartilage is performed (d).

RoI defines whole articular cartilage area of lateral condyle, tibia and part of lateral meniscus. In the central part, there is a fat body in joint and on the left side of RoI is partially captured cartilage of medial joint's segment. After taking segmentation, we observe a defect of cartilage's part lateral condyle. In the concluding step of segmentation, adjacent structures are suppressed cartilage is represented by yellow color. In the middle part of cartilage, chondromalacia defect is clearly observable.

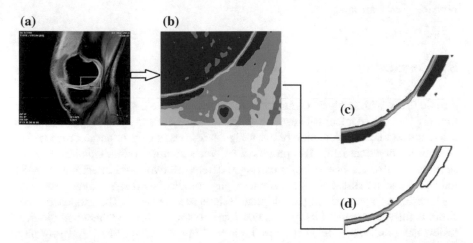

Fig. 5 Original MRI image with RoI (**a**), segmentation result (**b**), cartilage classification (**c**), detected cartilage's area (**d**)

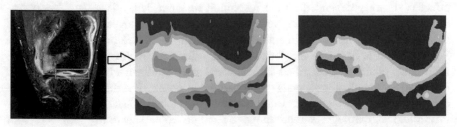

Fig. 6 Original MRI image with RoI (*left*), segmentation result (*middle*), cartilage classification (*right*)

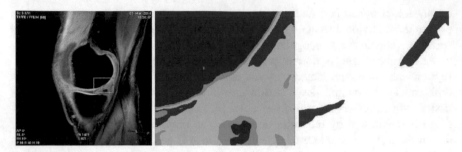

Fig. 7 The loss (*delimitation*) of dorsal part articular cartilage (*left*), segmentation result (*middle*), cartilage classification (*right*)

The last case deals with delimitation of dorsal part articular cartilage. There is a weak cartilage interruption, on Fig. 7. This lesion is not clearly observable due small contrast change from cartilage surface. Physiological part of cartilage is coded to red color. Segmentation model clearly recognizes chondromalacia lesion from physiological cartilage.

5 Conclusion

Article deals with analysis of articular cartilage structures. In the clinical practise, there is a severe problem with assessment of areas where early pathological changes are manifested. Early detection of pathological lesions is very important for making plan of further treatment. The proposed software solution offers suitable way for segmentation of surface articular cartilage. There are two main intentions of MRI data analysis: specification of areas where physiological cartilage is manifested and discovering of spots where pathological lesions are appeared. The proposed algorithm is fully automatic. Firstly, it is necessary to specify image where pathological lesion are observable. In next steps region of interest is specified and low pass filtering is applied. Filtration suppresses higher frequencies with are not important for cartilage analysis. By interpolation technique, image resolution is increased. This step is important for increasing image quality. It is obvious that, segmentation effect is strongly depended on the resolution of input data. If we used lower resolution, we would obtain badly recognizable segmented images. The main core of algorithm is consists segmentation approach with color pixel mapping. Segmentation is based on the penalized fuzzy C means with color transformation. Individual pixels are classified into isolated classes. Each class should contain information which corresponds with particular tissue. By this way, we obtain contrast image which precisely differentiates individual cartilage tissues. Physiological cartilage is represented by single color spectrum, gaps represent pathological interruptions. In the coming time we are going to focus on extraction of geometrical parameters segmented structures. This analysis is advantageous for clinical practise as well. It allows quantification of parameters cartilage structure for precise classification.

Acknowledgment This article has been supported by financial support of TA ČR PRE SEED: TG01010137 GAMA PP1. The work and the contributions were supported by the project SP2015/179 'Biomedicínské inženýrské systémy XI' and This work is partially supported by the Science and Research Fund 2014 of the Moravia-Silesian Region, Czech Republic and this paper has been elaborated in the framework of the project "Support research and development in the Moravian-Silesian Region 2014 DT 1—Research Teams" (RRC/07/2014). Financed from the budget of the Moravian-Silesian Region.

References

1. Yang, Y., Zhang, F., Zheng, C., Lin, P.: Unsupervised image segmentation using penalized fuzzy clustering algorithm. Lect. Notes Comput. Sci. **3578**, 71–77 (2005)
2. Kubicek, J., Penhaker, M., Bryjova, I., Kodaj, M.: Articular cartilage defect detection based on image segmentation with colour mapping. Lec. Notes Comput. Sci. (including subseries Lecture Notes in Artificial Intelligence and Lecture Notes in Bioinformatics) **8733**, 214–222 (2014)
3. Kaur, P., Soni, A.K., Gosain, A.: Image segmentation of noisy digital images using extended Fuzzy C-means clustering algorithm. Int. J. Comput. Appl. Technol. **47**(2–3), 198–205 (2013)
4. Kubicek, J., Penhaker, M.: Fuzzy algorithm for segmentation of images in extraction of objects from MRI. In: Proceedings of the 2014 International Conference on Advances in Computing, Communications and Informatics, ICACCI 2014, art. no. 6968264, pp. 1422–1427 (2014)
5. Yang, Y., Zheng, C., Lin, P.: Fuzzy c-means clustering algorithm with a novel penalty term for image segmentation. Opto-electronics Rev. **13**(4), 309–315 (2005)
6. Beevi, Z., Sathik, M.: A robust segmentation approach for noisy medical images using fuzzy clustering with spatial probability. Int. Arab J. Inf. Technol. **9**(1), 75–83 (2012)
7. Nie, F., Li, J., Tu, T., Pan, M.: Image thresholding using fuzzy correlation criterion and harmony search algorithm. Int. J. Computat. Intell. Appl. **13**(1), (2014). doi:10.1142/S1469026814500035
8. Augustynek, M., Penhaker, M.: Non invasive measurement and visualizations of blood pressure. Elektronika Ir Elektrotechnika 55–58 (2011)
9. Augustynek, M., Penhaker, M.: Finger plethysmography classification by orthogonal transformatios. In: 2010 Second International Conference on Computer Engineering and Applications (ICCEA), pp. 173–177. IEEE (2010)

A Formal Passive Testing Approach to Control the Integrity of Private Information in eHealth Systems

Azahara Camacho, Mercedes G. Merayo and Manuel Núñez

Abstract Intelligent Information is generated with each click that we do. The systems that are responsible of this tend to have a log where they record all the accesses to them. Unfortunately, the integrity and anonymity of this vast amount of information is not always ensured. There exist many threats that can change or even delete part of these logs. One of the purposes of the Internet of Things (IoT) is to generate Intelligent Information from different sources. Therefore, if there are some irregularities, then the generated information is not useful. In this paper we present a proposal to take into account some of these threats and prevent their consequences in some eHealth systems. In order to increase the applicability of our approach, we have implemented a tool that allow us to manage communications and check whether they are performed as intended.

1 Introduction

The appropriate control of data integrity is one of the open issues where more research areas involves. Different reasons, mainly security and privacy, have provoked this trend. In fact, this is currently a matter of public and professional areas because huge amounts of data are generated by the countless systems used in our

Research partially supported by the Spanish MEC projects ESTuDIo and DArDOS (TIN2012-36812-C02-01 and TIN2015-65845-C3-1-R) and the Comunidad de Madrid project SICOMORo-CM (S2013/ICE-3006).

A. Camacho (✉) · M.G. Merayo · M. Núñez
Departamento de Sistemas Informáticos y Computación, Universidad Complutense de Madrid, Madrid, Spain
e-mail: mariaazc@ucm.es

M.G. Merayo
e-mail: mgmerayo@fdi.ucm.es

M. Núñez
e-mail: mn@sip.ucm.es

© Springer International Publishing Switzerland 2016   221
D. Król et al. (eds.), *Recent Developments in Intelligent Information and Database Systems*, Studies in Computational Intelligence 642,
DOI 10.1007/978-3-319-31277-4_19

daily routine. There are many hazards that surround us and turn these devices on vulnerable systems. With the aim of preventing these risks, it is necessary to apply techniques to ensure the safety of our privacy. If the used technique has a formal basis then it will be possible to prove that they behave as expected. According to different features, we can find different testing methods, but a classical distinction is between *active* and *passive* [9]. In *active testing*, the tester needs to interact with the system under test (SUT). Therefore, it is necessary to stop the usual activities of the SUT in order to carry out testing activities. In addition, the tester needs to introduce some data (inputs) to decide whether the observed behaviour (outputs) is the expected one. These *fictitious* inputs can modify the data in the system (for example, the tester may require to modify the balance of a current account in order to test some features). Taking into account that systems in use are required to run 24/7 and that sensible data should not be modified in the real system, active testing, even though more powerful to find errors in a system, cannot be used in many situations. In *passive testing* the tester only observes the behaviour of the SUT, without influencing its performance by providing inputs. Thus, in order to perform the testing process, it is not necessary to change, stop or modify either the system or its associated data. Unfortunately, we lose the ability to *guide* the system through situations where it is more likely to find an error (this can be done in active testing by providing specific inputs targeting certain components of the SUT). In any case, although less powerful, passive testing techniques are being used more frequently and there are many proposals with a formal basis [4, 5].

In this paper, we adapt a formal approach to perform passive testing in systems where communications are asynchronous [16] to the control of the elements involved in the exchanging of information between two remote systems. We focus on eHealth computerized remote systems, more specifically, we are interested in checking the integrity of the transmitted data and the privacy of the process. In order to show a concrete case study, we consider the communication between a pacemaker and its monitor. The rest of the paper is structured as follows. In Sect. 2 we introduce the basic concepts of the Internet of Things (IoT) that are related to this paper. Section 3 explains the concept of eHealth and how it is related to IoT issues. Section 4 explains the scenario for the application of our approach and in Sect. 5 we present our proposal. Finally, in Sect. 6 we present our conclusions and provide some lines for future work.

2 The Internet of Things (IoT)

Technology makes life easier with original improvements. These could be in the form of better materials, new devices or healthier products. In this line, the Internet of Things is able to ease the management of information from different sources. Domotics, health-care, smart buildings and surveillance are only a few of them where IoT can be applied. According to the National Intelligence Council, "*by 2025 Internet nodes may reside in everyday things - food packages, furniture, paper documents, and more*" [18]. That is, each element of our daily routine will have a sensor able to send information about weather conditions, environment, its user, etc. Therefore, the evolution of IoT is an inevitable consequence and its ubiquitous application is only a matter of time. In this section we review the main characteristics of IoT, some open issues, and the main application areas.

2.1 Characteristics

In order to have a clearer idea about IoT, a possible definition is: "*Internet of Things is the combination of supporting resources to interconnect smart objects by means of extended Internet technologies, obtaining as a result the ensemble of applications and services leveraging them to open new business and market opportunities*" [17]. These *smart objects* are the *things* of the Internet of Things and for being one of them, there are some requirements that should be fulfilled [19]. In a nutshell, an IoT element has the following characteristics:

- *Existence*: It could be in the physical world such as computers, cars or houses, but it can be also the virtual data provided by the physical ones.
- *Sense of self*: Due to the huge number of elements to interconnect, it is essential to have an identification that let the other elements to find it in the network.
- *Connectivity*: Using the previous identification, the entity should be able to communicate with its surroundings and locate it.
- *Interactivity*: If the communication is established, then the element has to manage the sending and reception of information with other elements. Provoking a wide range of services as a result.
- *Dynamicity*: All the elements have to be available at any time, any place and in any way.

2.2 Open Issues

The previously mentioned characteristics generate new issues to support. The network where all the elements are connected is bigger everyday and this causes a greater difficulty for their performance [2]. The current issues to control are:

- *Authentication*: Objects need to be uniquely identified with an ID. The most used technique is IP addresses. The problem is the number of IPs in IPv4 are not enough to identify all the elements. Therefore, it will be necessary to use IPv6 to control all the elements [10].
- *Heterogeneity*: IoT has to be able to manage many different types of devices, without difficulty and being only aware of the IP for the identification. The

management of such a high number of differences between systems provokes the necessity of better protocol levels.

- *Scalability*: One of the unavoidable problems in a global information structure, such as IoT, is scalability. Due to the continuous growth of the network, it is essential that all the elements are able to adapt themselves to the new architecture.
- *Energy-optimization*: The dynamicity of IoT elements is the main reason why the optimization of energy is an open research issue. The devices depend on batteries and the minimization of the consumption is a trend that developers should adopt.
- *Localization and tracking*: The continuous tracking of IoT elements confronts the problem of the current state of the availability of wireless networks. Some applications such as health-care control or weather management could be affected by the lack of wireless resources in some areas.
- *Data management*: IoT generates massive amounts of data and its storage and exchanging require a proper architecture. Besides this, the design of the architecture should be the most efficient in order to manage data at the same time that other systems.
- *Security and privacy*: In this case, there are three key sub-issues which require some solutions. These are data confidentiality, privacy and trust. The guarantee that only authorized entities can access and modify the generated data is quite weak. Therefore, it is necessary to use control techniques to avoid the leakage of information.

2.3 Applications

Although the number of issues to resolve is large, there are a big amount of fields where IoT is starting to be successfully applied. Its potentialities offer the possibility of developing many applications and services that were unthinkable a few years ago. They could be divided in four domains:

1. *Transportation and logistics*: Thanks to the use of sensors and actuators, the application of IoT in roads and vehicles is gaining momentum. Some of the benefits in this area could be: transportation information, assisted driving and monitoring of transportation of hazardous materials.
2. *Smart environment*: This field is oriented to make easier and more comfortable our surroundings such as the house, the office or the supermarket. The appropriate distribution of sensors can facilitate the automatic regulation of room lighting, heating or energy consumption. In addition, a security option can be used to detecting natural phenomena. In the later case, real and current information can be provided and a rapid response can save human lives, mitigate the damage and reduce the level of disaster.

3. *Personal and social domain*: People tend to look for the faster and better social network in order to receive updates about news, friends and events. With the compilation of historical queries about websites and devices data, social networks can improve the information to show to users, according to their searches and preferences. In terms of security, this area can be benefited in case of losses or thefts.

4. *eHealth*: This is a sensitive area where any benefit can have an important impact and increase the number of lives saved. Considering the previous characteristics of IoT elements, there are many applications that can improve the current state of healthcare. Some of them are the automatic data collection from Implantable Medical Devices (IMD), patient identification and alarm actuators. This can be helpful for controlling its problems and either react in a faster way or inform about a critical situation of the patient.

We focus on eHealth and IMD. Therefore, in order to put our research in context, we need to explain some of the threats and open issues in the IMD area and what is our proposal to improve the current situation.

3 eHealth: Implantable Medical Devices

The healthcare industry has been affected by numerous (software) errors during the last years [20]. Since 2001, with the emerging of the *eHealth* concept [11], some approaches seem to be the improvement that this field needs. Around 3.337 recalls, out of 8.320 manufactured IMD, were reported to the US Food and Drug Administration (FDA) between 2010 and 2012, a 40 % from the total of elements [21]. In this case, the enhancements are been motivated by the errors related to software, hardware, data management and energy supplies. Some of them have been identified by the consequences of the open issues with IoT elements previously enumerated. Thanks to eHealth research, they might be farther reduced by implementing some of the following proposals:

- *Heterogeneity*: IMD need to be connected with different elements such as PCs and monitors. For the problem of connecting to all these types of heterogeneous devices, it is necessary to apply specific protocols managing all the necessary layers [8].
- *Scalability*: IMD developers need to be consistent with the continuous evolution of materials and software. A clear solution is the application of fault tolerance techniques, with the aim of diminishing the impact of this situation [1].
- *Localization and tracking*: Some IMD are provided with sensors that inform about the state, position and patient situation. In the case of insulin pumps, pacemakers and defibrillators, this technology allows physicians to check their data in case of emergency [7].

- *Data management*: The continuous service of these devices generates massive amounts of data, being some databases systems available for their management. We can mention the Electronic Health Records (EHR), a collection composed by the healthcare information of patients, and the Picture Archiving & Communication Systems (PACS), registering digital images in order to post-process and collect them for future medical cases [6].
- *Security and privacy*: In order to control the actors who access a device, it is necessary the implantation of an authorization system. In relation to the previously mentioned heterogeneity problem, specific medical protocols could include this feature with the aim of fulfilling two issues with one solution [13].

4 Scenario

There exists a wide range of vulnerable IMD and each of them has their own hazards. In this paper, we focus on pacemakers due to the fact that, alongside infusion pumps, they are the most implanted devices which require a software for the storing of data and the interaction with external elements [3]. For this group of IMD the main threats are related to data access, device identification and configurability [13]. Although their software is private, this is not a problem for hackers when they try to violate their security. Unfortunately, and difficult to understand because human lives are at risk, these devices tend to have an extremely low level of security measures. Developers do not pay too much attention to the possibility that an external element tries to capture data or alter its configuration. In order to prevent some of these situations, we present a scenario where our proposal can be applied in order to avoid the capture of private data by unauthorized actors.

The scenario is a patient whose pacemaker tries to send to the monitor, located in his house, the data captured during a certain period of time. This monitor, at the end of the transmission, sends this information to the computer of the physician, placed in the hospital. With this sequence of actions, the patient is completely tracked in terms of how the pacemaker works with the heart. A graphical representation of this scenario appears in Fig. 1.

Since the communication between these elements is by wireless connections, the level of danger is quite high. The risk that we seek to prevent is the intrusion of third parties inside this connection with the aim of capturing, modifying or deleting information from the pacemaker. As a solution, we propose controlling this threat by always checking the origin and destiny of messages. Monitoring this exchange of information can ensure that only the authorized actors are involved in the transmission. In order to achieve this goal, it is necessary the management of a list including all the users and devices that are authorized, so that during the communication, it should be possible to review the active actors. This identification will be possible thanks to the IP address of each source of the message.

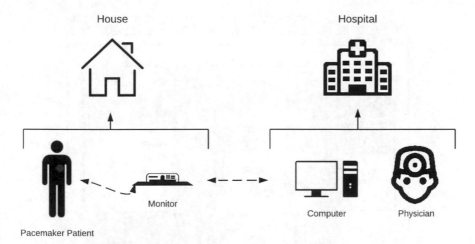

Fig. 1 Sending information from the pacemaker

5 Our Proposal

We propose a software tool to control the actors previously mentioned. The underlying methodology is based on a formal approach to perform passive testing with asynchronous communications [16]. The main reason why we decided to use a passive testing methodology is, as we explained in the introduction of the paper, is because we do not need to stop the system to check its correctness. In our case, due to the fact that we are working with the scenario of an implanted pacemaker, this kind of techniques are the most suitable. In addition, although this complicates the framework, we need to take into account that the communication between devices will be asynchronous.

The proposed software will be based on a graphical user interface (GUI) which is designed to connect it to the network where the pacemaker and the monitor communicate between them. The user is in charge of defining the elements conforming the list of authorized actors. This list is also made from the definition of properties where the user specify how messages should be transmitted and what actors will be allowed to participate in the exchange of information. The tool will be divided into different layers, each of them focusing on the specific feature for what it has been designed. Next we explain the main components of the tool.

5.1 Administration of Actors

This will be the most active part of the tool and it will allow the users to define the behaviour of the communications by using a certain set of properties. These

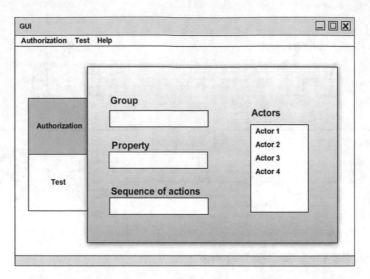

Fig. 2 Layer for the definition of the actions and actors of a property

properties will include the actors participating in the conversation and the messages that will be transmitted. The appropriate use of this layer of the GUI is essential in order to ensure correct communications. In Fig. 2 we show an example of this layer. Its main parts are:

- *Group*: This is the set of properties where the new one will be grouped.
- *Property*: This is the (unique) name of the property.
- *Sequence of actions*: This is the set of actions taking part in the property. Each of them includes information about the actor which performs it.
- *List of actors*: This is the group of authorized actors that can appear during the communication. It is not necessary to include the actors that defined in the property. This list is updated with the content of the sequence of actions.

Next we describe a typical property that we should ensure when a pacemaker is programmed to send its data to the monitor located in the house of the patient. In the property, actions prefixed by ? denote inputs while properties prefixed by *!* denote outputs.

- Group: Communication.
- Property: Sending records.
- Sequence of actions: (*?send*,pacemaker) (*!ready*,monitor) (*?ip*,pacemaker) (*! info*,monitor), where:

  - *?send* is the request of the pacemaker to send the data.
  - *!ready* is the confirmation of availability from the monitor to receive the records.

- – *?ip* is the request of the IP of the monitor to check against the list of actors that it is an authorized device.
- – *!info* is the IP of the monitor.

- List of actors: pacemaker and monitor.

5.2 Authorization Control

This will be the *passive* component of the tool. In order to verify the correctness of the transmission of information between the pacemaker and the monitor, an online checker tests the current transmission against the previously defined messages and actors. In Fig. 3 there is an example of this layer.

This component is the core of the GUI because it is in charge of checking the actors involved in the communication. This process will consist in the analysis of every message sent between the actors. However, this examination will concentrate on their features, not in their content, that is, the origin and destiny of the message, the type of the message and the order of the sequence. The result of checking the property against the observed trace will appear on the list of the right side of the GUI, showing the actors and their actions. If one of these sequences is invalid, then the process will stop and inform the user of the problem found in the transmission of the data.

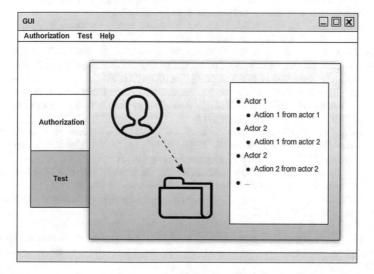

Fig. 3 Layer for the test of the current communication

6 Conclusions and Future Work

In this paper we have presented the main features of a tool which would be able to check the actors involved in the transmission of data from a pacemaker. By using passive techniques, it is possible to perform online checking of the performance against the properties previously defined without stopping the involved systems. If an irregularity is detected, then the user can apply the appropriate measures in order to avoid the alteration of private information.

As future work, we plan to analyse and study the complete medical protocol used in this type of communications. This will allow us to reflect its behaviour in our application and we will be able to ensure that the transmission of information is correct. In addition, we will study the impact of explicitly considering probabilities and time [4, 12, 14, 15].

References

1. Alemzadeh, H., Iyer, R.K., Kalbarczyk, Z., Raman, J.: Analysis of safety-critical computer failures in medical devices. IEEE Secur. Priv. **11**(4), 14–26 (2013)
2. Atzori, L., Iera, A., Morabito, G.: The internet of things: a survey. Comput. Netw. **54**(15), 2787–2805 (2010)
3. Allen, B.: The Eleven Most Implanted Medical Devices in America. http://247wallst.com/healthcare-economy/2011/07/18/the-eleven-most-implanted-medical-devices-in-america/. (2011)
4. Andrés, C., Merayo, M.G., Núñez, M.: Formal passive testing of timed systems: theory and tools. Softw Test. Verification Reliab. **22**(6), 365–405 (2012)
5. Bayse, E., Cavalli, A., Núñez, M,. Zadi, F.: A passive testing approach based on invariants: application to the WAP. Comput. Netw. **48**(2), 247–266 (2005)
6. Black, A.D., Car, J., Pagliari, C., Anandan, C., Cresswell, K., Bokun, T., McKinstry, B., Procter, R., Majeed, A., Sheikh, A.: The impact of eHealth on the quality and safety of health care: a systematic overview. PLoS Med. **8**(1), e1000387 (2011)
7. Burleson, W., Clark, S.S., Ransford, B., Fu, K.: Design challenges for secure implantable medical devices. In: ACM/EDAC/IEEE 49th Design Automation Conference, DAC'12, pp. 12–17. IEEE Computer Society (2012)
8. Chadwick, P.E.: Regulations and standards for wireless applications in eHealth. In: 29th Annual International Conference of the IEEE Engineering in Medicine and Biology Society, EMBC'07, pp. 6170–6173. IEEE Computer Society (2007)
9. Cavalli, A.R., Higashino, T., Núñez, M.: A survey on formal active and passive testing with applications to the cloud. Ann. Telecommun. **70**(3–4), 85–93 (2015)
10. Cooper, J., James, A.: Challenges for database management in the Internet of things. IETE Tech. Rev. **26**(5), 320–329 (2009)
11. Eysenbach, G.: What is e-Health? J. Med. Internet Res. **3**(2), e20 (2001)
12. Gregorio, C., Núñez, M.: Denotational semantics for probabilistic refusal testing. In: Workshop on Probabilistic Methods in Verification, PROBMIV'98, ENTCS 22. Elsevier (1999)
13. Halperin, D., Kohno, T., Heydt-Benjamin, T.S., Fu, K., Maisel, W.H.: Security and privacy for implantable medical devices. Pervasive Comput. **7**(1), 30–39 (2008)

14. Hierons, R.M., Merayo, M.G.: Mutation testing from probabilistic and stochastic finite state machines. J. Syst. Softw. **82**(11), 1804–1818 (2009)
15. Hierons, R.M., Merayo, M.G., Núñez, M.: Testing from a stochastic timed system with a fault model. J. Logic Algebraic Prog. **78**(2), 98–115 (2009)
16. Hierons, R.M., Merayo, M.G., Núñez, M.: Passive testing with asynchronous communications. In: IFIP 33rd International Conference on Formal Techniques for Distributed Systems, FMOODS/FORTE'13, LNCS, vol. 7892, pp. 99–113. Springer (2013)
17. Miorandi, D., Sicari, S., de Pellegrini, F., Chlamtac, I.: Internet of things: vision, applications and research challenges. Ad Hoc Netw. **10**(7), 1497–1516 (2012)
18. National Intelligence Council. Six Technologies with Potential Impacts on US Interests Out to 2025. http://fas.org/irp/nic/disruptive.pdf. (2008)
19. Roman, R., Najera, P., Lopez, J.: Securing the internet of things. Computer **44**(9), 51–58 (2011)
20. Sandler, K., Ohrstrom, L., Moy, L., McVay, R.: Killed by code: software transparency in implantable medical devices. Software Freedom Law Center (2010)
21. U.S. Food and Drug Administration. Medical Device Recall Report—FY2003 to FY2012, 2014

Part IV
Knowledge Management and Language Processing

Experience-Oriented Knowledge Management for Internet of Things

Haoxi Zhang, Cesar Sanin and Edward Szczerbicki

Abstract In this paper, we propose a novel approach for knowledge management in Internet of Things. By utilizing Decisional DNA and deep learning technologies, our approach enables Internet of Things of experiential knowledge discovery, representation, reuse, and sharing among each other. Rather than using traditional machine learning and knowledge discovery methods, this approach focuses on capturing domain's decisional events via Decisional DNA, and abstracting knowledge through deep learning process based on captured events data. The Decisional DNA is a flexible, domain-independent, and standard experiential knowledge repository solution that allows knowledge to be represented, reused, and easily shared. The main features, architecture, and an initial experiment of this approach are introduced. The presented conceptual approach demonstrates how knowledge can be discovered through its domain's experiences, and stored and shared as Decisional DNA.

Keywords Knowledge representation · Decisional DNA · Deep learning · Experience-Oriented Smart Things · Internet of Things

H. Zhang (✉)
Chengdu University of Information Technology, No. 24 Block 1,
Xuefu Road, Chengdu 610225, China
e-mail: Haoxi@cuit.edu.cn

C. Sanin
School of Engineering, The University of Newcastle, Callaghan,
NSW 2308, Australia
e-mail: Cesar.Sanin@newcastle.edu.au

E. Szczerbicki
Gdansk University of Technology, Gdansk, Poland
e-mail: Edward.Szczerbicki@zie.pg.gda.pl

© Springer International Publishing Switzerland 2016   235
D. Król et al. (eds.), *Recent Developments in Intelligent Information
and Database Systems*, Studies in Computational Intelligence 642,
DOI 10.1007/978-3-319-31277-4_20

1 Introduction

Finding valuable information from data produced by Internet of Things (IoT), and transforming the information into knowledge to make an intelligent world is the most challenging part and the essential goal of the concept of IoT [1–3]. By utilizing knowledge discovery technologies, such as data mining and machine learning, intelligence and smartness can be added to IoT [4, 5]. This paper describes the continuation of the development of the Experience-Oriented Smart Things (EOST) introduced in our previous work [6]. This research extension explores the possibilities and the practical implementations for using Decisional DNA and deep learning [7] technologies to acquire, represent, reuse, and share knowledge among IoT.

This paper is organized as follows: section two describes an academic background on basic concepts related to our work; section three presents the features, architecture and experiments for the experience-oriented knowledge management approach. Finally, in section four, concluding remarks are drawn.

2 Background

2.1 Knowledge Management

Knowledge Management (KM) is a term and a concept that arose approximately twenty years ago. Davenport [8] created the still widely cited definition at KM's early stage: "*Knowledge management is the process of capturing, distributing, and effectively using knowledge*". A few years later, another most frequently quoted definitions of KM was built by the Gartner Group [9] as "*Knowledge management is a discipline that promotes an integrated approach to identifying, capturing, evaluating, retrieving, and sharing all of an enterprise's information assets. These assets may include databases, documents, policies, procedures, and previously un-captured expertise and experience in individual workers*".

Both definitions share a very similar idea, which is that KM is a systematic effort of capturing, storing, distributing, and reusing information of an organization to make it available, actionable, and valuable to others. The discipline of KM is about building up and administrating the processes to deliver the right information to the right people at the right time, and help people act on information and share such information in order to improve the performance of organizations [10]. And the access to information should be considered more than just documents and databases, but also the experiences of individuals and teams through their day-to-day work, collaboration and communication.

From these views it is obvious that KM would involve many different organizational processes, and KM can be in various forms. A survey offered by Liao [11] says that there are generally seven categories of KM technologies and applications developed until 2002. In another study [12], after analyzed 30 published articles between 2003 and 2010 from high quality journals, nine core theories are found in KM area. These KM technologies, applications, and theories enable enterprises to manage their knowledge in different perspectives. However, there are limitations from these technologies: Most of them are designed for one specific kind of product, as well as not standard knowledge presentation; most of systems lack of the capability on information sharing and exchange; also, most of these systems only focused on supporting a particular stage of a product lifecycle [13]. Moreover, as the advances on IT and cyber technique, the working and living environment have been changed by innovations. For instance, tablets and smart-phones are changing the way of how we live, study, and work. People now are switching from desktops to these small computers. These small devices are researching into every corner of our lives—even restaurants start to get rid of menus and use tablets to order dishes). And traditional KM technologies are not designed to this situation. Specifically, Expert Systems are too slow for such devices [14].

Therefore, it is very urgent and necessary to have some new tools coping with these issues in order to establish a knowledge-rich environment that makes seamless knowledge connection among individuals, organizational processes, and enterprises possible.

2.2   Experience and the Experience-Oriented Smart Things

Experience, as a general concept, comprises previous knowledge or skill one obtained through daily life [15]. Usually experience is understood as a type of knowledge that one has gained from rather than books but from practice [16]. In this way, experience or experiential knowledge can be regarded as a specialization of knowledge carrying information of the tasks handled with the results of different handling strategies in the past.

From a knowledge management perspective, knowledge is useful only if it is accessible to all users, and can be used to solve problems and make decisions [17]. Additionally, the active KM is required to cover all organizational processes. The wide use of IoT has thus turned IoT to a physical basic for most of the knowledge-related activities, because IoT are largely used in organizational processes. By learning from practical processes such as problem solving and decision-making, experiential knowledge can be obtained. Subsequently, through proper reusing of knowledge, related IoT can help or complete these processes more efficiently or even automatically. Hence IoT can be built smart. Most importantly, the acquiring of knowledge is the creating of asset for such smart IoT's domain; which allows the domain/organization to know about itself from a systematic view, and keep its asset (i.e. the knowledge) once employees retired or machines replaced by new ones.

Therefore, the Experience-Oriented Smart Things (EOST) was proposed in order to allow experiential knowledge discovery, storage, involving, and sharing for IoT. It utilizes the Decisional DNA [18] as its knowledge representation method. More detailed information about EOST can be found in our previous publication [6].

3 The Experience-Oriented Knowledge Management

The Experience-oriented knowledge management framework, named Latte, is the core component of EOST. It is designed for knowledge discovery, representation, reuse, and sharing in smart things [6]. This section presents the main features, architecture, and the initial experiment of Latte.

3.1 Main Features

The Latte is designed and proposed to allow experiential knowledge discovery, presentation, and sharing among EOST (i.e. smart IoT). The Latte shall be able to handle different scenarios; in our case, we link each experience with a certain scenario describing the circumstance under which the decision was made, plus the outcome of the decision. Moreover, it shall be compatible with a range of different things, so that acquired knowledge can be shared and re-used in other things. In order to achieve these goals, the three key features of Latte are: Experience-oriented, deep learning based, and compatible.

(a) *Experience-oriented*: experience, as one kind of knowledge learned from practice, is the ideal source for improving performance of processes, in which a lot of practical activities involve. By reusing experiential knowledge, decision makers can make decisions faster, and more efficiently base their current decisions on experiences obtained from previous similar situations. Therefore, capturing the experience instead of all data produced by IoT is more preferred way for EOST to deal with the big data issue [3]. Like the man learns from experience, the Latte captures experience from data of things first, and then it abstracts knowledge from such captured experience.

(b) *Deep learning based*: the main mission of Latte is to abstract knowledge from data. However, early AI systems are based on expert knowledge, which is not universal [12, 19]. Deep learning algorithm can be universally applied to all learning [20, 7]; hence deep learning technology comes into play in Latte.

(c) *Compatible*: as the goal, the EOST is designed to be an open platform for all things. In other words, the Latte is expected to work in different domains, and deal with data from various IoT. Since most IoT are customized, the hardware and software for each of them could be distinctly different; thus, compatibility is essential for the Latte.

3.2   System Architecture

The Latte mainly consists of Prognoser, Knowledge Repository, and Deep Learning Engine (see Fig. 1).

The Prognoser is the control central of the Latte. It is in charge of experience capturing, knowledge abstraction, knowledge creating, knowledge retrieval, and knowledge reusing. For experience capturing, the Prognoser catches the scenario information when a decisional event is occurred, and send it to Knowledge Repository for store. Once it captures enough experience, the Prognoser abstracts knowledge based on captured experience by utilizing the Deep Learning Engine. Finally, it creates the knowledge and sends the knowledge to Knowledge Repository for store. While for supporting decision-making, it retrievals and reuses the knowledge. There are basically three situations for that matter:

(a) The Prognoser gets the same scenario in the knowledge repository, and then it will reuse the decision made previously on its current circumstance;
(b) The Prognoser does not get the same scenario, but it gets some similar ones. For this situation, the Prognoser will perform some analyzes and then come out with a solution to the current circumstance, apply it and save it as an experience into the knowledge repository;
(c) c) The Prognoser does not get neither the same nor similar scenario from the knowledge repository, then it will ask other similar systems for help if a connection is available, otherwise, it will return a default result.

The Deep Learning Engine runs deep learning algorithms, abstracts knowledge from experience, and reuses abstracted knowledge. Deep learning allows computational models that are composed of multiple processing layers to learn representations of data with multiple levels of abstraction [7], and it can automatically discover the representations of input data for detection or classification [20]. By

Fig. 1 System architecture of the Latte

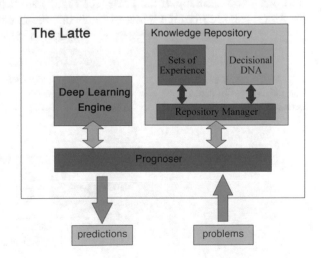

using deep learning, the knowledge can be abstracted from sets of experience, and stored as Decisional DNA. For knowledge reuse, the Deep Learning Engine loads the Decisional DNA, re-construct the deep learning network, and gives predictions according to learnt knowledge.

The Knowledge Repository is where the Latte's most valuable sources, experience and knowledge, stored and managed. In the Latte, a single decision event is captured and represented as an experience, stored as one Set of Experience (SOE) [18], and a set of SOE is organized as the Decisional DNA carrying the decisional fingerprint of the EOST. The Knowledge Repository provides functionality of query, store, insertion, editing and deletion of experience and knowledge.

3.3   Initial Experiments

In order to exam our conceptual knowledge management framework, we tested the Latte in classification tasks based on some third-party data that comes with the book '*Machine Learning in Action*' [21].

In the initial experiments, the data are collected from dating activities [21], and stored as sets of experience of each dating. Basically, there are three types of date: bad, OK, and good; which is based on the date mate's personal data in: (1) number of frequent flyer miles earned per year, (2) percentage of time spent playing video games, and (3) liters of ice cream consumed per week.

First, we constructed a five-layer network with thirty hidden neurons in the Latte; and then we train the network by using the training data; finally, we stored the acquired knowledge as Decisional DNA, and reuse it to classify date types in testing. We compared the performance of the Latte with the kNN algorithm provided in the book: the results show that both computation performance (see Fig. 2) and prediction accuracy of the Latte (see Fig. 3) are better than kNN.

As we can see from the initial experiments, the Latte is faster than kNN at any stage, most importantly, the kNN's computation time increases notably when

Fig. 2 The result of computation performance between kNN and the Latte

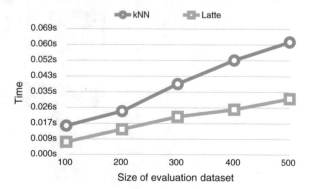

Fig. 3 The classification accuracy between kNN and the Latte

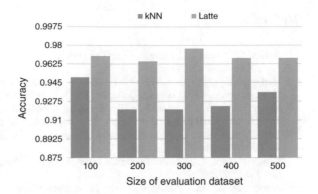

evaluation dataset becomes bigger. Moreover, the classification accuracy of the Latte is more stable and better than that of kNN. For the Latte, the classification accuracies are all higher than 0.965 in five different datasets, which is even higher than the best case of kNN (0.950), and the best case of the Latte is 0.975. The experiment results show that the combination of Decisional DNA and deep learning is very promising for knowledge management in IoT.

4 Conclusions and Future Work

In this paper, we introduced the Latte, a conceptual knowledge management platform for Internet of Things. By utilizing Decisional DNA and deep learning technologies, the Latte enables Internet of Things of experiential knowledge discovery, representation, reuse, and sharing among each other. Rather than using traditional machine learning and knowledge discovery methods, this approach focuses on capturing domain's decisional events via Decisional DNA, and abstracting knowledge through deep learning process based on captured events data. An initial experiment was made at the end of this paper, and the experiment results show that the Latte is very promising for knowledge discovery, representation, reuse, and sharing in EOST, and providing the smartness services for IoT.

Giving sense to IoT is a very challenging job, and our conceptual research is just at its first stage. There are many further work remaining to be done, some of them are:

- Further development of the Deep Learning Engine.
- Refinement of the Decisional DNA for working with deep learning.
- Further design and development of the Knowledge Repository.

Acknowledgement The authors would like to thank the editors and anonymous reviewer for their valuable comments and suggestions on this paper. This work was supported as part of the Project KYTZ201422 by the Scientific Research Foundation of CUIT.

References

1. Atzori, L., Antonio, I., Morabito, G.: The internet of things: a survey. Comput. Networks **54** (15), 2787–2805 (2010)
2. Bandyopadhyay, D., Sen, J.: Internet of things: applications and challenges in technology and standardization. Wireless Pers. Commun. **58**(1), 49–69 (2011)
3. Zhang Haoxi, S.C., Szczerbicki, E.: Applying decisional DNA to Internet of Things: the concept and initial case study. Cybern. Syst. **46**, 84–93(10) (2015)
4. Tsai, C., et al.: Data mining for Internet of Things: a survey. IEEE Commun. Surveys Tutorials **16**(1), 77–97 (2015)
5. Sánchez López, T. et al.: Adding sense to the Internet of Things. Pers. Ubiquitous Comput. **16** (3), 291–308 (2012)
6. Zhang Haoxi, S.C., Szczerbicki, E.: Experience-oriented enhancement of smartness for Internet of Things. Intelligent Information and Database Systems, pp. 506–515.Springer (2015)
7. LeCun, Y., Bengio, Y., Hinton, G.: Deep learning. Nature **521**(7553), 436–444 (2015)
8. Davenport, T.H.: Saving IT's Soul: human centered information management. Harvard Bus. Rev. **72**(2), 119–131 (1994)
9. Duhon, B.: It's all in our Heads. Inform 12(8) (1998)
10. O'Dell, C., Hubert, C.: The new edge in knowledge: how knowledge management is changing the way we do business. Wiley (2011)
11. Liao, S.H.: Knowledge management technologies and applications—literature review from 1995 to 2002. Expert Syst. Appl. **25**, 155–164 (2003)
12. Matayong, S., Mahmood, A.K.: The studies of Knowledge Management System in organization: a systematic review. In: Computer & Information Science (ICCIS), 2012 International Conference, IEEE, vol. 1, pp. 221–226 (2012)
13. Li, B.M., Xie, S.Q., Xu, X.: Recent development of knowledge-based systems, methods and tools for one-of-a-kind production. Knowl. Based Syst. **24**(7), 1108–1119 (2011)
14. Mazilescu, V.: A real-time algorithm for intelligent control embedded in knowledge based systems. Risk Contemp. Econ. **1**, 144–152 (2011)
15. Sun, Z., Finnie, G.: Brain-like architecture and experience based reasoning. In: Proceedings of the 7th JCIS, Cary, North Carolina, USA, pp. 1735–1738 (2003)
16. Sharma, N., Singh, K., Goyal, D.P.: Is technology universal panacea for knowledge and experience management? Answers from Indian IT Sector. Information Systems, Technology and Management, pp. 187–198. (2012)
17. Lao, G. et al.: Research on organizational knowledge sharing framework based on CAS theory. International Conference on Service Systems and Service Management, pp. 1–6 (2008)
18. Sanin, C., Toro, C., Zhang, H. et al.: Decisional DNA: a multi-technology shareable knowledge structure for decisional experience. Neurocomputing **88**(7), 42–53 (2012)
19. Haoxi, Z.: Experience-Oriented Smart Embedded System. The University of Newcastle's Digital Repository, NOVA (2013)
20. Ranzato, M., Hinton, G., Lecun, Y.: Guest editorial: deep learning. Int. J. Comput. Vision **2015**(113), 1–2 (2015)
21. Harrington, P.: Machine Learning in Action. Manning Publications Co. (2012)

Solutions of Creating Large Data Resources in Natural Language Processing

Huỳnh Công Pháp

Abstract Data resources in natural language processing (NLP) are mainly two kinds in terms of dictionaries and corpora. Their quality plays a crucial role in the quality and performance improvement of NLP tasks and systems. Indeed, the quality of NLP systems such as machine translation (MT) systems, search engines, text analyzing systems, etc., depends very much on the quality of data resources serving them Boitet (Revue française de linguistique appliquée XII:25–38, 2007 [1]). A data resource in NLP is evaluated good quality if it contains not only good data quality, but also various domains and numerous language pairs. Therefore, apart from the data quality factor, other factors regarding volume, covered domains and number of language pairs are also very important for the data resources in NLP. In this paper, we focus on proposing solutions to unify existing data resources in NLP for creating larger ones with a common structure and format, containing various domains and numerous language pairs. The results of the experiments are encouraging as we've created a very large corpus unified from many given corpora.

Keywords Natural language processing · Corpus · Dictionary · Unifying data resources · Large NLP data resources

1 Introduction

NLP has been emerging as all tasks and tools in this domain aims to support the need of human–human and human–machine communication. Hence, there are currently numerous NLP systems such as search engines, MT systems, dictionaries, textanalyzing systems, etc., that have been developed very fast in terms of amount as well as supported languages.

Nevertheless, the quality of the NLP tasks and systems, especially those used for under resourced languages such as Vietnamese, ethnic minority languages, is still

H.C. Pháp (✉)
The University of Danang, Da Nang, Vietnam
e-mail: hcphap@gmail.com

© Springer International Publishing Switzerland 2016 243
D. Król et al. (eds.), *Recent Developments in Intelligent Information and Database Systems*, Studies in Computational Intelligence 642,
DOI 10.1007/978-3-319-31277-4_21

Fig. 1 Unification illustration
of NLP data resources

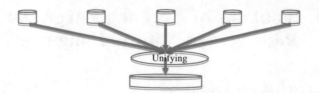

fairly low from the practical perspective. One of the factors strongly impacting the
quality of NLP tasks and systems is the quality of data resources serving them. This
can be proved by an example of quality of MT systems: several MT evaluation
campaigns, such as CSTAR, NESPOLE, IWSLT [2], which have been organized
on a large-scale using various subjective and objective evaluation methods, showed
that the quality of MT on common language pairs is only good for domains with
good quality corpus and very bad for domains with poor quality corpus.

Therefore, in addition to solutions of improving quality of NLP data resources
such as post-editing, enriching, etc., solutions of creating large volume of NLP data
resources are also very important and crucial to enhance the quality of NLP systems. Indeed, despite existing data resources being quite large, like corpora:
EuroParl, BTEC, ANC, ICE [3–6]; or dictionaries: Deutsches Wörterbuch, Oxford
English, Gregg Cox [7], their number of languages and covered domains are still
modest for the practical application needs.

Hence, the problem to be solved in this paper is how to create NLP data
resources with very large volume, various covered domains and numerous language
pairs to be used for enhancing quality of NLP systems. Our proposed solution in
this paper is to unify existing NLP data resources to acquire a larger one with a
common structure and format, containing various domains and languages (Fig. 1).

The existing NLP data resources we take into account in this paper are NLP
corpora and dictionaries. Unifying NLP data resources is thus to try to merge
different NLP corpora or dictionaries into a larger and uniformed one.

2 Related Works

There are a number of previous researches relating to create NLP data resources,
most of them focused on proposing solutions to automatically acquire corpora or
lexical dictionaries from multilingual resources. Among them, we can mention the
following common approaches:

- Building parallel corpora from multiple websites or aligned documents: many
  previous work proposed methods and algorithms for automatically extracting
  aligned sentences from multilingual websites or aligned documents to create
  parallel corpora [3, 8–12].
- Extending parallel corpora by calling MT systems: some researches proposed
  solutions of creating parallel corpora by calling MT systems to translate source

sentences of existing corpora into respective ones in other languages, then post-editing the translated sentences [2].

- Building lexical dictionaries from parallel corpora or websites: many researches worked on creating lexical dictionaries by using approaches based on word alignment on parallel corpora or multilingual websites [1, 13].

These approaches have allowed creating quite large corpora and dictionaries as mentioned in earlier section such as corpora: BNC, EuroParl, BTEC, ANC, ICE; dictionaries: Deutsches Wörterbuch, Oxford English, Gregg Cox. However, as we also mentioned above, these NLP data resources are still limited regarding their size, covered domains and languages for the practical needs. Furthermore, they were divergently constructed by various individuals or organizations, and they are located in dispersed places so that it's very inefficient and inconvenient to exploit, use and share them. So, the NLP data resources will become more useful and valuable if they are being unified into a larger one with a common structure and format.

Concerning unification of data resources, there are several researches and tools enabling a merge of a database into another or to convert some data format to another [14, 15]. However, until now, there is almost no research about creating large NLP data resources by unifying existing ones.

3 Solutions of Creating Large NLP Data Resources

As mentioned in previous section, if there are NLP data resources with large enough volume and good enough quality, then we can create very good quality NLP systems. Many previous researches as well as author's works proposed solutions for enhancing NLP data resources such as post-editing, enriching, getting rid of ambiguities/noises [2, 11, 16, 17]. In this paper, we only focus on proposing solutions to create large volume of NLP data resources. Before going through our solutions in detail, we need to clarify the concept of unifying NLP data resources. In [1, 2], showing that there are currently corpora and dictionaries with various size, language pairs, covered domains, as well as formats and structures. Unifying corpora or dictionaries is to merge them into a larger one with a common format and structure, containing various covered domains and numerous language pairs.

In order to address the problem of unifying NLP data resources, we propose three solutions as follows: Unifying data of NLP data resources; unifying languages of NLP data resources; and unifying formats and structures of NLP data resources.

3.1 Unifying Data of NLP Data Resources

For two given NLP data resources (corpora or dictionaries) to be unified, there may
be three cases for their language pairs: two NLP data resources with the same
language pairs (ex. Vn-En, Vn-En); two NLP data resources with a middle language
(ex. Vn-En, En-Fr); and two NLP data resources with totally different language
pairs (ex. Vn-En, Fr-Ja).

In this section, we focus on the first case, and the other two cases will be tackled
in the following sections.

To visualize it we provide an example, assume that we have two parallel corpora
with the same language pair Vn-En. The first one contains 20,000 sentences and the
second one contains 5000 sentences. After unifying these two corpora we have a
larger one containing from 20,000 to 25,000 sentences. From this example, we can
conclude that while unifying two NLP data resources with the same language pairs,
we must take two cases of data into account, data similarity and data difference.
Namely, data similarity is that a data unit exists simultaneously in both of the two
NLP resources and vice versa data difference is that a data unit exists only in one
NLP resource. Therefore, unifying data of two NLP resources is the process of
comparing and matching data units between two source data resources to select
relevant data units for creating the unified one. In general, suppose that each corpus
or dictionary is a set of data units (sentences or words), the unified NLP data
resource is then the result of union calculation of source NLP data resources, as
presented by the following formula and illustrative figure.

$$R_u = \bigcup_{i=1}^{n} R_i \tag{1}$$

where, R_u is the destination resource created from unifying various source data
resources, R_i, with the same language pairs. The proposed algorithm of unifying
two NLP data resources is as follows:

Input: $R_1 = (X_{L1}, Y_{L2})$, $R_2 = (M_{L1}, N_{L2})$
Output: $R_u = R_1 \cup R_2 = ((X + M)_{L1}, (Y + N)_{L2})$
1: $R_u \leftarrow \max(R_1, R_2)$
2: **for** i : $m_i \in M_{L1}$ **do**
3:    **for** j : $x_j \in X_{L1}$ **do**
4:       **if** $f(m_i) = f(x_j)$ **then**
5:          $R_u \leftarrow (m_i, n_i)$
6:       **end if**
7:    **end for**
8: **end for**

3.2 Unifying Languages of NLP Data Resources

In this section, we address the problems of unifying NLP data resources with different language pairs, in which there are two cases to be addressed: two NLP data resources with a middle language (ex. Vn-En, En-Fr); and two NLP data resources with totally different language pairs (ex. Vn-En, Fr-Ja).

The solution to the addressed problem we call "Unifying languages of NLP data resources".

- Unifying NLP data resources with a middle language.

Before presenting in details our solutions for this problem, we take an example, assuming that we have two corpora with a middle language: the first one (R_1) contains a language pair Vn-Fr; the second one (R_2) contains a language pair Vn-En.

After unifying these two corpora, we may have a corpus (R_u) with 3 language pairs Vn-Fr, Vn-En, and En-Fr (Fig. 2).

Thus, unifying NLP data resources with a middle language is indeed the process of matching data units in the middle language of two data resources to bridge the alignment of data units in the two remaining languages. If we see each data resource as a set of language pairs, then the unified data resource is resulted by the Descartes's calculation of sets of language pairs of the source data resources as presented by the following formula and figure:

$$R_u = R_1 \times R_2 = \{(L_i, L_j) \mid L_i \in R_1, \ L_j \in R_2\} \qquad (2)$$

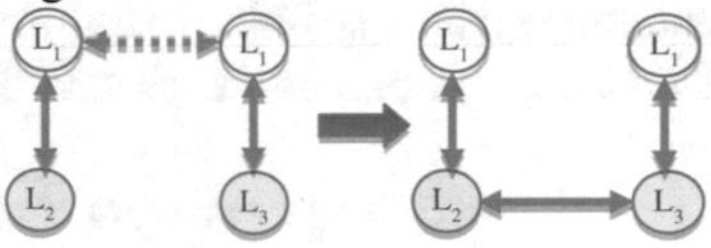

Our proposed algorithm for unifying two NLP data resources with a middle language is as follows:

```
Input:    R₁ = (X_L1, Y_L2), R₂ = (M_L1, N_L3)
Output:   R_u = R₁ x R₂ = { (X_L1, Y_L2), (M_L1, N_L3), (Y_L2, N_L3) }
1: R_u ← R₁
2: R_u ← R₂
3: for i : x_i ∈ X_L1 do
4:   for j : m_j ∈ M_L1 do
5:     if f(x_i) = f(m_j) then
6:        R_u ← (y_i, n_j) // y_i ∈ Y_L2, n_j ∈ N_L3
7:     end if
8:   end for
9: end for
```

Vn – En	Vn – Fr		Vn – Fr	Vn – En	En – Fr
.....	.....		.....	.....	.....
.....	.....		.....	.....	.....

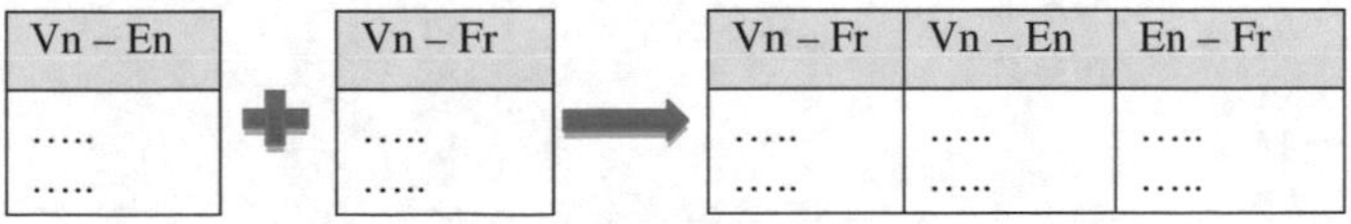

Fig. 2 Illustration of unifying languages of two NLP data resources

- Unifying NLP Data Resources with Totally Different Language Pairs.

Compared to the problems of unifying NLP data resources with the same language pairs or with a middle language, the problem of unifying NLP data resources with totally different language pairs is much more complex and challenging.

Suppose that there are two corpora with totally different language pairs: the first one (R_1) contains a language pair Vn-Fr, the second one (R_2) contains a language pair En-Ja. After unifying them, we may have a corpus (R_u) with 6 language pairs: Vn-Fr, En-Ja, En-Vn, Vn-Ja, En-Fr, Fr-Ja. From this example, we can state that if we see each data resource as a set of language pairs, then the unified data resource is also resulted by the Descartes's calculation of sets of language pairs of the source data resources as presented by the following formula and figure:

$$R_u = R_1 \times R_2 = \{(L_i, L_j) \mid L_i \in R_1,\ L_j \in R_2\}$$

(3)

Thus, in order to unify two NLP data resources with totally different language pairs, first we select two any languages from two data resources to align, then from this alignment we align the remaining language pairs. As the figure example above, we first align data in language L_1 and L_3 and then we align data between language pairs L_1-L_4, L_2-L_3, L_2-L_4. Therefore, unifying NLP data resources with totally different language pairs is indeed the problem of aligning data units in two languages of two data resources which can be presented by the following formula:

$$R_u = \{(x, y) \mid x \in R_{1L1} \land y \in R_{2L2} \land f(x) \approx f(y)\} \tag{4}$$

where, x is a data unit in language L_1 of NLP data resource R_{1L1}, y is a data unit in language L_2 of NLP data resource R_{2L2} and f is a function calculating the similarity between x and y.

Our proposed algorithm for unifying two NLP data resources with totally different language pairs is as follows:

Input: $R_1 = (X_{L1}, Y_{L2})$, $R_2 = (M_{L3}, N_{L4})$
Output: $R_u = R_1 \times R_2 = \{(L_i, L_j) \mid L_i \in R_1,\ L_j \in R_2\}$
1: $R_u \leftarrow R_1$
2: $R_u \leftarrow R_2$
3: **for** i : $x_i \in X_{L1}$ **do**
4: **for** j : $m_j \in M_{L3}$ **do**
5:   **if** $f(x_i) = f(m_j)$ **then**
6:     $R_u \leftarrow (x_i, m_j)$
7:     $R_u \leftarrow (y_i, m_j)$ // $y_i \in Y_{L2}$
8:     $R_u \leftarrow (x_i, n_j)$ // $n_j \in N_{L4}$
9:     $R_u \leftarrow (y_i, n_j)$ // $y_i \in Y_{L2}$, $n_j \in N_{L4}$
10:    **end if**
11: **end for**
12: **end for**

3.3 Unifying Formats and Structures of NLP Data Resources

One of the big problems to be solved with NLP data resources unification is to unify their formats and structures. Indeed, NLP data resources have been divergently created by various individuals and organizations, their formats and structures are hence very different and heterogeneous. Therefore, unifying formats and structures of NLP data resources is to transform existing NLP data resources into a larger one with a common structure and format.

Hence, in order to solve this problem, the solution we propose is to create a common format and structure that is able to represent as many NLP data resources as possible. From the common format and structure, we then build tools converting existing NLP data resources into a unified one with the common structure and format.

4 Experiment Results

In order to implement and experiment the solutions proposed in the previous sections, in the first phase we focus on unifying MT corpora to create a larger one with a common structure and format, various covered domains and numerous languages so as to improve the quality of NLP systems. The input corpora used for our experiments are EuroParl, BTEC, OPUS, JRC-AQUIS, ERIM, EOLSS, and DATIC. The first task we had to do is to analyze the given corpora and build a

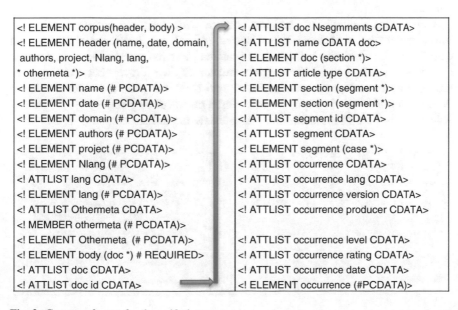

Fig. 3 Common format for the unified corpus

Corpus name	EOLSS ▼
Document name	[ ] - Existing docs - ▼
Language pair	- Source - ▼ - Translation - ▼
Source file	Choisissez un fichier Aucun f...choisi
Translation file	Choisissez un fichier Aucun f...choisi MT name [ ] - Common MT - ▼
UNL_FR file	Choisissez un fichier Aucun f...choisi

Import   Reset

Fig. 4 Corpus converter integrated into the import function of corpus exploiting system

common structure and format of unified corpus. This common structure and format must be able to enable the conversion of existing corpora into the unified corpus without losing data. We have used XML format to define the common format for the unified corpus as follows (Fig. 3).

The proposed common format of unified corpus consists of two parts: header containing information about corpus, languages, created date, … ; and body containing content of documents: <doc>, <dialogue>,…Each document containing the corpus hierarchical structure: chapters, pages, sections…and description of segments (<seg>, <TP>, , …), in which, the description of segments contains information: source, translation, context, sound attached, graph,…

From the proposed format of unified corpus, we developed a converting tool and integrated it into the import function of the system of exploiting corpora we've constructed as Fig. 4. This tool analyses and converts input corpora into the unified corpus.

Concerning solutions of unifying corpora with the same language pairs, we implemented the algorithms Edit distance, BLEU and NIST for matching and calculating the similarity between two sentences in two corpora. The Edit distance algorithm calculates the difference in character and word levels between two sentences from two corpora, whereas BLEU and NIST calculate the difference based on n-gram precision of various lengths. Based on calculating values, the program is able to unify data of the input corpora to create the unified corpus (Fig. 5).

French (Corpus 1)	Edit Dis	BLEU	NIST	French (Corpus 2)
Cela a frit du poisson, une saucisse avec les pois verts, s'il vous plaît.	Dc=25,Dw=8 D=11.4	0.39	2.77	Ce poisson Cela frit. a frit du poisson, une saucisse avec les des pois petits verts, pois. s'il vous plaît.
Steak avec un os en T et choucroute et a frit des pommes de terre, s'il vous plaît.	Dc=33,Dw=11 D=15.4	0.33	2.45	Du bifteck Steak à avec l'os un et os de en la T et choucroute et a frit des pommes de terre, terre frites. s'il vous plaît.
Poulet du rôti et deux tranches de jambon sur ce côté et épinards, s'il vous plaît.	Dc=8,Dw=2 D=3.2	0.81	4.08	Du Poulet du rôti et deux tranches de jambon sur ce côté et des épinards, s'il vous plaît.
J'aimerais petit déjeuner, s'il vous plaît.	Dc=3,Dw=1 D=1.4	0.77	2.99	J'aimerais un J'aimerais petit déjeuner, s'il vous plaît.
Café, s'il vous plaît.	Dc=0,Dw=0 D=0.0	1.0	2.58	Café, s'il vous plaît.

Fig. 5 Calculated results of Edit distance, BLEU and NIST between sentences in two corpora

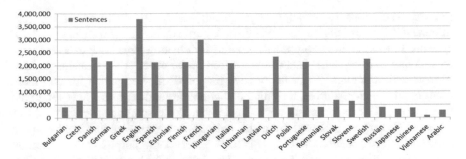

Fig. 6 The result corpus achieved from unifying existing corpora

However, the drawback of these algorithms is that they can only compare the similarity between two sentences at character and word levels. Whereas, two sentences from two corpora might be not similar in terms of characters but they have the similar meaning. Therefore, we're taking into account matching data units in the same language of two corpora also in terms of semantic aspect. However, this problem is very challenging, because to do this we must first annotate every concept of the corpora.

In relation to the solutions of unifying corpora with different language pairs, we also apply the algorithms Edit distance, BLEU and NIST to identify alignments of data units in the middle language of two corpora, and then align data units in the remaining languages. In case two corpora have totally different language pairs, we can apply existing alignment algorithms and tools such as GIZA+, GMA, A.Berger, Vanilla, Uplug, and Hunalign [6] with the corresponding language pairs. However, the experiment results of this solution are still modest, we hence try to improve the efficiency of this solution in next time.

The gained result of unifying the input corpora EuroParl, BTEC, ERIM, EOLSS, and DATIC is a very large corpus with 26 languages showed as Fig. 6.

Another problem, which is also very challenging and urgent, is how to enable large data resources to be convenient to use, exploit and share. A data resource will have a double value if it's always available and convenient to be used by NLP systems. Thus, apart from proposing solutions to unify NLP data resources, we have been building a system enabling the effective and convenient exploitation of NLP data resources. This system not only manages the unified NLP data resources, but also enables to integrate functions for exploiting data resources such as post-editing, evaluating, translating, segmenting, aligning, etc.

5 Conclusion

The quality of NLP systems, such as MT systems, search engines, evaluating systems, etc., highly depends on the data resources serving them in two aspects: quality and quantity. In addition to solutions for improving the quality of NLP data resources, solutions for creating large quantity of NLP data resources are very important to contribute to the quality and performance improvements of NLP systems. In this paper, we proposed solutions for creating large NLP data resources by unifying existing NLP data resources from three aspects: data, languages, and formats and structures. The solutions we proposed can be applied for two main kinds of NLP data resources, including in dictionaries and corpora. However, in the first phase of our experiments we've implemented the solutions for unifying MT corpora. The result achieved from the unification process of the input corpora *EuroParl, BTEC, ERIM, EOLSS, DATIC* is a very large corpus with a common structure and format, containing various domains and 26 languages, which is very convenient and useful for NLP systems to exploit and use. Based on the proposed solutions, we will continue to unify other existing corpora and dictionaries with the goal to create a very large corpus and dictionary with numerous of language pairs. Eventually, the unified NLP data resources are centrally located and effectively exploited by various NLP systems and functions such as post-editing, evaluating, translating, segmenting, aligning, etc.

References

1. Boitet, C.: Corpus pour la TA: types, tailles, et problèmes associés, selon leur usage et le type de système. Revue française de linguistique appliquée. **XII**(2007), 25–38 (2007)
2. Huynh, C.-P.: Des suites de test pour la TA à un système d'exploitation de corpus alignés de documents et métadocuments multilingues, multiannotés et multimedia. PhD thesis-National Polytechnic Institute of Grenoble, 228 p. (2010)
3. Koehn, P.: Europarl: A parallel corpus for statistical machine translation. In: Proceeding of the 10th Machine Translation Summit, Phuket, Thaïland, pp. 79–86 (2005)
4. Europarl corpus: http://www.statmt.org/europarl/
5. BTEC corpus: http://iwslt2010.fbk.eu/node/32
6. Alignment tools: http://web.eecs.umich.edu/~mihalcea/wa/
7. Largest dictionaries: http://www.worldslargestdictionary.com/
8. Munteanu, D.S., Marcu, D.: Extracting parallel sub-sentential fragments from non-parallel corpora. In: Proceedings of the 21st International Conference on Computational Linguistics and 44th Annual Meeting of the Association for Computational Linguistics, pp. 81–88 (2006)
9. Ying, Z., et al.: Automatic acquisition of Chinese-English parallel corpus from the web. Advances in Information Retrieval, pp. 420–431. Springer, Berlin, Heidelberg (2006)
10. Brunning, J.: Alignment Models and Algorithms for Statistical Machine Translation, Ph.D. Thesis. Cambridge University, 191 p. (2010)
11. Huynh, C.-P.: New approach for collecting high quality parallel corpora from multilingual Websites. In: iiWAS11 Conference. Proceedings of the 13th International Conference on Information Integration and Web-based Applications & Services (2011)
12. Amel, F., Paroubek, P.: Twitter as a comparable corpus to build multilingual affective lexicons. In: The 7th Workshop on Building and Using Comparable Corpora (2014)
13. Dosam, H.: A dictionary development system based on web. Int. Inf. Inst. (Tokyo) Inf. **14**, 11 (2011)
14. Tanmay, A.: Convert a Web Page into PDF File Format Online (2010)
15. Jun-min C.H.E.N.: Convert powerpoint presentations to word document based on VB6. 0. Comput. Knowl. Technol. **15** (2011)

16. Liu, P. et al.: Approaches to improving corpus quality for statistical machine translation. In: International Conference on Machine Learning and Cybernetics (ICMLC), 2010, vol. 6. IEEE (2010)
17. Maheshwari, S., Himanshu, S.: Corpus quality improvements for statistical machine translation. IJAICT **1**(3), 351–353 (2014)

Conference Calls' News Analysis and Stock Price Volatility

Jia-Lang Seng, Yi-Hui Wu and Hsiao-Fang Yang

Abstract This paper focuses on the relationship between news content of conference calls and stock price volatility for TWSE corporations and GTSM corporations. Our sample contains companies including listed in a Taiwan Stock Exchange companies and listed in Over-the-Counter companies that holding their conference calls on March during five-year period 2010–2014. We build rules of scoring to read the news of conference calls and use dictionary to objectively count the positive words and negative words in the news. This study's hypothesis is that (1) the less of the SCORE and TONE the more stock price volatility will produce; (2) the more QUANTITY of news, the more stock price volatility will produce. We use ordinary least squares regression to test our hypothesis. The empirical results show that there is a significantly negative relation between SCORE of news content and stock price volatility and there is a negative relation between TONE of news and stock price volatility. In addition, there is a significantly positive relation between QUANTITY of news and stock price volatility.

Keywords Conference call · Stock volatility · Content analysis · News

1 Introduction

The conference calls are a useful tool that business can use to convey their information to update the newest information to the professional analysts. The Taiwan Stock Exchange Corporation (TWSE) and the Gre Tai Securities Market (GTSM)

J.-L. Seng (✉) · Y.-H. Wu · H.-F. Yang
Department of Accounting, College of Commerce, National Chengechi University,
11605 Taipei City, Taiwan, ROC
e-mail: jia.lang.seng@gmail.com

Y.-H. Wu
e-mail: jses427@gmail.com

H.-F. Yang
e-mail: hfyang.wang@gmail.com

© Springer International Publishing Switzerland 2016
D. Król et al. (eds.), *Recent Developments in Intelligent Information
and Database Systems*, Studies in Computational Intelligence 642,
DOI 10.1007/978-3-319-31277-4_22

are usually holding company's conference calls every season or half a year. In short conference calls is indeed a good opportunity for companies to enhance the notability and pass on their complete information to professional analysts and media which make them make investment choice, so the conference calls is a very important activities between every company and their investors.

Ball and Brown [3] find evidence that stock prices continue to drift upward (downward) after initial positive (negative) earnings announcements, rendering the initial stock price reaction to the earnings news incomplete and raising questions of market efficiency. Recent studies such as [5, 10, 13, 19, 21, 22] indicate that language used in firms' earnings disclosures or in news media reports significantly affects stock returns. All studies find that the statistical significance for the linguistic tone of disclosure documents. Consequently, suggests that conference calls may provide a better setting in which to explore the relation between linguistic content and firm performance.

We focus on the how the financial news report conference calls and use our content analysis techniques to extract information from our data and analyze its relationship between stock market volatility and conference calls. This research develops a customized word list of financially oriented words based on an analysis of commonly used words in conference calls. The results show that conference calls' content conveys information to market participants and affect stock price volatility.

The remainder of this paper is organized as follows. Section 2 presents relevant literatures. Research method is described in Sect. 3. In Sect. 4, the study presents the empirical results and findings. Finally, conclusion is provided in Sect. 5.

2 Conference Calls, Content Analysis, and Stock Price Volatility

Conference calls are often used to supplement mandated disclosures, in particular, quarterly earnings releases. Franke et al. [9] find that conference calls provide information to the market beyond that which is found in the press release alone. Kimbrough [12] contends that conference calls are a voluntary disclosure mechanism of increasing importance that provide corporate managers with a forum in which they can emphasize specific aspects of recent performance and highlight their implications for future financial performance. Price et al. [18] find that the earnings-specific dictionary is much more powerful in detecting relevant conference call tone; conference call discussion tone has highly significant explanatory power for initial reaction window abnormal returns as well as the post-earnings-announcement drift.

Content analysis[1] is a wide and heterogeneous set of manual or computer-assisted techniques   for   contextualized   interpretations   of   documents   produced   by

---

[1]https://en.wikipedia.org/wiki/Content_analysis.

communication processes in the strict sense of that phrase or signification processes, having as ultimate goal the production of valid and trustworthy inferences. Tetlock [21] examines investor sentiment by measuring the pessimism index from the GI dictionary. Tetlock [22] find that negative words in firm-specific news stories predict low firm earnings. Henry [10] use computer-based analytical tools to measure the frequency of positive and negative words found in earnings press releases. Mayew and Venkatachalam [15] incorporate the linguistic content of conference calls as a control in their analysis. Liu [14] aims to classify an opinion document as expressing a positive or negative opinion or sentiment. Past studies is also commonly known as the document-level sentiment classification because it considers the whole document as a basic information unit. The application of content analysis, in general, for example, [5, 7, 10, 19] extract the tone of the wording of quarterly earnings press releases and relate it to things such as stock returns, volatility, and firm performance. All of these textual analysis studies find statistical significance for the linguistic tone of disclosure documents, suggesting that relevant information is conveyed by managers in their word choices.

In finance, volatility[2] is the degree of variation of a trading price series over time. Schwert [20] has computed sample standard deviations as volatility from intra-daily returns. Frankel et al. [9] empirically examine conference calls as a voluntary disclosure medium by analyzing stock volatility and trading volume levels. Antweiler and Frank [2] find evidence of a relationship between message activity and both trading volume and return volatility. Johnson and Marietta-Westberg [11] document a positive relationship between firm level news and firm level volatility contrary to the negative relationship between idiosyncratic news and firm volatility predicted by current theories. Sadique and Veeraraghavan [19] find that positive tone is directly related to returns, and negatively related to volatility. Similarly, [2010] find evidence that voluntary disclosure following the introduction of the Regulation Fair Disclosure.

3 Research Method

In this paper, we propose a research flowchart shown in Fig. 1.

3.1 Hypothesis Development

The research question of whether the score and tone of the conference calls news content affect stock-price volatility. Abarbanell and Bernard [1] state that the literature on both under reaction and overreaction provides evidence suggesting that

---

[2]https://en.wikipedia.org/wiki/Volatility_(finance).

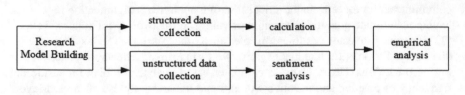

Fig. 1 Research flowchart

stock price behavior around earnings announcements may be caused by the failure of market participants to fully appreciate the information content of current earnings. The investor's expectations about future cash flows become more sensitive to new information, which causes greater stock price volatility [23]. This implies that we need to further our understanding of contemporaneous stock price reactions to the information content associated with earnings announcements. Firm issued conference calls and their news coverage give investors some idea about the company's past performance and future prospects. Accordingly, it is logical to expect that a report's SCORE and TONE affects volatility. Since positive (negative) tone reduces (increases) uncertainty about a firm's future performance, volatility is expected to fall for reports with positive tone and to rise for reports with negative tone. The quantity of news is deemed as whether the market have a lively discussion, so the QUANTITY of news are expected that the more quantity of conference calls news reported, the more stock price volatility will produce. Past research and above discussion leads to our two hypotheses: Hypothesis 1. The score and tone of conference calls news affects stock price volatility. Hypothesis 2. The quantity of news has impact on stock price volatility.

3.2 Data Collection

This paper observed that most enterprise uploaded their financial statements on March. Then, our sample contains companies including listed in TWSE and listed in Over-the-counter companies that holding their conference calls on March during 2010–2014. After our collection, this paper find many related news missed, so this paper extends our event window to 5 trading days (t − 1 to t + 3). It uses a time horizon [4, 17] beginning the day prior to the event day (for the company's conference calls) because that date can reflect the date of print media based on a press release issued on the previous day. We collect the news talking about firm's conference calls from news database.[3] In this paper, we got 435 conference calls news. We eliminate banking and insurance business, not in event windows' company, foreign Company, repeated company, not holding conference calls' company, and

---

[3]Knowledge Management Winner, http://kmw.chinatimes.com/.

missing value and get our final sample 198 companies. Most our sample companies are high-tech industry.

3.3   Research Design

Pastor and Veronesi [16] provide evidence of an inverse link between ROE (Return on Shareholders' Equity) and volatility. This paper also uses control variables as indicative of the size, value and level of indebtedness. The variables for size and value are included because they are factors that significantly affect the expected returns of stock prices in financial markets (see [8]). The size is measured by the number of OSHARE. The coefficient of the variable is expected to be negative [6]. The value is measured by the BtoM, and the Firm's indebtedness is determined by the LEV. This study uses Ordinary Least Squares (OLS) regression to test the relation between the news content of conference calls and stock price volatility. The regression model as follow:

$$
\begin{aligned}
STDEV = {} & \beta_0 + \beta_1 QUANTITY + \beta_2 SCORE + \beta_3 TONE \\
& + \beta_4 ROE + \beta_5 LEV + \beta_6 OSHARES + \beta_7 BtoM + \varepsilon
\end{aligned}
\tag{1}
$$

where STDEV is stock price volatility. The most commonly used measure of stock return volatility is standard deviation [20]. The methodology used to test our hypothesis follows the work of [11]. In this study, we use a daily average stock volatility measure (STDEV) expressed as follows:

$$
\sigma = \sqrt{\sum_1^T (R_{j\_}\overline{R}_j)^2}
\tag{2}
$$

where T is the number of trading days in our time window, Rj is the one-day closing price and $\overline{R}_j$ is the one-day average stock price. The QUANTITY is the quantity of news that are reported company's conference call. The SCORE is score of company's conference calls, and its range from −10 to +10. Follow [14] concept, we use numeric score expressing the strength/intensity scores. If the news make mention of company's stock price limit up, the company will get +8 to +10 scores; if the news refer to the company's net income is significantly increasing, the company will get +4 to +7 scores; if the news refer to the company operating achievements is not good as expectation, the company will get +1 to +3 scores; If the news refer to a company's operating performance is under expectation, their operating income is negative, the company will get −1 to −3 scores; if the news mention that a company's revenue is significantly decreasing, the company will get −4 to −7 scores; if the news make mention of company's stock price limit down, the company will get -8 to -10 scores. The TONE is the affect or feeling of news that refer to conference calls of a company. The formula of TONE is as following:

$$TONE_j = \frac{PositiveWords_j - NegativeWords_j}{PostiveWords_j + NegativeWords_j} \tag{3}$$

The ROE is a company's net income divided by its average stockholder's equity, the LEV is Leverage that is computed as net debt/total asset, the OSHARE (Outstanding shares) is the tradable shares available on the market at the end of our research year, and the BtoM (Book to Market) is the ratio between book value and market capitalization.

4   Research Results and Findings

This study applies the descriptive statistics to analyze the data from sample companies. The mean value of TONE is 0.667. The means of SCORE is 4.38, which indicate that news are mostly report optimistic operating condition. The max of SCORE is 9 and the min of SCORE is -4, indicating when company are in a good operating condition, news will give enormous publicity to the company, but when company are in a bad operating condition, news will report the company with retention of the phrase. The means of QUANTITY is 1.8, indicating that when company held conference calls, there are average of 2 news report the conference calls. The standard deviation for OSHARE is quite high reflecting the different operating structure and dimension of the companies in different industry. The LEV is generally very low, with an average net debt/asset ratio close to zero. The correlation is above 0.5 for ROE and BtoM.

The finding indicates that STDEV has a significantly negative association with news that are quantified by the SCORE which are given by reading the news that refer to company's conference calls; this result is consistent with the hypothesis that the score of conference calls news affects volatility; it is expected that positive score will reduce volatility and negative score will increase volatility. It is an evidence that conference calls' news indeed influences the stock market in the time window. However, the TONE is not significant but the negative coefficient is as this research expected. The QUANTITY of news that reported company's conference call is significant that prove the stock market are indeed influenced not only by the news content but also the frequency of the news reported. As expected, the LEV coefficient is also positive, coherent with the hypothesis, higher leverage results in higher financial risk and ultimately higher stock price volatility, but LEV is not statistically significant. The BtoM enters the regression is in our expectation with a negative coefficient, but it is not statistically significant. The negative coefficient of OSHARE variable is significant as our expectation that a larger quantity of tradable outstanding shares should be correlated to greater stability in stock prices. They are less likely to be influenced by volatility resulting from a scarce availability of shares on the market. The coefficient of ROE is positive as this research expect, and it is statistically significant.

5 Conclusion and Discussion

This study focus on the relationship between news content of conference calls and stock price volatility. We empirically examine the content of conference calls' news and its relation to stock price volatility. The results show that conference calls' content conveys information to market participants and affect stock price volatility. In addition, SCORE has significant explanatory power for stock price volatility and QUANTITY has significant explanatory power for stock price volatility. However, the TONE does not have significance for stock price volatility. Conference calls' news indeed have impact on investors and meaningfully impact capital allocation. Although rules has been built up, and its empirical result is significant, the result may be affected by the subjective judgment of readers. The specified dictionary used to count positive words and negative words has limitation to express what the words want to convey. Consequently, the quantity and the year of sample news can be expanded and there may have additional insights this research can research.

Acknowledgement This research is supported by NSC 102-2627-E-004-001, MOST 103-2627-E-004-001, MOST 104-2627-E-004-001.

References

1. Abarbanell, J.S., Bernard, V.L.: Tests of analysts overreaction underreaction to earnings information as an explanation for anomalous stock-price behavior. J. Finance **47**(3), 1181–1207 (1992)
2. Antweiler, W., Frank, M.Z.: Is all that talk just noise? The information content of internet stock message boards. J. Finance **59**(3), 1259–1294 (2004)
3. Ball, R., Brown, P.: An empirical evaluation of accounting income numbers. J. Account. Res. **6**, 159–178 (1968)
4. Ball, R., Kothari, S.P.: Security returns around earnings announcements. Account. Rev. **66**(4), 718–738 (1991)
5. Davis, A.K., Piger, J.M., Sedor, L.M.: Beyond the numbers: an analysis of optimistic and pessimistic language in earnings press releases. Federal Reserve Bank of St. Louis Working Paper Series. https://m.research.stlouisfed.org/wp/2006/2006-005.pdf. (2006)
6. Dell'Acqua, A., Perrini, F., Caselli, S.: Conference calls and stock price volatility in the post-reg FD era. Eur. Financ. Manage. **16**(2), 256–270 (2010)
7. Demers, E., Vega, C.: Linguistic tone in earnings announcements: news or noise? (2008)
8. Fama, E.F., French, K.R.: The cross-section of expected stock returns. J. Financ. **47**(2), 427–465 (1992)
9. Frankel, R., Johnson, M., Skinner, D.J.: An empirical examination of conference calls as a voluntary disclosure medium. J. Acc. Res. **37**(1), 133–150 (1999)

10. Henry, E.: Are investors influenced by how earnings press releases are written? J. Bus. Commun. **45**(4), 363–407 (2008)
11. Johnson, W.C., Marietta-Westberg, J.: The effect of news on volatility: a study of IPOs. Available at SSRN: http://ssrn.com/abstract=556786 (2004)
12. Kimbrough, M.D.: The effect of conference calls on analyst and market underreaction to earnings announcements. Account. Rev. **80**(1), 189–219 (2005)
13. Li, F.: Annual report readability, current earnings, and earnings persistence. J. Account. Econ. **45**(2–3), 221–247 (2008)
14. Liu, B.: Sentiment analysis and opinion mining. Synth. Lect. Hum. Lang. Technol. **5**(1), 1–167 (2012)
15. Mayew, W.J., Venkatachalam, M.: The power of voice: managerial affective states and future firm performance. J. Finance **67**(1), 1–43 (2012)
16. Pastor, L., Veronesi, P.: Stock valuation and learning about profitability. J. Finance **58**, 1749–1789 (2003)
17. Patell, J.M., Wolfson, M.A.: The ex ante and ex post price effects of quarterly earnings announcements reflected in option and stock-prices. J. Account. Res. **19**(2), 434–458 (1981)
18. Price, S.M., Doran, J.S., Peterson, D.R., Bliss, B.A.: Earnings conference calls and stock returns: the incremental informativeness of textual tone. J. Bank. Finance **36**(4), 992–1011 (2012)
19. Sadique, S., In, F.H., Veeraraghavan, M.: The impact of spin and tone on stock returns and volatility: evidence from firm-issued earnings announcements and the related press coverage. Available at SSRN: http://ssrn.com/abstract=1121231 (2008)
20. Schwert, G.W. Stock market volatility. Financ. Anal. J. **46**(3), 23–34 (1990)
21. Tetlock, P.C.: Giving content to investor sentiment: the role of media in the stock market. J. Finance **62**(3), 1139–1168 (2007)
22. Tetlock, P.C., Saar-Tsechansky, M., Macskassy, S.: More than words: quantifying language to measure firms' fundamentals. J. Finance **63**(3), 1437–1467 (2008)
23. Veronesi, P.: Stock market overreaction to bad news in good times: a rational expectations equilibrium model. Rev. Financ. Stud. **12**(5), 975–1007 (1999)

Lexicon-Based Sentiment Analysis of Facebook Comments in Vietnamese Language

Son Trinh, Luu Nguyen, Minh Vo and Phuc Do

Abstract Social media websites like Twitter, Facebook etc. are a major hub for users to express their opinions online. Sentiment analysis which is also called opinion mining, involves in building a system to collect and examine opinions about the product made in blog posts, comments, or reviews. Sentiment analysis can be useful in real life. In this paper, we propose a lexicon based method for sentiment analysis with Facebook data for Vietnamese language by focus on two core component in a sentiment system. That is to build Vietnamese emotional dictionary (VED) including 5 sub-dictionaries: noun, verb, adjective, and adverb and propose features which based-on the English emotional analysis method and adaptive with traditional Vietnamese language and then support vector machine classification method to be use to identify the emotional of the user's message. The experimental show that our system has very good performance.

Keywords Lexicon-based sentiment analysis · Vietnamese · Text analytics · Vietnamese emotional dictionary · Proposing features · Facebook

S. Trinh (✉) · L. Nguyen · M. Vo · P. Do
University of Information Technology, Ho Chi Minh City,
Ho Chi Minh City, Vietnam
e-mail: sontq@uit.edu.vn

L. Nguyen
e-mail: luut.ng@gmail.com

M. Vo
e-mail: voleminh10t2@gmail.com

P. Do
e-mail: phucdo@uit.edu.vn

© Springer International Publishing Switzerland 2016
D. Król et al. (eds.), *Recent Developments in Intelligent Information and Database Systems*, Studies in Computational Intelligence 642,
DOI 10.1007/978-3-319-31277-4_23

263

1 Introduction

Social media websites like Twitter, Facebook etc. are a major hub for users to express their opinions online. On these social media sites, users post comments and opinions on various topics. Hence these sites become rich sources of information to mine for opinions and analyze user behavior and provide in-sights for user behavior, product feedback, user intentions, lead generation. Businesses spend an enormous amount of time and money to understand their customer opinions about their products and services.

Sentiment analysis is a type of natural language processing for tracking the mood of the public about a particular product or topic. Sentiment analysis, which is also called opinion mining, involves in building a system to collect and examine opinions about the product made in blog posts, comments, or reviews. Sentiment analysis can be useful in several ways. For example, in marketing it helps in judging the success of an ad campaign or new product launch, determine which versions of a product or service are popular and even identify which demo graphics like or dislike particular features. Thus Sentiment Analysis has become a hot research area since 2002. Sentiment Analysis is used to determine sentiments, emotions and attitudes of the user. The text used for analysis can range from big document (e.g. Product reviews from Amazon, blogs) to small status message (e.g. Tweets, Facebook comments).

Lexicon-based approaches to sentiment analysis differ from the more common machine-learning based approaches in that the former rely solely on previously generated lexical resources that store polarity information for lexical items, which are then identified in the texts, assigned a polarity tag, and finally weighed, to come up with an overall score for the text. Such sentiment analysis systems have been proved to perform on par with supervised, statistical systems, with the added benefit of not requiring a training set. In this paper, we implemented a system with lexicon-based approaches for analysis of Facebook message in Vietnamese language.

The rest of the paper is organized as follows. In Sect. 2, we present related work on sentiment analysis and then present sentiment analysis for Vietnamese language system in details in Sect. 3. In Sect. 4, we experiment with the training and evaluate the final result obtained from the test data. In Sect. 5, we present our conclusions and outline our future work.

2 Related Work

Sentiment Analysis on raw text is a well known problem. The Liu [1] book covers the entire field of sentiment analysis. Sentiment analysis can be done using machine learning, lexicon-based approach or combined.

The machine learning approach applicable to sentiment analysis mostly belongs to supervised classification in general and text classification techniques in particular [2]. However, their obvious disadvantage in terms of functionality is their limited applicability to subject domains other than the one they were designed for. In a machine learning based classification, two sets of documents are required: training and a test set. A training set is used by an automatic classifier to learn the differentiating characteristics of documents, and a test set is used to validate the performance of the automatic classifier. A number of machine learning techniques have been adopted to classify the reviews. Machine learning techniques as Naive Bayes, maximum entropy and support vector machines (SVM) [2]. Although interesting research has been done aimed at extending domain applicability [3], such efforts have shown limited success. An important variable for these approaches is the amount of labeled text available for training the classifier, although they perform well in terms of recall even with relatively small training sets [4]. On the other hand, a growing number of initiatives in the area have explored the possibilities of employing unsupervised lexicon-based approaches.

The semantic orientation approach to sentiment analysis is unsupervised learning because it does not require prior training in order to mine the data. Instead, it measures how far a word is inclined towards positive and negative. Much of the research in unsupervised sentiment classification makes use of lexical resources [2]. The lexicon based approach is based on the assumption that the contextual sentiment orientation is the sum of the sentiment orientation of each word or phrase. Turney [5] identifies sentiments based on the semantic orientation of reviews. Taboada et al. [6], Mclville et al. [7], Ding et al. [8] use lexicon based approach to extract sentiments.

These rely on dictionaries where lexical items have been assigned either polarity or a valence, which has been extracted either automatically from other dictionaries, or, more uncommonly, manually. The works by Hatzivassiloglou and McKewon [9] and Turney [5] are perhaps classical examples of such an approach. The most salient work in this category is Taboada [6], whose dictionaries were created manually and use an adaptation of Polanyi and Zaenen's [10] concept of Contextual Valence Shifters to produce a system for measuring the semantic orientation of texts, which they call SOCAL(culator).

Combining both methods (machine learning and lexicon-based techniques) has been explored by Kennedy and Inkpen [11], who also employed contextual valence shifters, although they limited their study to one particular subject domain (the traditional movie reviews), using a "traditional" sentiment lexicon (the General Inquirer), which resulted in the "term-counting" (in their own words) approach. The degree of success of knowledge based approaches varies depending on a number of variables, of which the most relevant is no doubt the quality and coverage of the lexical resources employed, since the actual algorithms employed to weigh positive against negative segments are in fact quite simple.

3 Sentiment Analysis for Vietnamese Language

Currently, the research of sentiment problem in Vietnamese language has some results. In particular, in [12] the author show up the problem in building a sentiment dictionary for less popular languages, such as Vietnamese, is difficult and time consuming. An approach to mining public opinions from Vietnamese text using a domain specific sentiment dictionary in order to improve the accuracy. The sentiment dictionary is built incrementally using statistical methods for a specific domain such as sentiment classification from hotel reviews [13], computer products reviews [14].

In another way, in [15, 16] authors have explored different methods of improving the accuracy of sentiment classification. The sentiment orientation of a document can be positive (+), negative (−), or neutral (0). Dictionary has many verbs, adverbs, phrases and idioms. The author based on the combination of Term-counting method and enhanced contextual valence shifters method has improved the accuracy of sentiment classification. The combined method has accuracy 68.9 % on the testing dataset, and 69.2 % on the training dataset. All of these methods are implemented to classify the reviews based on our new dictionary and the internet movie data set.

Comparing with previous researches related to our topic, our propose method has different points, that are features which were selected adaptation in Vietnamese language, build a Vietnamese emotional language with more words consistent with the Vietnamese grammar based on spelling that people are using on social network.

Our proposed system has 3 components: data collection from the Facebook, preprocessing and extracting features and the third component is analysis emotional for comments from the data (Fig. 1).

In more details:

- Data collection: we collect all comments (sentences) from the Facebook via API which provided to get automatic data in this component.

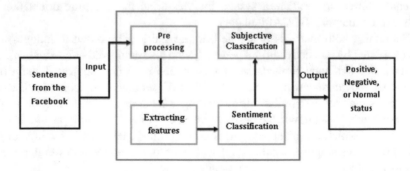

Fig. 1 Our proposed system

- Preprocessing and extracting features: This is a component step for pre processing data where we have a lots of processing for text including: Removing stopword and foreign words, icon and then pos tagging for all of sentence from the data which has been collected in the previous step and then extract features.
- Analysis emotional: Evaluating the sentence has emotion or non-emotion based on features which were selected adaptation in Vietnamese language and Vietnamese emotional dictionary, and then sentiment classification algorithm has been processed to evaluate the emotion for sentence is in positive, negative or normal status based on support vector machine (SVM) classification method.

3.1 Building Emotional Dictionary

The calculation of sentiment based on dictionary begins with two assumptions: that individual words have what is referred to as prior polarity, that is, a semantic orientation that is independent of context and that said semantic orientation can be expressed as a numerical value. Several lexicon based approaches have adopted these assumptions.

Adjectives

Much of researches in sentiment focused on adjectives or adjective phrases as the primary source of subjective content in a document. In general, the semantic orientation of an entire document is the combined effect of the adjectives or relevant words found within, based upon a dictionary of word rankings (scores) [17].

Nouns, Verbs, and Adverbs

Although the sentences have comparable literal meanings, the plus-marked nouns, verbs, and adverbs indicate the positive orientation of the speaker towards the situation, whereas the minus-marked words have the opposite effect. It is the combination of these words in each of the sentences that conveys the semantic orientation for the entire sentence [17].

In order to make use of this additional information, we created separate noun, verb and adverb dictionaries, hand-ranked using the same +5 to −5 scale as our adjective dictionary.

We created Vietnamese emotional dictionary (VED) which contains 5 sub-dictionaries: noun, verb, adjective, and adverb dictionary. Our dictionary is essentially based on the English SO-CAL (Dictionaries for the Semantic Orientation CALculator) dictionary. We choose SO-CAL, because this dictionary is the best in overall for a lots of topic in experiments as shown in Fig. 2 in which

Emotional Dictionaries	Efficiency				
	Topic Epinion 1	Topic Epinion 2	Topic Movie	Topic Camera	Overall
Google-Full	62.00	58.50	66.31	61.25	62.98
Google-Basic	53.25	53.50	67.42	51.40	59.25
Maryland-Full-NoW	58.00	63.75	67.42	59.46	62.65
Maryland-Basic	56.50	56.00	62.26	53.79	58.16
General Inquirer-Full	68.00	70.50	64.21	72.33	68.02
General Inquirer-Basic	62.50	59.00	65.68	63.87	64.23
SentiWordNet-Full	66.50	66.50	61.89	67.00	65.02
SentiWordNet-Basic	59.25	62.50	62.89	59.92	61.47
Subjective-Full	72.75	71.75	65.42	77.21	72.04
Subjective-Basic	64.75	63.50	68.63	64.83	66.51
SO-CAL-Full	80.25	80.00	76.37	80.16	78.74
SO-CAL-Basic	65.50	65.25	68.05	64.70	66.04

Fig. 2 List of emotional dictionaries

topic Epinion 1, 2 has 400 documents for each about book, car, computer, cooking, hotel, music and phone. Movie has 1900 documents about films. Camera has 2400 documents about printers and cameras.

In addition, we added some words to our dictionary to make consistent with the Vietnamese grammar and concise spelling that people are using on social network. The number of words in each dictionary of noun, verb, adjective and adverb are 1546 words, 1108 words, 2357 words, 749 words respectively and each word is paired with an integer which describes the corresponding emotional value from the most negative (-5) to the most positive ($+5$). Notice that no word has SO emotional value at zero value (0) (Tables 1, 2, 3, 4 and 5).

Table 1 Some words from
dictionary of noun

Noun	
Noun	Emotional value
hoàn hảo (perfection)	5
lộng lẫy (luxury)	4
chiến thắng (victory)	3
phước lành (blessing)	2
độc lập (liberty)	1
tội phạm (crime)	−1
điểm yếu (weakness)	−2

Table 2 Some words from
dictionary of verb

Verb	
Verb	Emotional value
tôn kính (respect)	4
hoan hỉ (delight)	4
thành công (succeed)	3
sáng tạo (create)	2
tăng (increase)	1
vùi dập (ruin)	−1
xấu hổ (shame)	−2

Table 3 Some words from
dictionary of adjective

Adjective	
Adjective	Emotional value
tuyệt vời (perfect)	5
cao cấp (high-grade)	4
bổ ích (helpful)	3
chặt chẽ (close)	2
hợp lý (agreed)	1
cũ (old)	−1
đần độn (silly)	−2

Table 4 Some words from dictionary of adverb

Adverb	
Adverb	Emotional value
thú vị (interestingly)	5
huy hoàng (splendidly)	4
giỏi (well)	3
Tươi (freshly)	2
sạch (clean)	1
kỳ quặc (weirdly)	−1
thô (crudely)	−2

Table 5 Some words from intensification dictionary

Intensification	Emotional value
Ít (Slenderly)	−1.5
chút ít (Slightly)	−0.9
Hơi (a little)	−0.5
Khá (rather)	−0.2
chắc (surely)	0.2
Siêu (super)	0.4
hoàn toàn (completely)	0.5

The intensification dictionary has 185 special words in Vietnamese language and each word also has a accompanied decimal to demonstrate the increase or decrease of its emotional value.

Example: If emotional value for word "nhếch nhác" (messy) is (−3) then word "khá nhếch nhác" (rather messy) has emotional value (−3) * (1 − 0.1) = (−2.7). On the same, if emotional value for word "xuất sắc" (excellent) is (5) then word "xuất sắc nhất" (the most excellent) has emotional value 5 * (1 + 1) = 10.

Table 6 Result of subjective manual classification

Number	Topic	Training data	
		Subjective sentences	Objective sentences
1	Education	173	99
2	Movie	194	95
3	Sport	248	76
4	All	615	270

Table 7 Result of sentiment manual classification

Number	Topic	Training data	
		Positive sentences	Negative sentences
1	Education	133	40
2	Movie	115	79
3	Sport	201	47
4	All	449	166

3.2 Training

As we know, emotional is extremely complicated. Hence to build a manageable data, we conducted collecting comments and opinions of the social network user and labeled those each sentence in comment to analyze them. Each sentence has subjective nature of every person. The first task is to classify which comment is emotional or non-emotional (also known as subjectivity classification) and the second task is to classify which comment is negative or positive (also known as sentiment classification).

Our data source was chosen from 3 topics: education, movie and sport. Each database contains from 250 to 350 comments of those topics, and then we created a bigger synthetic database from 3 topics which includes 885 comments. In the next step, we divide manually the synthetic database into 2 parts: subjective and objective sentences. After that, the subjective sentences were classified manually into 2 parts: negative and positive sentences (Tables 6 and 7).

4 Experimental Model and Result

4.1 Subjective Classification

This method uses 6 features to classify which sentence is emotional or non-emotional:

- Feature 1st: The amount of word in the sentence. It partly displays what the users want to express through the comments. If a lots of number of words are appeared, the user is really interested in this topic.

- Features 2nd, 3rd, 4th and 5th: The total of emotional value of noun, adjective, verb, adverb in the comments. The emotional value of a sentence depends on the type of word which was compared with the VED dictionary.
- Feature 6th: The total of emotional value of a sentence is basically total of 4 attributes that is 2nd, 3rd, 4th, 5th.
- Moreover, the emotional value of a sentence also depend on the type of the sentence. The emotional value of a sentence will be 0 point if this sentence is a condition or a question sentence.

Algorithm

- Input: Sentence has been preprocessing, VED emotional dictionary.
- Output: Feature vectors
- Steps:
  - Count number of words in sentence
  - Find and calculate sum of emotional value of adjective in sentence. (2)
  - Find and calculate sum of emotional value of adverb in sentence. (3)
  - Find and calculate sum of emotional value of noun in sentence. (4)
  - Find and calculate sum of emotional value of verb in sentence. (5)
  - Sum of emotional value of sentence = sum of all values in 2, 3, 4, 5
  - If sentence is question or conditional sentence return 0
    Otherwise, return sum of emotional value of sentence.
  - Return feature vector.

From the feature vector, we use SVM method to classify sentence into subjective (emotional) or objective class (non-emotional).

4.2   Sentiment Classification

After the subjective classification has been processed, we continued to apply the sentiment classification on these sentences. We proposed features which presented in below was based-on the English emotional analysis method and the consistent with traditional Vietnamese language.

- Firstly, emotional value of a sentence depends on the emotional value of each emotional word or phrase. The most basic attributes inherited from subjective analysis. The summary of emotional value of a sentence is total in value of all features above.
- Secondly, emotional value of a sentence which depends on the emotional value of the intensification will be calculated by: Emotional value = value of intensification * value of emotional word

The total of these values will be the new value of the emotion after review intensification. In the absence of intensification in sentence, this value is the total value of all kinds of emotional words in a sentence.

- Thirdly, emotional value of a sentence also depends on the negative words in the sentence: "không" (no),"không có" (without), ... will be calculated by: Emotional value = (−1) * value of emotional word
- Fourthly, emotional value of a sentence which depends on the imperfect words: "nên" (should), "phải" (must have), "có thể" (maybe), ... will be calculated by: Emotional value = (0.5) * total value of all imperfect words in a sentence
- Fifthly, emotional value of a positive sentence: In fact, traditional vietnamese culture, people avoid using negative words to express their opinions so that the positive words are commonly used. Hence, the emotional value of a positive word will be calculated by: Emotional value = (1 + 0.5) * value of positive word
- Lastly, emotional value of a sentence which has a contrasting-linked word likes: "nhưng" (but), "tuy nhiên" (however), ... will be calculated by total of the emotional value of words that subtract the emotional value of the words before the contrasting-linked word by: Emotional value = Emotional value − total of emotional value of the words before the contrasting-linked word

Algorithm

- Input: Sentence has been preprocessing and VED emotional dictionary.
- Output: Feature vectors
- Steps:
For each sentence from the data do:

- Find and calculate sum of emotional value of adjective in sentence. (2)
- Find and calculate sum of emotional value of adverb in sentence. (3)
- Find and calculate sum of emotional value of noun in sentence. (4)
- Find and calculate sum of emotional value of verb in sentence. (5)
- Sum of emotional value of sentence = sum of all values in 2, 3, 4, 5
- Find intensification words in the sentence and update the value of emotional:
  *Emotional value = value of intensification * value of emotional word*
- Find negative words in the sentence and update the value of emotional:
  *Emotional value = (−1) * value of negative word*
- Find imperfect words in the sentence and update the value of emotional:
  *Emotional value = (0.5) * total value of all imperfect words in a sentence*
- Find positive words in the sentence and update the value of emotional:
  *Emotional value = (1 + 0.5) * value of positive word*
- Find linked word in the sentence and update the value of emotional:
  Emotional value = Emotional value − total of emotional value of the words before the contrasting-linked word
- Return feature vector.

4.3 Result

Testing data include 4 topics: education, movie, sport and all data combined which are shown in Table 8. We classified manually emotions through 2 steps: subjective classification and sentiment classification. The first step, data of each topic is classified into two categories: subjective and objective. In the second step, we classified subjective data into two categories as positive and negative.

By using our features which had consistent with traditional Vietnamese language and classify based on SVM classification method and then we calculated the precision measure of subjective classification according to the algorithm in previous step. Precision measure is calculated as the proportion of the true sentence classification against all the sentences. Results are presented in Table 9.

We saw the results of subjective classification for each topic, in education topic, precision value is 125/135 sentences which has been classified true (92.6 %), movie is 131/146 (89.7 %), sport is 145/162 (89.5 %) and ALL data are 398/443 (89.8 %). The average of precision value is 90.4 %.

We continue to assess the accuracy of the sentiment classification method. Results are presented in Table 10.

In Table 10 shows results of sentiment classification for each topic, in education topic, precision value is 79/87 sentences which has been classified true (90.8 %), movie is 84/106 (79.2 %), sport is 115/121(95 %) and ALL data are 281/314 (89.5 %). The average of precision value is 90.4 %. The average of precision value is 88.6 %. The experimental show that our system has very good performance, because of features which were selected adaptation in Vietnamese language and Vietnamese emotional language has been built with more words consistent with the Vietnamese grammar based on spelling that people are using on social network. These point confirmed to help improvement accuracy in sentiment analysis.

Table 8 Manually classification

ID	Topic	Testing data			
		Subjective sentences	Objective sentences	Positive sentences	Negative sentences
1	Education	87	48	76	11
2	Movie	106	40	80	26
3	Sport	121	41	113	8
4	All	314	129	269	45

Table 9 Results of subjective classification

ID	Topic	Number sentences	Result – precision (%)
1	Education	135	92.6
2	Movie	146	89.7
3	Sport	162	89.5
4	All	443	89.8

Table 10 Results of sentiment classification

ID	Topic	Number sentences	Result – precision (%)
1	Education	135	90.8
2	Movie	146	79.2
3	Sport	162	95.0
4	All	443	89.5

5 Conclusion

Sentiment detection has a wide variety of applications in information systems, including classifying reviews, summarizing review and other real time applications. In this paper, we proposed a lexicon based method for sentiment analysis with Facebook data for Vietnamese language by focus on two core component in a sentiment system. That is to built Vietnamese emotional dictionary (VED) which contains 5 sub-dictionaries: noun, verb, adjective, and adverb and proposed features which based-on the English emotional analysis method and adaptive with traditional Vietnamese language and then support vector machine classification method has been used to identify the emotional of the user's message. The experimental show that our system has very good performance. In future, we are continuing on improving the performance by building a larger emotional dictionary for Vietnamese language and integrating some technical nature language processing, and solve the big data problem.

References

1. Liu, B.: Sentiment analysis and opinion mining. Synth. Lect. Hum. Lang. Technol., 1–167 (2008)
2. Vinodhini, G., Chandrasekaran, R.M.: Sentiment analysis and opinion mining: a survey. In: Int. J. Adv. Res. Comput. Sci. Soft. Eng. 2(6) (2012)
3. Aue, A., Gamon, M.: Customizing sentiment classifiers to new domains: a case study. Presented at the Recent Advances in Natural Language Processing (RANLP), Borovets, Bulgaria (2005)
4. Andreevskaia, A., Bergler, S.: ClaC CLaC-NB: knowledge-based and corpus-based approaches to sentiment tagging. In: Proceedings of the 4th International Workshop on Semantic Evaluations. Association for Computational Linguistics, Prague, Czech Republic (2007)
5. Turney, P.D.: Thumbs up or thumbs down? Semantic orientation applied to unsupervised classification of reviews. In: Proceedings of the 40th Annual Meeting of the Association for Computational Linguistics (ACL), pp. 417–424 (2002)
6. Taboada, M., Brooks, J., Tofiloski, M., Voll, K., Stede, M.: Lexicon-based methods for sentiment analysis. Comput. Linguist. 37(2), 267–307 (2011)
7. Melville, P., Gryc, W., Lawrence, R.D.: Sentiment analysis of blogs by combining lexical knowledge with text classification. Proceedings (2011)
8. Ding, X., Liu, B., Yu, P.S.: A holistic lexicon-based approach to opinion mining. In: Proceedings of the International Conference on Web Search and Web Data Mining, pp. 231–240. ACM (2008)

9. Hatzivassiloglou, V., McKeown, K.R.: Predicting the semantic orientation of adjectives. In: Proceedings of the Eighth Conference on European Chapter of the Association for Computational Linguistics, pp. 174–181. Association for Computational Linguistics, Madrid, Spain (1997)

10. Polanyi, L., Zaenen, A.: Contextual valence shifters. In: Computing Attitude and Affect in Text: Theory and Applications, pp. 1–10. Springer, Dordrecht (2006)

11. Kennedy, A., Inkpen, D.: Sentiment classification of movie reviews using contextual valence shifters. Comput. Intel. **22**(2), 110–125 (2006)

12. Nguyen, H.N., Van Le, T., Le, H.S., Pham, T.V.: Domain specific sentiment dictionary for opinion mining of vietnamese text. In: The 8th International Workshop, MIWAI 2014, Bangalore, India (2014)

13. Duyen, N.T., Bach, N.X., Phuong, T.M.: An empirical study on sentiment analysis for Vietnamese. In: International Conference on Advanced Technologies for Communications (2014)

14. Kieu, B.T., Pham, S.B.: Sentiment analysis for vietnamese. In: KSE '10 Proceedings of the Second International Conference on Knowledge and Systems Engineering, pp. 152–157 (2010)

15. Nguyen N.D.: Document summarization based on sentiment classification. Master thesis in computer science (Vietnamese), University of Technology Hochiminh city (2014)

16. Phu, V.N., Tuoi, P.T.: Sentiment classification using enhanced contextual valence shifters. In: Proceedings of International Conference on Asian Language Processing, Malaysia (2014)

17. Taboada, M., Brooke, J., Tofiloski, M., Voll, K., Stede, M.: Lexicon-based methods for sentiment analysis. Comput. Linguist. J. **37**(2) (2011)

Scoring Explanatoriness of a Sentence and Ranking for Explanatory Opinion Summary

Trung Thien Vo, Bac Le and Minh Le Nguyen

Abstract On the online reviews, one of the important types of information is the sentiment explanation which expresses a content users generated. Sentiment explanation is a sentence that expresses detailed reason of sentiment (i.e., "explanatoriness") and plays an important role in opinion summarization. In this paper, we propose and study a method for scoring the explanatoriness of a sentence. A first method is to adapt an existing method and a second method based on a probabilistic model. Experimental results show that the proposed methods are effective, presenting a better value for a state of the art sentence ranking method for standard text summarization.

Keywords Explanatory scoring · Ranking explanatory sentence · Opinion mining

1 Introduction

One similar work with this paper is [1], the authors et al. [1] scored the explanatoriness for each sentence, and then ranked explanatory sentences for opinion summarization. Although their approach was unsupervised, their approach has advantage of not asking many manual efforts and applying into various domains. Their formulas were based on the assumption that existing technique can be used to classify review sentences into different aspects and specify sentiment polarity for

T.T. Vo (✉) · B. Le
Faculty of Information Technology, University of Science, VNU-HCM,
Ho Chi Minh City, Viet Nam
e-mail: vothientrung90@gmail.com

B. Le
e-mail: lhbac@fit.hcmus.edu.vn

M. Le Nguyen
School of Information Science, Japan Advanced Institute of Science
and Technology, Nomi, Japan
e-mail: nlminh2007@gmail.com

© Springer International Publishing Switzerland 2016
D. Król et al. (eds.), *Recent Developments in Intelligent Information
and Database Systems*, Studies in Computational Intelligence 642,
DOI 10.1007/978-3-319-31277-4_24

each sentence. Our approach takes advantage of the fundamental of the previous work [1] and is closer to pragmatic demands.

BM25 [2] is clearly one of the most important and widely used information retrieval functions. In [1], the authors study and focus on ranking sentences in opinionated text. In this paper, we improve the ranking explanatory sentence model which is introduced in [1]. Explanatory sentences could not only be relative to the mentioned topic, but also include details explaining reasons of sentiments. However, we consider that general positive or negative sentences are not explanatory.

In this paper, we focus on studying how to solve this problem in an unsupervised way. We propose three features for scoring explanatoriness of a sentence such as length of sentence, popularity, and distinctive of sentence. We also introduce a two general method for scoring explanatoriness of a sentence based on these features. They are methods which use TF-IDF weighting and probability model based on word-level frequency.

We use one data set [1] which is based on a collection of Amazon product reviews used in [3, 4] and a evaluation method named weighted Mean Average Precision (*wMAP*). Both of them are mentioned at [1]. Experiment results show that the proposed methods are effective in ranking explanatory sentences.

2 Framework

We assume that existing methods classify opinionated sentences into different aspects such as subtopics and determine the sentiment polarity of a review sentence. An opinionated sentence is either positive or negative. Consequently, to solve this problem, we assume the input which is a group of four sets as:

- A topic T is described by a phrase or word. Furthermore, a topic T has one or more aspects, e.g., a topic T is a phrase 'Nikon'.
- An aspect A is expressed by a phrase, e.g., a function 'camera' of Nikon.
- A sentiment polarity P on the specific aspect A of topic T. Moreover, a sentiment polarity P is either positive or negative.
- A set of review sentences $O = \{S_1,\ldots, S_n\}$ of the sentiment polarity P.

For example, if we want to ranking positive opinion sentences about Nikon camera, our input would be T = 'Nikon', A = 'camera', P = 'positive', and a set of positive review sentence about Nikon camera, O. The output result which we desire is a ranked list of all sentences in O based on their explanatory score and we denote $L = (S'_1,\ldots, S'_n)$. In L, each sentence S'_i belongs to a set O and we desire explanatory sentences would be ranked on top of non-explanatory ones. A ranked list L can be useful to help users know opinions carefully or combined with current summarization algorithm so as to construct an explanatory opinion summary.

We have an image to illustrate this framework as follows (Fig. 1).

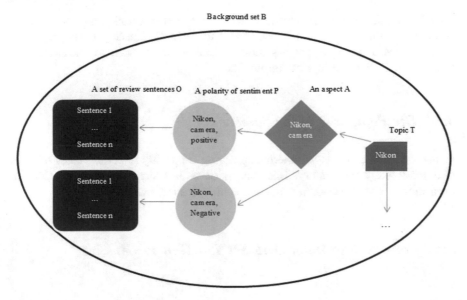

Fig. 1 The proposed framework

3 Basic Features and an Input Data Set

In this paper, we apply three sentence features to two explanatoriness scoring methods which we propose. The three features are presented as follows.

3.1 Sentence Features

The length in this feature means the number of words in the sentence. In general, a longer sentence is more probable explanatory than a shorter one since a longer sentence may contain more information. Longer sentences would probably be scored higher because they would contain more non-stop word terms.

3.2 The Distinctiveness of the Sentence to Background Data Set

We see that a sentence with more distinctive terms that can differentiate O from background information is more probable explanatory. Explanatory sentences should provide us more specific information about the given topic. Intuitively, an explanatory sentence would more likely contain terms that can help discriminate the set of sentences to be summarized O from more general background sets which contain opinions that are not as specific as those in O. Indeed, in this paper, we should focus a sentence that has more distinctive terms such as terms that are frequent in O, but rare in a background set.

3.3 The Popularity of the Sentence

A sentence that contains more terms appearing frequently in all sentences of a set O is more likely explanatory. This conception is basically the main idea which is used in the existing standard extractive summarization techniques.

3.4 Construct an Input Data Set and How to Combine Features

In this paper, we introduce the way to create the background set. In general, the background set is a superset of a set O. In our problem context, the set O contains sentences which satisfy the constraints that they belong to aspect A of topic T with sentiment polarity P. We construct a background data set B by gathering all sentences about topic T. This is easy to do that.

However, how to combine all the three features is a challenge problem. If we use any single feature to rank, the undesirable results will happen. For instance, if we only focus on the popularity feature, the result is that the non-informative will be ranked at higher level. Otherwise, if we only concentrate a distinctive feature, the rare opinions mentioned only a few times may be ranked at higher level. Therefore, in this paper, we propose two methods to combine these features for explanatoriness scoring.

4 Explanatoriness Scoring Models

4.1 $Bm25_{EX}$

The method is to adapt an existing ranking function of information retrieval such as $BM25$ [5], which is one of the most effective basic information retrieval function. $BM25$ [5] is also a function of term frequencies, document frequencies and the field length for the single field. This popularity feature which we mention above can be demonstrated by Term Frequency (TF) weighting and the distinctive feature can be presented through Inverse Document Frequency (IDF) weighting. Consequently, we propose the following modified $BM25$ for explanatoriness scoring. We name it $BM25_{Ex}$ function.

In explanatoriness ranking and computing an explanatory score, we can refer a sentence as a query and compute explanatoriness of each word in the sentence based on how frequent the word is in the background data set (B) and the input data set (O). With a given sentence $s = w_1, w_2,..., w_n$, the modified $BM25_{Ex}$ method is defined as:

$$BM25_{Ex}(S, O, B) = \sum_{w \in S} IDF(w, B) * \frac{TF(w, O, B)}{TF(w, O, B) + 1} \tag{1}$$

$$TF(w, O, B) = \frac{c(w, O)(k_1 + 1)}{c(w, 0) + k_1(1 - b + b\frac{|O|}{avgt})} \tag{2}$$

$$IDF(w, B) = \log \frac{|B| - c(w, B) + 0.5}{c(w, B) + 0.5} \tag{3}$$

$BM25_{Ex}$ is computed for a sentence s with description over a background data set B and an input data set O. The sum is over all terms w in a sentence s. The length feature is implicitly captured by this scoring function because the sum is computed over all the words in sentence s. The popularity feature is captured via Term Frequency (TF) weight, whereas the distinctiveness is also captured via Inverse Document Frequency (IDF) weight. $TF * IDF$ measures how frequent the words in a sentence occur relative to their occurrence in the background data set B and the input data set O. A sentence that has more terms which appear frequently in all the sentences in O and more distinctive terms which can distinguish O from the background data set B is more likely explanatory and is scored higher. In our $BM25_{Ex}$ formula, $c(w, O)$ is the count of word w in O, $c(w, O)$ is the count of word w in B, $|O|$ and $|B|$ is the total number of term occurrences in a data set O and B respectively, and $avgt$ is the average number of total term occurrences of sub-clusters in topic T which O is extracted from. The two parameters k_1 and b are parameters which users pass into. Moreover, k_1 and b can be used experimentally.

Our $BM25_{Ex}$ method has an advantage that it is flexible, and can be extended to include other fields in the document or sentence description as new fields become available. Moreover, the $BM25_{Ex}$ method has an additional advantage that it is language-independent model. The advantage of this approach is that all languages can be represented within the same method.

However, a disadvantage of our $BM25_{Ex}$ method is that they commonly contain a number of parameters which need to be tuned. In general, we can use machine-learning methods to learn unknown parameters of a ranking function or combine the outputs of different ranking functions.

4.2 Probabilities Explanatory Scoring

The second ranking explanatoriness sentence method we introduce is a method which based on a probabilities model. Here, we consider each word in sentence as an entity in scoring explanatoriness. We compute explanatoriness score of sentence as the sum of the explanatoriness of each word. Therefore, we note *ExplanatoryScore_s* to represent an explanatoriness score of sentence s and *ExplanatoryScore_w* to symbol an explanatoriness score of word w. We define *ExplanatoryScore_s* is a sum of *ExplanatoryScore_w* over all the word in the sentence s, i.e. $ExplanatoryScore_s = \sum_{w \in s} ExplanatoryScore_w$.

How to compute *ExplanatoryScore_w* is the problem. First, we assume that word w can be explanatory or not explanatory. We symbolize explanatoriness of word w is *Ex* and *Ex* is either 1 or 0. To compute the explanatoriness score of word w, we use the conditional probability model $p(Ex = 1|w)$ that can be explained as the posterior probability which the word w is explanatory. We can explain $p(Ex = 0|w)$ similarly.

We can see that the value of *ExplanatoryScore_w* is $p(Ex = 1|w)$ or we can rewrite *ExplanatoryScore_w* $= p(Ex = 1|w)$. We also use the idea in [1] which is $p(Ex = 1|w)$ that equal to $\frac{p(Ex=1|w)}{p(Ex=0|w)}$. We use Bayes rule to apply this formula as follows:

$$\frac{p(Ex = 1|w)}{p(Ex = 0|w)} = \frac{p(w) * p(w|Ex = 1)}{p(w) * p(w|Ex = 0)} = \frac{p(w|Ex = 1)}{p(w|Ex = 0)} \tag{4}$$

So, we have a formula to compute explanatoriness of word w like this:

$$ExplanatoryScore_w = p(Ex = 1|w) = \frac{p(w|Ex = 1)}{p(w|Ex = 0)} \tag{5}$$

There is a question which is how to compute $p(w|Ex = 1)$ and $p(w|Ex = 0)$. To do this, we will calculate a value of $p(w|Ex = 1)$ and $p(w|Ex = 0)$ based on a kind of word which we consider an explanatory or not and the set of sentences. We use the set of opinionated sentences O to approximate a sample of explanatory data sources and the background data set B to proximate a sample of non-explanatory data sources. Therefore, the likelihood $p(w|Ex = 1)$ and $p(w|Ex = 0)$ would be defined as:

$$p(w|Ex = 1) = \frac{num(w, 0)}{len(0)} \tag{6}$$

$$p(w|Ex = 0) = \frac{num(w, B)}{len(B)} \tag{7}$$

In two Eqs. 6 and 7, we denote *num(w, O)* and *num(w, B)* being the number of sentences which word w occur in a set O and a set B, respectively. We also denote *len(O)* and *len(B)* that mean the number of total sentences in a set O and a set B.

With this Eq. 6, if a word w appears in more sentences of a set O, an explanatoriness score of word w would have a higher value. This is corresponding to popularity feature. We call this probabilities explanatory scoring function is the Sum of Word Explanatory Probability (*SWEP*).

The *SWEP* method is also a language-independent model, do not have any parameter to tune, do not require any manual effort, and is applicable to many different domains. Moreover, the language-independent approach offers several advantages for extension to new languages. First, the computing time essentially remains the same when a new language is added. Second, when the language-independent model set has been built, there should be no need for corpus data for many languages.

5 Data Set and Evaluation

5.1 Data Set Preparation

One of the questions is whether the proposed methods can actually find explanatory sentences. In this paper, we use the data[1] for explanatory sentence extraction [1] which is based on a collection of Amazon product reviews used in [3, 4]. In this paper, we call it the 'Products' data set. The sentences have been clustered in each cluster based on their aspect and sentiment polarity. Each cluster contains a set of opinionated sentences O corresponding to an one (T, A, P). The authors et al. [1] required 4 labelers to make explanatoriness labels for each sentence. An explanatoriness label has three values: 0, 1 and 2. A value 0 means "no explanation", a value 1 means "weak explanation" and a value 2 means "strong explanation". Two values 1 and 2 mean that explanation while 0 means that no additional explanation. Moreover, a sentence which is labeled as explanation has an additional label that is assigned with 1 meaning 'half or less than half of the sentence provides explanations' or 2 meaning 'more than half or entire sentences provides explanations', while a sentence that are labeled as no explanation has an additional label that is assigned with 0 which is 'no additional explanation'. Because the test input of our methods is a review sentence and to make evaluations, we ensure that each sentence of the review for testing is labeled as sentiment explanation, disregarding review sentences with no sentence labeled as sentiment explanation.

5.2 Baseline

We consider that our proposed sentence ranking methods may work better than the previous methods which mainly implement a part of all three basic features or all three basic features. To evaluate this assumption, we use $BM25_E$ and *SumWordLR* [1] ranking explanatory methods and can generate ranking score of sentences.

---

[1]http://sifaka.cs.uiuc.edu/ ~ hkim277/expSum.

5.3   Result

Tables 1 and 2 represent the weighted MAP (*wMAP*) score of all the reviews and our approach attains a relative high performance compared with baselines. In this paper, we used two-fold cross validation so as to evaluate the performance of $BM25_{Ex}$ and $BM25_E$ [1]. The *wMAP* values shown in the table are the average over all the test instances in the 'Products' data set. *SWEP* and *SumWordLR* [1] do not have any parameter to tune, so we use one-cross validation to estimate. In Table 1, the parameter values chosen for $BM25_E$ and $BM25_{Ex}$ were $k_1 = 1.2$ [5] in the standard *BM25* method and b = 0.8 [6].

It can be seen that in Tables 1 and 2, our methods have slightly better performance over performance baselines. The Sum of Word Explanatory Probability model (*SWEP*) performs better than *SumWordLR* [1] on the 'Products' data set. Indeed, the gap between two performance values of *SWEP* and *SumWordLR* [1] does not show significance. In conclusion, all proposed methods show better performance than the baseline methods. Especially, $BM25_{Ex}$ and *SWEP* performed slightly better than $BM25_E$ and *SumWordLR* [1]. Beside superior methods, our proposal is Sum of Word Explanatory Probability model (*SWEP*) because it does not have any parameter.

We also have an illustrated example to describe our $BM25_{Ex}$ method and $BM25_E$ baseline method and try a range of values for k_1. We tuned k_1 from 1.2 to 2.0 in increments of 0.1 and the *b* value is 0.8. For a pair of parameter values k_1 and *b*, we show the average performance of two methods over the 'Products' data set in Table 3 and Fig. 2.

Furthermore, we perform a loop number of times to compare two-fold cross validation results of *SumWordLR* baseline method and our *SWEP* method. The results will be presented in Table 4 and Fig. 3 as follows.

5.4   Error Analysis

It is difficult to say that our $BM25_{Ex}$ method is better than a $BM25_E$ method [1] in all situations. By experimenting, we realize that the words which describe a sentiment are more likely to have a small appearance frequency in two sets *O* and *B* and are

Table 1 Comparison $BM25_{Ex}$ with $BM25_E$ baseline method in *wMAP*

Methods	Performance
$BM25_E$	0.7147
$BM25_{Ex}$	0.7167

Table 2 Comparison *SWEP* with *SumWordLR* baseline method in *wMAP*

Methods	Performance
SumWordLR	0.7377
SWEP	0.7430

Table 3 Comparison of scoring methods in *wMAP*

Parameter values	$BM25_E$ method	$BM25_{Ex}$ method
$k_1 = 1.2; b = 0.8$	0.7151	0.7165
$k_1 = 1.3; b = 0.8$	0.7151	0.7164
$k_1 = 1.4; b = 0.8$	0.7151	0.7164
$k_1 = 1.5; b = 0.8$	0.7154	0.7164
$k_1 = 1.6; b = 0.8$	0.7154	0.7165
$k_1 = 1.7; b = 0.8$	0.7153	0.7167
$k_1 = 1.8; b = 0.8$	0.7149	0.7168
$k_1 = 1.9; b = 0.8$	0.7151	0.7166
$k_1 = 2.0; b = 0.8$	0.715	0.7164

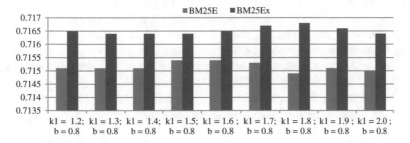

Fig. 2 A diagram illustrate results of our $BM25_{Ex}$ method and $BM25_E$ baseline method

Table 4 A comparison *SWEP* and *SumWordLR* scoring methods via a number of loop numbers

A loop number of times	*SWEP* method	*SumWordLR* method
10 times	0.6909	0.6850
20 times	0.6909	0.6860
50 times	0.6920	0.6859
100 times	0.6917	0.6854
200 times	0.6913	0.6855
500 times	0.6917	0.6858
1000 times	0.6921	0.6851

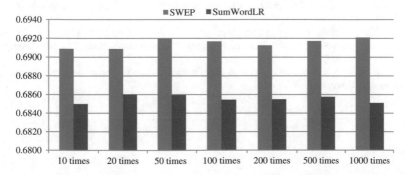

Fig. 3 A diagram illustrate results of our *SWEP* method and *SumWordLR* baseline method through a loop number of times

capable of the higher $BM25_{Ex}$ score. In summary, the more the words that appearance frequency in O has a small value in a sentence are, lower the $BM25_E$ score is and the higher $BM25_{Ex}$ score is. Nevertheless, the less the words that appearance frequency in O has a small value in a sentence are, the lower the $BM25_{Ex}$ score is and the higher $BM25_E$ score is.

6 Conclusions

In this paper, we modified two models for a novel sentence ranking problem called explanatory sentence extraction (*ESE*). We proposed two explanatory scoring methods which are modified *TF-IDF* weighting model for scoring explanatoriness and probabilistic explanatory scoring. Experimental results showed that proposed methods are more effective in ranking sentences than previous methods.

In this paper, our main work is studying and modifying the problem of measuring explanatoriness. For the future work, we can study, adjust more different ways of estimating the proposed probabilistic models and $BM25_{Ex}$ method effectively and improve semantic analysis of an opinionated sentence.

References

1. Kim, H.D., Castellanos, M.G., Hsu, M., Zhai, C., Dayal, U., Ghosh, R.: Ranking explanatory sentences for opinion summarization. In: Proceedings of the 36th International ACM SIGIR Conference on Research and Development in Information Retrieval, pp. 1069–1072. SIGIR '13 (2013)
2. Robertson, S., Walker, S.: Some simple effective approximations to the 2-Poisson model for probabilistic weighted retrieval. In: ACM SIGIR Conference on Research and Development in Information Retrieval (SIGIR), pp. 345–354 (1994)
3. Ding, X., Liu, B., Yu, P.S.: A holistic lexicon-based approach to opinion mining. In: Proceedings of the 2008 International Conference on Web Search and Data Mining, pp. 231–240. WSDM '08 (2008)
4. Hu, M., Liu, B.: Mining and summarizing customer reviews. In: Proceedings of the Tenth ACM SIGKDD International Conference on Knowledge Discovery and Data Mining, pp. 168–177. KDD '04 (2004)
5. Jones, K.S., Walker, S., Robertson, S.E.: A probabilistic model of information retrieval: development and comparative experiments. In: Information Processing and Management, pp. 779–808 (2000)
6. Okapi BM25: a non-binary model. http://nlp.stanford.edu/IR-book/html/htmledition/okapi-bm25-a-non-binary-model-1.html

Using Content-Based Features for Author Profiling of Vietnamese Forum Posts

Duc Tran Duong, Son Bao Pham and Hanh Tan

Abstract This paper reports the results of author profiling task for Vietnamese forum posts to identify the personal traits, such as gender, age, occupation, and location of the author using content-based features. Experiments were conducted on the different types of features, including stylometric features (such as lexical, syntactic, structural features) as well as content-based features (the most important words) to compare the performance and on the data sets we collected from the various forums in Vietnamese. Three learning methods, consisting of Decision Tree, Bayes Network, Support Vector Machine (SVM), were tested and the SVM achieved the best results. The results show that these kinds of features work well on such a kind of short and free style messages as forum posts, in which, content-based features yielded much better results than stylometric features.

1 Introduction

The rapid growth of World Wide Web has created a lot of online channels for people to communicate, such as email, blogs, social networks, etc. However, online forums are still among the most popular channels for people to share the opinions and discuss about the topics which are interested in common. Forum posts created by users can be considered as informal and personal writings. Authors of these posts can indicate their profiles for other people to view as a function of forum. But not many users reveal their personal information, because of information privacy

D.T. Duong (✉) · H. Tan
Posts and Telecommunications Institute of Technology, Hanoi, Vietnam
e-mail: ducdt@ptit.edu.vn

H. Tan
e-mail: tanhanh@ptit.edu.vn

S.B. Pham
Faculty of Information Technology, University of Engineering and Technology, Vietnam National University, Hanoi, Vietnam
e-mail: sonpb@vnu.edu.vn

© Springer International Publishing Switzerland 2016   287
D. Król et al. (eds.), *Recent Developments in Intelligent Information and Database Systems*, Studies in Computational Intelligence 642, DOI 10.1007/978-3-319-31277-4_25

issues on the online systems. Moreover, personal information of users is not mandatory to input when they register as a user of forums. Therefore, most of people do not provide their personal information or input the incorrect/unclear data.

As a result, the task of automatically classifying the author's properties such as gender, age, location, occupation, etc. becomes important and essential. Applications of this task can be in commercial field, in which providers can know which types of users like or do not like their products/services (for targeted marketing and product development). For the social research domain, researchers also want to know the profile of people who have a specific opinion about some social issues (when doing a social survey). It can also be used to support the court, in term of identifying if a text was created by a criminal or not.

Profiling the author of forum posts is also a challenging task when compared to doing this on other formal types of text such as article or novel or even the other types of online texts such as blog posts or email. Forum posts are often short and written in free style, which may contain grammar errors or informal sentence structures.

Most of earlier works in author profiling were conducted on other types of text (blog posts, email) and focused on using the stylometric features (or only small part of content-based features). This work presents a study in which we applied the machine learning algorithms to predict profiles of authors of forum posts using both types of features. Motivations for this work are:

- Only few previous works (e.g. [12]) on author profiling were done on forum posts, especially none of them was tested on Vietnamese. The work of Abbasi and Chen [1] was conducted on forum posts, but for author attribution, not author profiling task.
- Only one research in author profiling was done in Vietnamese [6], but was tested on blog posts, and used the stylometric features only. Our work is not only conducted on a more informal and noisier type of document, but also explored the use of content-based features.

The organization of the paper is as follows. In Sect. 2, we present the related work on the author analysis problem. Section 3 describes the methods and the system. Section 4 presents the result and discussion. In Sect. 5, we draw a conclusion and future work.

2 Related Work

The problem of authorship analysis has been studied for decades, mostly on English and some other languages (Dutch, French, Greek, Arabia etc.). In the early stage, it was often conducted on the long and formal documents such as article or novel. However, since 1990s, when the WWW grew and created a large amount of online text, the task of author analysis has moved the focus to this type of text.

According to Zheng et al. [22], the authorship analysis studies can be classified into three major fields, including authorship attribution, authorship profiling, and similarity detection.

Authorship attribution is the task of determining if a text is likely written by a particular author or not. It also is the technique to identify which one from a set of infinite authors is the real author of a disputed document. Therefore, it is also called authorship identification. The first study in this field dates back to 19th century when Mendenhall (1887) investigated the Shakespeare's plays. But the work which was considered the most thorough study in this field was conducted by Mosteller and Wallace (1964) when they analyzed the authorship of FederalList Papers. From that point, a number of works have been conducted by various researchers, including [2, 5, 7, 10, 17, 19, 22].

Authorship profiling, also known as authorship characterization, detects the characteristics of an author (e.g. gender, age, educational background, etc.) by analyzing the texts created by him/her. This technique is different from the former in that it is often used to examine the anonymous text, which is created by an unknown author, and generates the profile of the author of that text. For this reason, the author profiling task is often conducted on the online documents rather than literary texts. Therefore, this field is only more concerned by researchers from the late of 1990s, when more and more online documents are created by Internet's users. The most typical studies in this fields are from [2–4, 6, 8–11, 12–14, 16, 18, 20].

Similarity detection, on the other hand, doesn't focus on determining the author or his/her characteristics, but analyzes two or more documents to find out if they are all created by the same author or not. This technique is also used to verify if a piece of text is written by the author himself/herself or copied from the product of other authors. This task is mostly used for plagiarism detection. Some of the most convincing studies in this field were conducted by [2, 5, 7, 10].

Regarding the process of authorship analysis, there are two main issues that may significantly affect the performance, namely features set and analytical techniques [22].

Features set can be considered as a way to represent a document in term of writing style. With a chosen features set, a document can be represented as a features vector in which entries represent the frequency of each feature in the text [11]. Although various types of features have been examined, there is no features set that is the best to all the cases. According to Argamon et al. [4], there are two types of features that often can be used for authorship profiling: stylometric features and content-based features.

Stylometric features can be grouped into three types, including lexical, syntactic, and structural features. Lexical features are used to measure the habit of using characters and words in the text. The commonly used features in this kind consist of the number of characters, word, frequency of each kind of characters, frequency of each kind of words, word length, sentence length [7], and also the frequency of individual alphabets, special characters, and vocabulary richness [10]. Syntactic features include the use of punctuations, part-of-speeches, and function words. Function words feature is the interesting kind of features, which is examined in a number of studies and yielded very good results [10, 18, 22]. The set of function words used is also varying, from 122 to 650 words. Structural features show how the author organizes his/her documents (sentences, paragraphs, etc.) or other special structures such as greetings or signatures [5, 10].

Content-based features are often specific words or special content which are used more frequent in that domain than in other domains [21]. These words can be chosen by correlating the meaning of words with the domain [2, 10, 21] or selecting from corpus by frequency or by other feature selection methods [4].

Also the investigation of Zheng et al. [21] showed that, in early studies most authorship analytical techniques were statistical methods, in which the probability distribution of word usage in the texts of each author was examined. Although these methods achieved good results in authorship analysis, there are still some limitations, such as the ability to deal with multiple features or the stability over multiple domains. To overcome those limitations, the extensive use of machine learning techniques has been investigated. Fortunately, the advent of powerful computers allows researchers to conduct the experiments on complicated machine learning algorithms, in which Support Vector Machine (SVM) shows the better results in many cases [1, 2, 5–7, 10, 11, 14, 16, 18, 22]. Some other machine learning algorithms also have been examined and yielded good results, including Bayesian Network, Neural Networks, Decision Tree [4, 10, 18, 21]. In general, machine learning methods have advantages over statistical methods because they can handle the large features sets and the experiments also shown that they achieved the better results.

In this report, we investigated the use of machine learning techniques for the task of author profiling of online forum posts, using both stylometric and content-based features. We have found that content-based features outperformed stylometric features on this kind of text, and the combination of both features yielded the best result.

3 System Description

3.1 System Overview

In this work, we built a system which can take sample texts from web crawlers, then used text and linguistic processing components to extract features to create the data sets for the purpose of training the classifier. The classifier then can be used to predict the profile of the author of an anonymous forum post.

In the data processing step, data is cleaned and grouped by author profiles. Unlike the gender and location trait, which can be divided into two groups (male/female, north/south), the other traits are grouped by more than 2 classes. For age trait, we categorized our data into 3 subclasses (less than 22/24–27/more than 32). Age is categorized according to the life stages of a person (students or pupils/young working adults/middle-age people) and age periods are not continuous because distinguishing two contiguous ages is almost impossible. With the occupation trait, we tried to identify three occupations which are the most popular (business, sale, and administration/technical and technology/education and healthcare).

Linguistic processing is the task of tokenizing the text into sentences or word and the tagging for part-of-speeches. These tasks are important for extracting the word and syntactic features in the next step. In this work, we used existing tools from [15].

In the next sections, we describe the features and techniques which were used for classification in detail.

3.2   Features

As mentioned earlier, various features can be used to identify the characteristics of an author. In this work, we used both stylometric and content-based features.

Stylometric features include character-based, word-based, structural, and syntactic features. Character-based features include the number of characters in total and the ratio of each type of characters (number, letter, special, etc.) or each individual character (letters from a to z, and the special characters such as @, #, etc.) to the total number of characters. Some other features related to character are the average number of characters per word, per sentence, the number of upper case letters or how the author uses upper case letters in a word, etc. Word-based features group consists of the total number of words of a post, the average number of words per sentence, and the ratio of some types of word to the total number of words, such as words with a specific length, special words, the vocabulary richness (hapax legomena, hapax dis legomena etc.). Syntactic features indicate the use of punctuations such as "!", "?", function words, and part-of-speech tags. Function words chosen are the words which have little lexical meaning and express the grammatical relationship with other words in a sentence (212 Vietnamese function words). Part-of-speech tags include 18 word types, such as noun, verb, preposition, etc. Structural features present the structure of a post, such as the number of paragraphs, number of lines, etc.

Content-based features used in our work were chosen from the corpus, which are the words that can discriminate best between classes of each trait. Firstly, these words were selected based on the frequency of them in the corpus (separately by classes of each trait). Then the Information Gain method was applied to select the best features. Information Gain is one of the most popular feature selection methods, which attempts to measure the significance of each feature in distinguishing between classes. This method was tested on various previous works and yielded the good result.

For gender trait, we selected 3000 words which were used most frequently by male/female separately. After eliminating the identical words and applied the Information Gain method, we chose 1000 words which have highest significance.

Using the similar process, we chose about 1000 most significant words to use as content-based features for discriminating the age, occupation, and location traits.

All of these features are extracted from the text and store in a numeric vector. For features which need some kinds of linguistic processing activities, such as the word segmentation or the part-of-speech tagging, we used existing tools available for Vietnamese. Extracted features are stored in the features containers (ARFF files), then are sent to classifiers for training purposes and prediction models are built for classifying the new data.

We also conducted experiments on subsets of features, including stylometric features, content-based features, and all features for analysis of performance of each type.

3.3 Learning Methods

In this work, we used 3 machine learning algorithms to build the classifiers for input messages, namely Decision Tree J4.8, Bayesian Network, and Support Vector Machine.

Support Vector Machine is a learning method having an advantage that it does not require a reduction in the number of features to avoid the problem of over-fitting. This property is very useful when dealing with large dimensions as encountered in the area of text categorization [5]. SVM has been used in many previous works in author analysis and in most case yielded the better result than other classifiers.

Decision Tree and Bayesian Network are also popular learning algorithms. Although, they are not shown the better results than SVM in the earlier works, we still tried them in our experiments to compare the performance.

For each algorithm, 3 subsets of features were experimented to find out the best classifier and the feature set (Stylometric, Content-based, All).

4 Experiments

4.1 Data

There are a number of Vietnamese forums, of which we can collect the data. However, each of them often serves for a specific type of user only (e.g. for ladies or gentlemen) or for a specific subject of interest such as technology, automobile etc. Therefore, we selected three forums to collect data to ensure that the data collected will cover a wide range of users and subjects.

- Webtretho forum (www.webtretho.com/forum): A forum for girls and ladies to discuss about the variety of subjects in life and work.

Table 1 The statistic of data in corpus

Trait	Total posts	Class	Percent in corpus
Gender	4474	Male	54
		Female	46
Age	3017	Less than 22	21
		From 24 to 27	27
		More than 32	52
Location	3960	North	57
		South	43
Occupation	3453	Business, Sale, Admin	36
		Technical, Technology	31
		Education, Healthcare	33

- Otofun forum (www.otofun.net/forum): A forum for mostly the men to exchange about issues of automobile and related subjects.
- Tinhte forum (www.tinte.vn/forum): A forum for young people to exchange the topics about technological devices and interests.

Users of these forums can indicate the personal information such as name, age, gender, interest, job etc. in their profiles. However, none of them is the explicit field in the user's profile. As a result, we must use both of methods, automatic and manual, to collect and annotate the data.

After the last step, we obtained a collection of 6831 forum posts from 104 users (736,252 words in total), for which we also received at least one of the information about age, gender, location, occupation of the author of each post. The length of each post is also restricted in the range from 250 to 1500 characters to eliminate the too long or too short posts (too long post may contain the text copied from other sources) (Table 1).

The cleaned data then is analyzed by NLP tools, including word segmentation and part-of-speech tagging as mentioned earlier.

4.2 Results and Discussion

We conducted experiments on 4 traits of authors as mentioned earlier (gender, age, location, occupation) using the Weka[1] toolkit. The results were verified through a ten-fold cross validation process.

Table 2 shows the results of author profiling experiments of 4 traits.

As the results shown in Table 2, we can observe that content-based features outperformed stylometric features. Although content-based features are often

---

[1]http://www.cs.waikato.ac.nz/ml/weka/.

Table 2 The results of author profiling experiments

Trait	Feature	J48	SVM	BayesNet
Gender	All features	83.35	90.47	87.35
	Stylometric	73.31	82.94	77.17
	Content-based	83.36	89.97	87.58
Age	All features	55.76	63.96	63.92
	Stylometric	52.03	62.14	56.17
	Content-based	55.24	61.74	62.55
Location	All features	69.32	80.06	74.54
	Stylometric	65.73	70.39	66.99
	Content-based	69.23	79.39	75.01
Occupation	All features	43.41	56.98	50.65
	Stylometric	43.97	51.77	46.44
	Content-based	43.32	55.38	51.34

considered domain-specific and may be less accurate when moving the other domains, the results in this task are still promising. Firstly, the data in corpus was collected from various source, therefore it is not so domain-specific. Secondly, even the results are domain-specific to some extent, it is still useful when we conduct the research or apply the results in that domain. Besides, the results of stylometric features are also good, especially for gender and location.

Regarding the learning methods, the SVM outperformed the other two methods, in which Bayesian Network gave better results than Decision Tree. This is a reasonable result and again proves that SVM is a good algorithm for classifying the author characteristics.

In comparison to the results of previous works, although forum posts are shorter and noisier than other types of online messages such as blog posts or emails, but the results can be considered as promising, especially for gender and location traits. The accuracy of 90.47 % when predicting the gender is even better than the results of most of previous works which were conducted on blogs or emails (which had base-line about 80 %). The percentage of age prediction (63.96 %) is not as good as the results conducted on blog posts or emails (which had the base-line around 77 % for blog posts), but much better compared to the result of a research on forum posts conducted by [12], which is only 53 %. The same evaluation can be used when saying about the location trait, but the occupation prediction is not so good. The main reason is that occupation information is very noisy and subtle. For example, a person who studied about technical but then works as a sale person is not an easy case when predict his/her job. This needs to be investigated further in later researches.

When comparing with the only previous work on author profiling in Vietnamese by [6], for the gender trait, we achieved the better result (90.47 and 83.3 %) when using content-based features, and the same result (82.94 and 83.3 %) without content-based features. It showed that our approach when adding the content-based features has improved the results significantly. The same evaluation can be said when comparing the results of location trait. But for other traits, our results are less accurate, but it is understandable and still promising, because our experiments were conducted on a shorter and more informal type of text than blog posts.

5 Conclusion and Future Work

In this study, we showed that it is feasible to classify authorial characteristics of the informal online messages as forum posts based on linguistic features, in which using content-based features improved the results significantly. Experiments conducted show the promising results, although some aspects still need to be improved such as the solutions for noisy information in occupation trait or the result for age prediction should be better and so on. This also showed that the SVM algorithm outperformed the other classifiers, while Decision Tree gave the poor results.

In the future, this study can be expanded to other domains, such as social networks or user comments/product reviews. The data in these domains is even shorter and noisier than forum posts, so it is more challenging task. But the results of such kind of works have promising applications in commercial fields, such as analyzing market trends or user behaviors prediction etc.

We also have planned to investigate more about the use of content-based features in this kind of task. We have conducted experiments and found that content-based features work very well on the author profiling task for Vietnamese text. However, more insightful analytics should be investigated to show why they are better than stylometric features and which kinds of content are more significant.

References

1. Abbasi, A., Chen, H.: Applying authorship analysis to extremist-group Web forum messages. IEEE Intel. Syst. (2005)
2. Abbasi, A., Chen, H.: Writeprints: a stylometric approach to identity-level identification and similarity detection in cyberspace. ACM Trans. Inf. Syst. **26**(2), 1–29 (2008)
3. Argamon, S., Koppel, M., Fine, J., Shimoni, A.: Gender, genre, and writing style in formal written texts. Text **23**(3) (2003)
4. Argamon, S., Koppel, M., Pennebaker, J., Schler, J.: Automatically profiling the author of an anonymous text. Commun. ACM **52**(2), 119–123 (2009)
5. Corney, M., DeVel, O., Anderson, A., Mohay, G.: Gender-preferential text mining of e-mail discourse. In: ACSAC'02: Proceedings of the 18th Annual Computer Security Applications Conference, Washington, DC, pp. 21–27 (2002)
6. Dang, P., Giang, T., Son, P.: Author profiling for Vietnamese blogs. In: International Conference on Asian Language Processing (2009)
7. De Vel, O., Anderson, A., Corney, M., Mohay, G.M.: Mining e-mail content for author identification forensics. SIGMOD Rec. **30**(4), 55–64 (2001)
8. Goswami, S., Sarkar, S., Rustagi, M.: Stylometric analysis of bloggers' age and gender. In: Adar, E., Hurst, M., Finin, T., Glance, N.S., Nicolov, N., Tseng, B.L. (eds.) ICWSM. The AAAI Press (2009)

9. Gressel, G., Hrudya, P., Surendran, K., Thara, S., Aravind, A., Prabaharan, P.: Ensemble learning approach for author profiling. Notebook for PAN at CLEF (2014)
10. Iqbal, F.: Messaging forensic framework for cybercrime investigation. A Thesis in the Department of Computer Science and Software Engineering, Concordia University Montréal, Canada (2010)
11. Koppel, M., Argamon, S., Shimoni, A.R.: Automatically categorizing written texts by author gender. Literary Linguist. Comput. 17(4), 401–412 (2002)
12. Nguyen, D., Smith, N.A., Rosé, C.P.: Author age prediction from text using linear regression. In: Proceedings of the 5th ACL-HLT Workshop on Language Technology for Cultural Heritage, Social Sciences, and Humanities, LaTeCH '11, pp. 115–123, Stroudsburg, PA, USA. Association for Computational Linguistics (2011)
13. Nguyen, D., Gravel, R., Trieschnigg, D., Meder, T: How old do you think I am?; a study of language and age in twitter. In: Proceedings of the Seventh International AAAI Conference on Weblogs and Social Media (2013)
14. Peersman, C., Daelemans, W., Vaerenbergh. L.V.: Predicting age and gender in online social networks. In: Proceedings of the 3rd International Workshop on Search and Mining User-Generated Contents, SMUC '11, pp. 37–44, New York, NY, USA, 2011. ACM (2007)
15. Le-Hong, P., Roussanaly, A., Nguyen, T.M.H., Rossignol, M.: An empirical study of maximum entropy approach for part-of-speech tagging of Vietnamese texts. In: Proceedings of Traitement Automatique des Langues Naturelles (TALN-2010), Montreal, Canada (2010)
16. Rangel, F., Rosso, P.: Use of language and author profiling: Identification of gender and age. In: Natural Language Processing and Cognitive Science, pp. 177 (2013)
17. Savoy, J.: Authorship attribution based on specific vocabulary. ACM Trans. Inf. Syst. 30, 2 (2012)
18. Schler, J., Koppel, M., Argamon, S., Pennebaker, J.: Effects of age and gender on blogging. In: 43rd Proceedings of AAAI Spring Symposium on Computational Approaches for Analyzing Weblogs (2006)
19. Stamatatos, E., Fakotakis, N., Kokkinakis, G.: Automatic text categorization in terms of genre and author. Comput. Linguist. 26(4), 471–495 (2000)
20. Zhang, C., Zhang, P.: Predicting gender from blog posts. Technical report, University of Massachusetts Amherst, USA (2010)
21. Zheng, R., Chen, H., Huang, Z., Qin, Y.: Authorship analysis in cybercrime investigation. In: ISI. LNCS, vol. 2665, pp. 59–73 (2003)
22. Zheng, R., Li, J., Chen, H., Huang, Z.: A framework for authorship identification of online messages: writing-style features and classification techniques. J. Am. Soc. Inform. Sci. Technol. 57(3), 378–393 (2006)

Trigram-Based Vietnamese Text Compression

Vu H. Nguyen, Hien T. Nguyen, Hieu N. Duong and Vaclav Snasel

Abstract This paper presents a new and efficient method for text compression using tri-grams dictionary. There have been many methods proposed to text compression such as: run length coding, Huffman coding, Lempel-Ziv-Welch (LZW) coding. Most of them have based on frequency of occurrence of letters in the text. In this paper, we propose a method to compress text using tri-grams dictionary. Our method firstly splits text to tri-gram then we encode it based on tri-grams dictionary, with each tri-gram, we use 4 bytes to encode. In this paper, we use Vietnamese text to evaluate our method. We collect text corpus from internet to build tri-grams dictionary. The size of text corpus is around 2.15 GB and the number of tri-grams in dictionary is more than 74,400,000 tri-grams. To evaluate our method, we collect a testing set of 10 different text files with different sizes to test our system. Experimental results show that our method achieves better results with compression ratio around 82 %. In comparison with WinZIP version 19.5 (http://www.winzip.com/win/en/index.htm) (the software combines LZ77 (Ziv and Lempel in IEEE Trans Inf Theory 24(5), 530–536, 1978 [20]) and Huffman coding) and WinRAR version 5.21 (http://www.rarlab.com/download.htm) (the software

V.H. Nguyen · H.T. Nguyen (✉)
Faculty of Information Technology, Ton Duc Thang University,
Ho Chi Minh City, Vietnam
e-mail: hien@tdt.edu.vn

V.H. Nguyen
e-mail: nguyenhongvu@tdt.edu.vn

H.N. Duong
Faculty of Computer Science and Engineering, Ho Chi Minh City
University of Technology, Ho Chi Minh City, Vietnam
e-mail: dnhieu@cse.hcmut.edu.vn

V. Snasel
Faculty of Electrical Engineering and Computer Science,
VSB-Technical University of Ostrava, Ostrava, Czech Republic
e-mail: vaclav.snasel@vsb.cz

© Springer International Publishing Switzerland 2016   297
D. Król et al. (eds.), *Recent Developments in Intelligent Information and Database Systems*, Studies in Computational Intelligence 642, DOI 10.1007/978-3-319-31277-4_26

combines LZSS (Storer and Szymanski in J ACM 29(4), 928–951, 1982 [17]) and Prediction by Partial Matching [2]), our method achieves a higher compression ratio applied for any size of text in our test cases.

Keywords Text compression · Tri-grams dictionary · Vietnamese text compression · Dictionary-based compression

1 Introduction

In 2012, every day we create 2.5 EB of data and in 2015, every minute we have nearly 1750 TB of data transferring over the internet according to IBM[1] and the forecast of Cisco[2] respectively. Reducing size of data is one of solutions to increase the data transfer rate and save storage space to enhance the performance of the system. A simple way to do that is data compression which involves two main phases: compression and decompression. In compression phase, an input X will be encoded to generate an output Y that requires fewer bit than X. In contrast, decompression phase decodes the output of compression phase to get the original input file. According to [15], data compression has two main broad classes: lossless and lossy. With lossless compression, encoded data and decoded data are identical whereas in lossy compression methods, the compressed data cannot be completely recovered.

There are several lossy and lossless techniques proposed in the past decades. These techniques can be further classified into three major types: substitution, statistical and dictionary [13]. The substitution data compression techniques replace a certain longer repeating of characters by a shorter one, the remarkable method of this technique is run length encoding.[3] The statistical techniques usually calculate the probability of characters to generate shortest average code length, such as Shannon-Fano coding [6, 16], Huffman coding [8]. The last type is dictionary data compression techniques, such as Lempel-Ziv-Welch (LZW), which involves substitution of sub-string of text by indices or pointer code relating to a position in dictionary of the substring [18–20]. Every method has own strength, weakness and applied to a specific field, none of the above methods has been able to achieve best case compression ratio.

Normally, users will decide to choose the appropriate method based on their purposes. With systems that allow reconstruction information from output which are not as same as the input, we can use lossy methods, such as systems to compress images, compress audio files. With systems that require the original data must be recovered exactly from the compressed data, we should use lossless methods such

---

[1]http://www-01.ibm.com/software/data/bigdata/what-is-big-data.html.

[2]http://www.cisco.com/c/en/us/solutions/collateral/service-provider/ip-ngn-ip-next-generation-network/white_paper_c11-481360.html.

[3]https://en.wikipedia.org/wiki/Run-length_encoding.

as text compression systems. Regarding to text compression, there have been several approaches in recent years, most of them based on dictionary or word level or character level [3–5, 7, 9, 14]. In these researches, they do not consider for the structure of word or morpheme in the text. There are some approaches for text compression based on syllables. These approaches involve to some languages that have the morphology in the structure of word or morpheme (German, Arabic, Turkish, Czech, etc.) such as [1, 10–12].

In this paper, we propose a new and efficient method for text compression using tri-grams dictionary. Our method firstly splits text to tri-grams then encode them based on tri-grams dictionary that we build from a text corpus collected from internet. We use 4 bytes to encode for index of every tri-gram that occurs in dictionary. For tri-gram that does not occur in dictionary and for other cases (uni-gram, bi-gram), we encode it using Unicode UTF8 encoding.

This paper presents the first attempt to text compression using tri-gram dictionary and the contribution is three-folds: (1) a method for text compression using tri-grams dictionary, (2) collect text corpus of Vietnamese language from internet and build a tri-grams dictionary with more than 74,400,000 tri-grams, and (3) a testing set of 10 different text files with different sizes to evaluate our system and compare with other systems such as: WinRAR and WinZIP. The rest of this paper is organized as follows: Sect. 2 presents our proposed method, our experimental results are shown and analysed in Sect. 3. Finally, we draw conclusions in Sect. 4.

2   Proposed Method

In this section, we present a text compression using tri-grams dictionary model for our proposed method. This model has two main parts. The first part is used for text compression and the second part for decompression. Figure 1 describes our method model. In our model, the tri-grams dictionary are used for both compression and decompression phases. We will describe more details in following subsections.

2.1   n-Gram Theory and Dictionary

2.1.1   n-Gram Theory

In this paper, we employ n-gram theory from Wikipedia[4]: in the fields of computational linguistics and probability, **an n-gram is a contiguous sequence of n items from a given sequence of text or speech**. The items can be phonemes, syllables, letters, words or base pairs according to the application. The n-grams typically are

---

[4]https://en.wikipedia.org/wiki/N-gram.

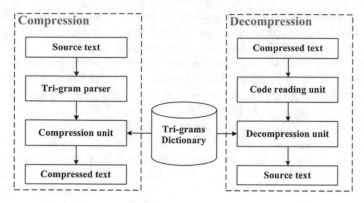

Fig. 1 Trigram-based Vietnamese text compression

collected from a text or speech corpus. An n-gram of size 1 is referred to as a "uni-gram"; size 2 is a "bi-gram"; size 3 is a "tri-gram". Larger sizes are referred to the value of n, e.g., "four-gram", "five-gram", and so on. Particularly, in our paper, we focus on tri-gram to build dictionary.

2.1.2 Dictionaries

In our paper, we use Vietnamese language to evaluate for our method. We build two tri-gram dictionaries with different sizes to evaluate the effects of size of dictionary with our method. Every dictionary has two columns, one contains tri-grams and one contains addresses of these tri-grams. These dictionaries have built based on a text corpus which is collected from open access databases. The size of text corpus for dictionary 1 is around 800 MB and for dictionary 2 is around 2.5 GB. We use SRILM[5] to generate the tri-gram data for these dictionaries. The data of tri-grams after using SRILM for dictionary 1 is around 1.15 GB with more than 40,514,000 tri-grams and for dictionary 2 is around 2.15 GB with more than 74,400,000 tri-grams. To reduce the time of searching in dictionaries, we arrange them according to alphabet. Table 1 describes the size and number of tri-grams of each dictionary.

2.2 Compression

According to Fig. 1, the compression phase has two main parts, the first part is tri-grams parser and the second is compression unit. In following subsections, we will focus to explain detail for them.

---

[5]http://www.speech.sri.com/projects/srilm/.

Table 1 Dictionaries

Dictionary	Number of tri-grams	Size (GB)
1	40,514,822	1.15
2	70,400,000	2.15

2.2.1 Tri-Gram Parser

Tri-gram parser is used to read the source text file, separate it to sentences based on newline and split all text of sentences to tri-grams. In the case of last tri-gram, maybe it is not a tri-gram (it just has uni-gram or bi-gram). Therefore, we must assign an attribute to it to distinguish tri-gram with uni-gram and bi-gram. In Table 2, this value of attribute is one.

For example, we have a file with two sentences: "**Tôi là sinh viên Trường Đại học Tôn Đức Thắng**" (*I am a student of Ton Duc Thang university*) and "**Tôi đang học môn Ngôn ngữ lập trình**" (*I am studying the subject Principles of Programming Languages*). In the first sentence, we have three standard tri-grams: "**Tôi là sinh**", "**viên Trường Đại**", "**học Tôn Đức**" and one uni-gram: "**Thắng**" and for the second sentence, we have following tri-grams: "**Tôi đang học**", "**môn Ngôn ngữ**", "**lập trình**". We have the output presented in Table 2.

2.2.2 Compression Unit

Compression unit uses the result from tri-gram parser, detect tri-grams in dictionary to find the corresponding codes for standard tri-grams. If a tri-gram occurs in dictionary, we encode it by four bytes otherwise we encode it with the exactly number of characters that it has. The compression task can be summarized as following:

Encoding for tri-grams occur in dictionary

When a tri-gram occurs in dictionary, we use four bytes to encode it. To distinguish with tri-grams which do not occur in dictionary and bi-gram, uni-gram, we set the

Table 2 Output of tri-gram parser

No.	Tri-gram	Attribute
1	Tôi là sinh	0
2	viên Trường Đại	0
3	học Tôn Đức	0
4	Thắng	1
5	Tôi đang học	0
6	môn Ngôn ngữ	0
7	lập trình	1

most significant bit of the first byte to zero. So the four bytes encoding has the structure like below:

$$0\ B_0^4\ B_0^3\ B_0^2\ B_0^1\ B_0^0\ B_1^7\ B_1^6\ B_1^5\ B_1^4\ B_1^3\ B_1^2\ B_1^1\ B_1^0\ B_2^7\ B_2^6\ B_2^5\ B_2^4\ B_2^3\ B_2^2\ B_2^1\ B_2^0\ B_3^7\ B_3^6\ B_3^5\ B_3^4$$
$$B_3^3\ B_3^2\ B_3^1\ B_3^0$$

Where:

- The most significant bit of the first byte is **0**: Encode for a tri-gram which occurs in dictionary.
- $B_0^6\ B_0^5\ B_0^4\ B_0^3\ B_0^2\ B_0^1\ B_0^0\ B_1^7\ B_1^6\ B_1^5\ B_1^4\ B_1^3\ B_1^2\ B_1^1\ B_1^0\ B_2^7\ B_2^6\ B_2^5\ B_2^4\ B_2^3\ B_2^2\ B_2^1\ B_2^0\ B_3^7$
  $B_3^6\ B_3^5\ B_3^4\ B_3^3\ B_3^2\ B_3^1\ B_3^0$: Encode for the index of tri-gram in dictionary.

Encoding for tri-grams do not occur in dictionary and for other cases

When a tri-gram does not occur in dictionary and for other cases (uni-gram, bi-gram) we encode it by exactly number of characters that it has. In this case, we set the most significant bit of the first byte to one. Next seven bits of this byte will present the number of characters of the tri-gram and the other cases. So, we have the encoding structure like below:

$$1\ B_0^6\ B_0^5\ B_0^4\ B_0^3\ B_0^2\ B_0^1\ B_0^0\ B_i^7\ B_i^6\ B_i^5\ B_i^4\ B_i^3\ B_i^2\ B_i^1\ B_i^0$$

Where:

- The most significant bit of the first byte is **1**: Encode for a tri-gram which does not occur in dictionary and for other cases.
- $B_0^6\ B_0^5\ B_0^4\ B_0^3\ B_0^2\ B_0^1\ B_0^0$: Number of encoding characters.
- $B_i^7\ B_i^6\ B_i^5\ B_i^4\ B_i^3\ B_i^2\ B_i^1\ B_i^0$: encoded bytes of characters of tri-gram that does not occur in dictionary and other cases. For our testing data, we use Vietnamese language and normally, it is presented by unicode. In encoding stream, we use Unicode UTF8 encoding. So, the value of i is the number of bytes that Unicode UTF8 uses to encode for this tri-gram or other cases.
- We set all values of $B_0^6\ B_0^5\ B_0^4\ B_0^3\ B_0^2\ B_0^1\ B_0^0$ to 1 to encode for newline (\r\n). So, to encode for newline we just use one byte.

In Table 3, we conduct the compression result of two sentences: "**Tôi là sinh viên Trường Đại học Tôn Đức Thắng**" (*I am a student of Ton Duc Thang university*). "**Tôi đang học môn Ngn ngữ lập trình**" (*I am studying the subject Principles of Programming Languages*). In Table 3, Att. means attribute. In this table, we use four bytes to encode for tri-grams that occur in the dictionary (from number one to number 3 and number six, seven). With tri-gram number four ("Thắng"), it does not occur in dictionary. So we turn the most significant bit of the first byte to 1. The three last bits of the first byte are "111". It means that we encode the tri-gram "Thắng" by seven bytes. We present "Thắng" byte Unicode UTF8 encoding. So the corresponding code of every character in "Thắng" is 'T': 01010100; 'h': 01101000; 'ắ': 11100001-10111010-10101111; 'n': 01101110; 'g': 01100111. For newline, we use one byte: 11111111 to encode for it.

Table 3 Compression result

No.	Tri-gram	Att.	Codeword
1	Tôi là sinh	0	00000010-00010110-10011101-10001110
2	viên Trường Đại	0	00000010-01010001-11001110-00100110
3	học Tôn Đức	0	00000000-11111100-00000100-00101110
4	Thắng	1	10000111-01010100-01101000-11100001-10111010-10101111-01101110-01100111
5	\r\n (newline)		11111111
6	Tôi đang học	0	00000010-00010110-00001001-11110000
7	môn Ngôn ngữ	0	00000001-01011000-01100011-00010000
8	lập trình	1	10001100-01101100-11100001-10111010-10101101-01110000-00100000-01110100-01110010-11000011-10101100-01101110-01101000

2.2.3 Compression Ratio

Compression ratio is used to measure the efficiency of compression method, the higher of compression ratio the higher quality of compression method. The compression ratio can be calculated by Eq. 1.

$$CR = \left(1 - \frac{compressed\_file\_size}{original\_file\_size} \right) \times 100 \, \%$$ (1)

where:

- **original_file_size**: Size of the original file.
- **compressed_file_size**: Size of file after encoding.

2.3 Decompression

Decompression is the inversion of compression phase. The decompression process has undergone in two steps and can be summarized as follows:

- **Code reading unit**: this unit reads the encoding stream from compressed text, separate it byte to byte.
- **Decompression unit:**
  This unit decodes one by one tri-gram or other cases (uni-gram, bi-gram or tri-gram does not occur in dictionary). The decompression unit decides decode for tri-gram or other cases based on the first bit of the first byte that it reads from the encoded stream. If the first bit is **0**, it decodes for standard tri-gram. Otherwise, it decodes for other cases. We describe the detail of decompression unit as following:

  - **The most significant bit of first byte is 0**: the decompression unit will read **three bytes** more, search the tri-gram corresponding the value of four bytes in dictionary and decodes it. This task is the inversion of tri-gram encoding.

- **The most significant bit of first byte is 1**:

  If all remaining bits of the first byte is not equal to 1: the decompression unit computes the value of the remaining bits of first byte. This value is the number of bytes that it will read from encoded stream. It will decode for these bytes based on Unicode UTF8 decoding.

  If all remaining bits of first byte is 1: this is the encoded of newline. The compression unit decodes a newline (\ r\ n) for it.

After finishing the decoding for one tri-gram or other cases, it will read one next byte, repeat decompression task to decode for other tri-gram or other cases until it reads to the last byte.

3 Experiments

3.1 Experiment Result

We conduct experiment to evaluate our method, we use a data set which randomize collect from Vietnamese text and online newspaper. The data set includes 10 files completely different from size and content.

In order to evaluate the effects of tri-grams in dictionary (the size of dictionary), we conduct two experiments with dictionary 1 and dictionary 2. According to Table 1. We show the result of two experiments in Table 4. According the Table 4, we find that the compression ratio from the dictionary 2 is better than the compression ration from the dictionary 1. So if we have an enough dictionary of all tri-grams, the compression ratio will be the best. In Tables 4 and 5, Figs. 2 and 3, we have some abbreviations and meanings as following: OFS: original file size, CFS: compressed file size, CR: compression ratio, D1: dictionary 1, D2: Dictionary 2.

Table 4 Compression ratio of dictionary 1 and dictionary 2

No.	OFS (Byte)	CFS-D2 (Byte)	CR of D2 (%)	CFS-D1 (Byte)	CR of D1 (%)
1	1166	185	84.13	305	73.84
2	2240	359	83.97	719	67.90
3	6628	1710	74.20	2404	63.73
4	12,224	2057	83.17	3321	72.83
5	22,692	3702	83.69	7469	67.09
6	49,428	7870	84.08	15,872	67.89
7	96,994	17,723	81.73	27,161	72.00
8	156,516	27,434	82.47	41,228	73.66
9	269,000	49,902	81.45	70,105	73.94
10	489,530	92,739	81.06	135,639	72.29

Table 5 Compression ratio of our method, WinRAR and WinZIP

No.	OFS	CFS (Byte) of our method	CR of our method (%)	CFS (Byte) of WinRAR	CR of WinRAR (%)	CFS (Byte) of WinZIP	CR of WinZIP (%)
1	1166	185	84.13	617	47.08	676	42.02
2	2240	359	83.97	887	60.40	946	57.77
3	6628	1710	74.20	2052	69.04	2111	68.15
4	12,224	2057	83.17	3378	72.37	3442	71.84
5	22,692	3702	83.69	6162	72.85	6150	72.90
6	49,428	7870	84.08	12,504	74.70	12,286	75.14
7	96,994	17,723	81.73	21,389	77.95	21,321	78.02
8	156,516	27,434	82.47	34,162	78.17	34,362	78.05
9	269,000	49,902	81.45	56,152	79.13	57,671	78.56
10	489,530	92,739	81.06	101,269	79.31	108,175	77.90

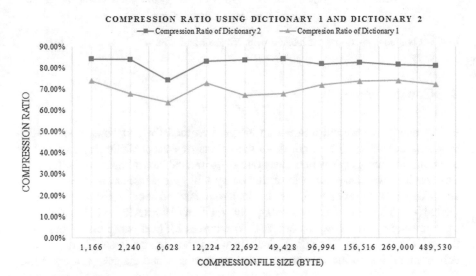

Fig. 2 Comparison between dictionary 1 and dictionary 2

In Fig. 2, the compression ratio when we use the dictionary 2 is higher than the compression ratio of dictionary 1.

In order to evaluate our method with other methods, we compress these input files using WinRAR version 5.21 and WinZIP to have a comparison. In Table 5, we show the result of our method in 10 cases above in comparison with WinRAR and WinZIP. According to Table 5 and Fig. 3, our compression ratio is better than WinRAR, WinZIP. Especially when file size is small. Our method achieves a higher compression ratio than WinRAR and WinZIP.

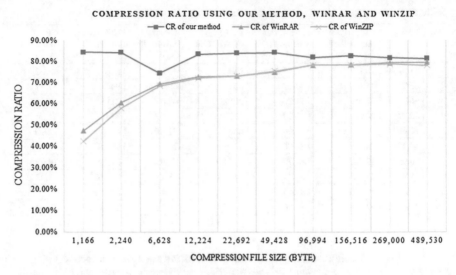

Fig. 3 Comparison between our method with WinRAR and WinZIP

4 Conclusion

In this paper, we present a new and efficient method for text compression using tri-grams dictionary. To demonstrate the effects of size of dictionary (number of tri-grams in dictionary) we conduct our experimental results on two dictionaries. The result shows that a big size of dictionary will give a higher compression ratio and vice versa. In order to further evaluate our method, our experiments are conducted on the same data set using our method, WinRAR and WinZIP. The experimental results show that our method achieves a higher compression ratio than that of WinRAR and WinZIP. Especially when the file size is small. We plan to collect more data to increase size of dictionary to enhance the compression ratio.

References

1. Ziv, J., Lempel, A.: Compression of individual sequences via variable-rate coding. IEEE Trans. Inf. Theory **24**(5), 530–536 (1978)
2. Storer, A.J., Szymanski, T.G.: Data compression via textual substitution. J. ACM **29**(4), 928–951 (1982)
3. Cleary, G.J., Witten, I.H.: Data compression using adaptive coding and partial string matching. IEEE Trans. Commun. **32**(4), 396–402 (1984)
4. Sayood, K.: Introduction to Data Compression, 3rd edn. Morgan Kaufmann Publishers (2000)
5. Pandya, M.K.: Data compression: efficiency of varied compression techniques. Technical report, University of Brunel, UK (2000)

6. Fano, R.M.: The transmission of information. Technical report, Massachusetts Institute of Technology, Research Laboratory of Electronics (1949)
7. Shannon, C.E.: A mathematical theory of communication. Mobile Comput. Commun. Rev. **5**(1), 3–55 (2001)
8. Huffman, D.A.: A method for the construction of minimum redundancy codes. Proc IRE **40** (9), 1098–1101 (1952)
9. Welch, T.A.: A technique for high-performance data compression. IEEE Comput. **17**(6), 8–19 (1984)
10. Ziv, J., Lempel, A.: A universal algorithm for sequential data compression. IEEE Trans. Inf. Theory **23**(3), 337–343 (1977)
11. Crochemore, M., Langiu, A., Mignosi, F.: Note on the greedy parsing optimality for dictionary-based text compression. Theor. Comput. Sci. **525**, 55–59 (2014)
12. Dvorský, J., Pokorný, J., Snásel, V.: Word-based compression methods and indexing for text retrieval systems. In: Advances in Databases and Information Systems, Third East European Conference, ADBIS'99, Maribor, Slovenia, Proceedings. pp. 75–84, 13–16 Sept 1999
13. Dvorský, J., Pokorný, J., Snásel, V.: Word-based compression methods for large text documents. In: Data Compression Conference, DCC 1999, Snowbird, Utah, USA, p. 523, 29–31 Mar 1999
14. Fariña, A., Navarro, G., Paramá, J.R.: Boosting text compression with word-based statistical encoding. Comput. J. **55**(1), 111–131 (2012)
15. Kalajdzic, K., Ali, S.H., Patel, A.: Rapid lossless compression of short text messages. Comput. Stand. Interfaces **37**, 53–59 (2015)
16. Platos, J., Dvorský, J.: Word-based text compression. CoRR abs/0804.3680 (2008)
17. Akman, I., Bayindir, H., Ozleme, S., Akin, Z., Misra, S.: A lossless text compression technique using syllable based morphology. Int. Arab J. Inf. Technol. **8**(1), 66–74 (2011)
18. Kuthan, T., Lansky, J.: Genetic algorithms in syllable-based text compression. In: Proceedings of the Dateso 2007 Annual International Workshop on DAtabases, TExts, Specifications and Objects, Desna, Czech Republic, 18–20 Apr 2007
19. Lansky, J., Zemlicka, M.: Text compression: syllables. In: Proceedings of the Dateso 2005 Annual International Workshop on DAtabases, TExts, Specifications and Objects, Desna, Czech Republic, pp. 32–45, 13–15 Apr 2005
20. Lansky, J., Zemlicka, M.: Compression of small text files using syllables. In: 2006 Data Compression Conference, Snowbird, UT, USA. p. 458, 28–30 Mar 2006

Big Data Analysis for Event Detection in Microblogs

Soumaya Cherichi and Rim Faiz

Abstract The growing complexity of the Twitter micro-blogging service in terms of size, number of users, and variety of bloggers relationships have generated a big data which requires innovative approaches in order to analyse, extract and detect non-obvious and popular events. Under such a circumstance, we aim, in this paper, to use big data analytics within twitter to allow real time event detection. These challenges present a big opportunity for Natural Language Processing (NLP) and Information Extraction (IE) technology to enable new large-scale data-analysis applications. Taking to account all the difficulties, this paper proposes a new metric to improve the results of the searches in microblogs. It combines content relevance, tweet relevance and author relevance, and develops a Natural Language Processing method for extracting temporal information of events from posts more specifically tweets. Our approach is based on a methodology of temporal markers classes and on a contextual exploration method. To evaluate our model, we built a knowledge management system. Actually, we used a collection of 10 thousand of tweets talking about the current events in 2014 and 2015.

Keywords Microblogs · Relevant information · NLP · Event detection · Big data

1 Introduction

One of the most important facts on twitter is that data varies in volume, velocity, veracity and variety (unstructured and structured). However, this creates an opportunity to improve effective event detection models by taking advantage of patterns that is created from big data analysis. A great deal of research has

S. Cherichi (✉)
LARODEC, ISG, University of Tunis, Tunis, Tunisia
e-mail: soumayacherichi@gmail.com

R. Faiz
LARODEC, IHEC, University of Carthage, Tunis, Tunisia
e-mail: Rim.Faiz@ihec.rnu.tn

© Springer International Publishing Switzerland 2016   309
D. Król et al. (eds.), *Recent Developments in Intelligent Information and Database Systems*, Studies in Computational Intelligence 642,
DOI 10.1007/978-3-319-31277-4_27

addressed the problem of trending news and hot topics in real time from microblogging messages (Tweets) to satisfy the user's information needs and provide hot events. However, most research works rely on traditional data warehouses (EDWs) and Traditional data analysis tools to analyse microblogs such as twitter. Nonetheless traditional EDWs can't handle unstructured data, which is the bulk of most blogger's interaction data being generated on twitter (i.e., #Hashtag,). Whereas Big data analysis can be deemed to analyse this special kind of chaotic data.

Automatic event detection is one of Big Data tasks that have emerged to gain big insight about popular topic and events on social media. More precisely, social media is greatly impacting the growth of big data; and big data is providing suitable tool to understand the huge volume of data generated on microblogs. Most importantly, big data enables to do iterative data discovery that leads to analyse, detect and extract useful knowledge such as popular events. With particular reference to this issue, we focus on this work on the problem of automatic online event detection on Twitter microblogs by combining a big data analytics environment with Twitter analytics to create a novel approach that can enhance events detection within the big data space. We aim to leverage an accurate and timely automatic identification of events in tweets.

In recent world events, social media data has been shown to be effective in detecting earthquakes [1, 2], rumors [3], and identifying characteristics of information propagation [4]. This incites us to study the problem of event detection, which is an interesting and important task in such circumstances. In fact Event detection approaches designed for documents cannot be strictly applied to tweets due to their specific characteristics. Our work consists in suggesting a new metric, which allows studying the impact of each feature on impact on the quality of search results. We also intend to develop a Natural Language Processing method to extract temporal information from tweets.

We gathered the features on three groups: those related to content, those related to tweet and those related to the author. We used the coefficient of correlation with human judgment to define our score. For processing the content of tweets, we intend to use resources and linguistic methods. To identify event information from tweets, we proposed to identify five classes of linguistic markers (key-words) namely temporal markers (calendar term, occurrence indicator, relative pronoun, transitive verb). Our experimental result uses a corpus of 10 thousand of subjective tweets, which are neither answers nor retweets.

The remainder of this paper is organized as follows. In Sect. 2, we give an overview of related works. In Sect. 3, we present our approach of Event information extraction and discuss experiments and obtained results in Sect. 4. Finally, Sect. 5 concludes this paper and outlines future work.

2 Related Works

Researches on methods for Big Data are of interest in social computing area, since there are millions of online social contents, with many more being generated daily. As long as we know, there are no any papers that study the effects of Big Data analytics for event detection on Twitter. As the primary goal of this paper is to provide an event extraction approach from tweets using big data analytics methods, we aim, on the following, to present a brief strong and comprehensive study of current research on Big Data. This study aims to underline and explain the importance of detecting events using Big Data techniques.

By definition, "Big data refers to data sets whose size is beyond the ability of typical database software tools to capture, store, manage and analyse" [5]. In fact, nowadays, we are in the dawn of the big data era, where people are concerned with how to rapidly get the desired information, and Big Data analytics present suitable methods and techniques to extract key information from massive data. Indeed, big data analytics is where advanced analytic techniques are applied on big data sets [6]. Thence, those analytics have as a goal to discover associations and understand patterns and trends within a huge volume of data. Big data analytics are described by three primary characteristics: volume, velocity and variety and it has the potential to take advantage of the explosion in data to extract insights for making better informed decisions and bring values for enterprises and individuals.

Owing by the increasing importance of Big Data analytics, many tools for big data mining and analysis are available. The most knowing offline analysis architecture are essentially based on Hadoop such as include Facebook's open source tool Scribe, LinkedIn's open source tool Kafka. They are using in order to reduce the cost of data format conversion and improve the efficiency of data acquisition. Twitter data volume is expected to grow dramatically in the years ahead. In addition, large volumes of high velocity, complex, and variable data are generated that require advanced techniques able to manage and analyse the information. Hence, it is vitally important for "social media analysis approaches" to acquire the available tools, infrastructure, and techniques to leverage big data effectively or else risk losing potentially their innovative nature.

Several works have focused on the analysis of data posted on microblogs, particularly in Twitter. References [7, 8] propose approaches for sentiment classification regarding Twitter messages i.e. determine whether tweets express a positive, negative or neutral feeling. Positive and negative polarities correspond respectively to a favourable and unfavourable opinion as well. To solve this task, the authors have used natural language processing and machine learning techniques. Reference [9] proposes an approach to measure user influence in twitter. Many studies have found that there is a high correlation between the information posted on the web and present results. Reference [10] have used tweets to analyze awareness and anxiety levels of Tokyo inhabitants during the events of earthquakes tsunami and the sites of nuclear emergencies in Japan in 2011. Reference [11] have presented a method to measure the prevalence of H1N1 disease in the population of

United Kingdom. They also sought in the tweets the symptoms related to the disease and obtained results, which were compared with real results from the Health Protection Agency. Besides [12, 13] analysed the tweets to predict public opinion and then compared the results with the surveys.

Tweets reflect useful event information for a variety of events of different types and scale. These event messages can provide a set of unique perspectives, regardless of the event type [14, 15], reflecting the points of view of users who are interested or who participate in an event. In case of unexpected events such as Earthquakes, Twitter users sometimes spread news prior to the traditional news media [16]. Previous work on event extraction [17, 18] have focused largely on news articles, as historically this genre of text has been the main source of information on current events. In the meantime, several research efforts have focused on identifying events in social media in general and on Twitter in particular [19, 20]. Recent work on Twitter has started to process data as a stream, as it is produced, but has mainly focused on identifying events of a particular type, e.g., news events [21, 22], earthquakes. Other works identify the first Twitter message associated to an event as soon as it happens [11].

In the context of event extraction from tweets, [23] have developed a framework that takes a keyword related to a particular event, returns a summary that responds to the request. The summary contains the time of the beginning that indicates when the event began to be discussed, a term that specifies how long the event was discussed, and a small number of posts during this time interval. In the same context, [24] proposed a method to generate summaries from tweets (in real time) covering an event e. Our work consists in examining the role and impact of social networks, in particular microblogs, on public opinion. We aim to analyze the behavior of users through the texts they post in order to extract the events that reflect the interests and opinions of a population. We introduce in this paper our approach for tweet search that integrates different criteria namely the social authority of micro-bloggers, the content relevance, the tweeting features as well as the hashtag's presence. Once we selected relevant tweets, we move to the step of identifying event information from these tweets. In the addition we want to identify classes of linguistic markers (key-words) namely temporal markers. This way our work can be seen different and unlike the work of [25, 26] which use only sets of keywords to detect events known in advance. In addition to the previous works we intend to detect events "not previously known" that can be stimulating for users at the same time.

There is no need to include page numbers or running heads; this will be done at our end. If your paper title is too long to serve as a running head, it will be shortened. Your suggestion as to how to shorten it would be most welcome.

3 Event Detection

To detect a target event from Twitter, we search from Twitter and find useful tweets. Our method of acquiring useful tweets for target event detection is portrayed in Fig. 1. After normalizing the feature scores, these three scores are combined linearly using the following formula:

$$Score(Ti, Q) = Scorecontent(Ti, Q) + \beta \, ScoreTweet(Ti, Q) + \gamma \, ScoreAuthor(a, Q)$$

(1)

with

- Scorecontent(Ti, Q) on [0, 1] because tweet content should deal with the topic of request. In fact, being able to measure the content relevance of a tweet is essential from a semantic perspective, since it enables distinguishing between noise and pertinent tweets: pertinent tweets must have a content score that goes beyond a threshold value which is the mean of the scores, otherwise it is considered non pertinent and can't be considered for the second filtering step. Once we selected the most relevant tweets according to the most reliable scorecontent, then we calculate the score of tweets according to scoretweet and scoreauthor mentioned above
- Scorecontent (Ti, Q) is the normalized score of the relevance of content;
- ScoreTweet (Ti, Q) is the normalized score of the specificity of the tweet Ti;
- ScoreAuthor (a, Q) is the normalized score of the importance of the author (a) corresponds to the blogger who published the tweet Ti; $\beta + \gamma = 1$;
  We note that:
- Scorecontent(Ti, Q) = Relevance(T, Q) + Lg(Ti) + Popularity(Ti, Tj, Q) + Quality(Ti);
- ScoreTweet(Ti, Q) = Url count(Ti) + Hashtag Count(Ti) + Retweet(Ti) + Reply (Ti) + Favor(Ti);
- ScoreAuthor(a, Q) = TwitterPageRank(a) + Audience(a) + Tweet Count (a) + Mention Count(a) + Expertise(a) + RetweetRank(a) + Follower (a) + Following(a);

Once we selected relevant tweets, we move to the step of identifying event information from these tweets. We automatically extract all information about events from tweets and specify more information about these events: associations, locations, temporal settings, etc. We propose an event extraction system that aims at automatic extracting of significant sentences (or paragraphs) bearing information with temporal knowledge from news articles as well as identifying the agent, the location, and the temporal setting of those events.

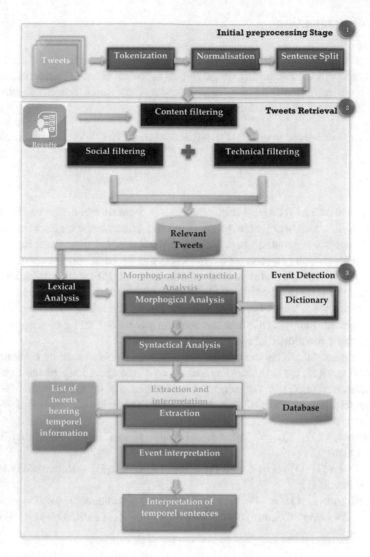

Fig. 1 Architecture of our method

Our system (cf. Fig. 1) is divided into five modules:

1. A lexical analysis module allowing the chunking of a tweet into words.
2. A morphological analysis module identifying words while triggering functions that deal with morphological inflexions and generate a morpho-syntactic code for each word.
3. A syntactic analysis module that re-establishes the order of the morpho-syntactic codes generated by the morphological analyser with the aim of building some morpho-syntactic structures.

4. An extraction module, which allows us to pick out markers in order to identify distinctive sentences, which represent events.
5. A module for interpretation of the extracted tweets to identify "Who did what?", "to whom?" and "where?".

3.1 Extraction and Interpretation of Event Information

Based on the results of the research of Faiz [17], we analyzed several tweets; we noticed that they might have one of the following forms:

1. Calendar term followed by an event. Example:
   #Tunisie: et maintenant, le tour de l'**élection**#présidentielle; blog de @GeopolisFTV http://geopolis.francetvinfo.fr/tunisie-la-democratie-en-marche/ 2014/11/17/tunisie-apres-les-legislatives-la-presidentielle.html …
2. Preposition followed by a calendar term. Example: Depuis le 25 mai, les Français de **Tunisie** sont représentés par 5 conseillers consulaires: leurs noms, leur mission. http://www.ambassadefrance-tn.org/Election-des-conseillers …
3. Event followed by a calendar term. Example: Officiel: Le deuxième tour de l'**élection** présidentielle en **#Tunisie** aura lieu avant le 31 décembre 2014. https://twitter.com/albawsalatn/status/4817
4. Subject followed by a relative pronoun, followed by a verb cause-consequence, and followed by event. Example: Jour historique en **#Tunisie** qui organise sa première **élection** présidentielle libre après 24 ans de benalisme
5. Subject followed by a verb cause-consequence, followed by event. Example: L'Union européenne déploie 100 observateurs pour l'**élection** présidentielle en **Tunisie** http://fb.me/1J6uq02NQ
6. Subject followed by a verb cause-consequence, followed by event. Example: En **Tunisie**, l'**élection** présidentielle s'achemine vers un second tour http://www. lemonde.fr/tunisie/article/2014/11/23/president

This representation has led us to draw the main **linguistic markers** and to sequence them according to their types.

1. The calendar term class

   (a) propo-num stands for preposition + number. Example: Depuis 2012
   (b) Cal-num stands for: calendar + number. Example: Janvier 2010.
   (c) Prepo stands for preposition. Example: maintenant,
   (d) Num-cal-num stands for number + calendar + number Example: 17 janvier 2014.

2. The occurrence indicator class

   (a) Adj_occ stands for adjective + occurrence. Example: une autre fois, la dernière fois, la première fois

(b) Adt_det_occ stands for tense adverb + determiner + occurrence. Example: encore une fois.

3. The relative pronoun class

(a) Prr_aux_ppa: relative pronoun + auxiliary + past participle. Example: #Tunisie qui a organisé

(b) Prr_aux_adv_ppa stands for relative pronoun + auxiliary + adverb + past participle. Example: qui a trop bu.

4. The cause-consequence verb

(a) Verbconsq_subject: event + verb + event
Example: mini-tornade a provoqué des dégâts

(b) Verbconsq_argument: subject + verb + event. Exemple: le Conseil de prévention et de lutte contre le dopage avait provoqué une petite crise avec l'Union cycliste.

As the temporal markers are independent from the language, our Twitterim system can also be applied to English corpus and Arabic corpus. Examples of temporal markers:

- maintenant (french), now (English), الآن (Arabic).
- depuis 2012 (french), since 2012 (English), 2012 عاممنذ (Arabic).
- une autre fois (french), another time (English), اخرىمرة (Arabic).
- avant (french), before (English), قبل (Arabic).

4 Experimental Evaluation

We built a search engine that we have called "TWEETRIM", which allows to calculate all scores and display the most relevant tweets according to these score. It has as input a query composed of three keywords and as output a set of relevant tweets relative to the query.

To collect 10 thousand articles from Twitter, we implemented a Java program that used the Twitter4 J library. This library provides access to data (tweets, user information...) Twitter via its programming interface, Twitter API. We mainly studied the content of tweets (their sizes, the most frequents words, words known by a French lexicon), the preoccupations users based on hashtags used, the behavior of users. …

To perform queries and to collect the human judgment of relevance followed the following steps:

- We collected 1000 queries on recent actualities in Tunisia from users,
- then, we used the system that we have built which allows us to view the relevant 10 results according to the score of the content,

- And then, we asked 450 users to judge the 10 first results of each query.

We suppose that the content relevance already exists and we will improve our search result by varying our two other scores; ScoreTweet and ScoreAuthor. We calculate the correlation coefficient between our scores and the corpus, which allowed us to find our weighting coefficients β and γ.

4.1 Results

4.1.1 Estimation of Weights

We make a comparison within the values of correlation coefficients and through the results, we observe that the best correlation coefficient between βScoreTweet + γScoreAuthor with human judgment score = 0.3842 when β = 0.8 and thus γ = 0.2 (Fig. 2).

4.1.2 Evaluation of the System

The reference model combines only the features linearly without weighting. This model gave us the correlation coefficient equal to 0.2459 and our model gave us the correlation coefficient of 0.3842. It can clearly be noticed that there is 56 % improvement in the satisfaction of our human judgment (Fig. 3).

The event information extraction was derived by running the system on tweets already selected through Twitterim from our dataset. The tweets covered different themes like weather reports, politics, statements of people and editorials.

On the whole, around 10 thousand tweets were used to ascertain that the extraction module (event information extraction) did work. We have conducted experiments to verify the effectiveness of our proposed approach to event information extraction. Firstly, human experts evaluated the extracted event sentences and over 80 % of them were deemed good event sentences. Secondly, in order to

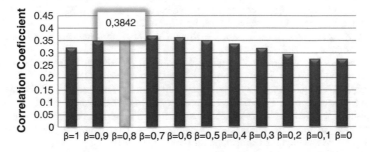

Fig. 2 Estimation of weights

Fig. 3 Comparing our model
with reference model

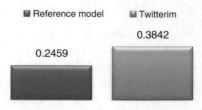

mcasure the performance of the system, the results for the testing of the event extraction system were measured using standard information extraction units recall and precision where:

$$\text{Recall} = \frac{No.\ of\ relevant\ event\ sentences\ identified}{No.\ of\ relevant\ event\ sentences} = 80\,\%$$

$$\text{Precision} = \frac{No.\ of\ relevant\ event\ sentences\ identified}{No.of\ event\ sentences\ identified} = 88.9\,\%$$

5 Conclusion

We have proposed in this paper a new metric for Social Research on twitter and event information detection. Through experiments we demonstrated that the proposed approach is efficient and is able to capture reasonable events in topic streams and random streams on Twitter.

References

1. Sakaki, T., Okazaki, M., Matsuo, Y.: Earthquake shakes Twitter users: real-time event detection by social sensors. In: WWW'10 (2010)
2. Mendoza, M., Poblete, B., Castillo, C.: Twitter under crisis: can we trust what we RT? In: Proceedings of the First Workshop on Social Media Analytics (2010)
3. Qu, Y., Huang, C., Zhang, P., Zhang, J.: Microblogging after a major disaster in China: a case study of the 2010 Yushu earthquake. In: Proceedings of the ACM 2011 Conference on Computer Supported Cooperative Work, pp. 25–34 (2011)
4. Zheng, X., Zeng, Z., Chen, Z., Yu, Y., Rong, C.: Detecting spammers on social networks. Neurocomputing 159(2), 27–34 (2015)
5. Manyika, J., Chui, M., Brown, B., Bughin, J., Dobbs, R., Roxburgh, C., Byers, A.H.: Big data: the next frontier for innovation, competition, and productivity (2011)
6. Elgendy, N., Elragal, A.: Big data analytics: a literature review paper. In: Advances in Data Mining. Applications and Theoretical Aspects, pp. 214–227. Springer, Berlin (2014)
7. Barbosa, L., Feng, J.: Robust sentiment detection on Twitter from biased and noisy data. In: Proceedings of the 23rd International Conference on Computational Linguistics: Posters, pp. 36–44. Association for Computational Linguistics (2010)

8. Jiang, L., Yu, M., Zhou, M., Liu, X., Zhao, T.: Target-dependent Twitter sentiment classification. In: Proceedings of 49th ACL: HLT, vol. 1, pp. 151–160 (2011)
9. Cha, M., Haddadi, H., Benevenuto Krishna, F., Gummadi, P.: Measuring user influence in twitter: the million follower fallacy. In: Proceedings of the Fourth International AAAI Conference on Weblogs and Social Media 2010, ICWSM (2010)
10. Doan, S., Vo, B.K.H., Collier, N.: An analysis of Twitter messages in the 2011 Toho earthquake. Arxiv preprint arXiv:1109.1618 (2011)
11. Lampos, V., Cristianini, N.: Tracking the flu pandemic by monitoring the social web. In: 2010 2nd International Workshop on Cognitive Information Processing (CIP), pp. 411–416. IEEE (2010)
12. OConnor, B., Balasubramanyan, R., Routledge, B.R., Smith, N.A.: From tweets to polls: linking text sentiment to public opinion time series. In: Proceedings of the International AAAI Conference on Weblogs and Social Media, pp. 122–129 (2010)
13. Duan, Y., Jiang, L., Qin, T., et al.: An empirical study on learning to rank of tweets. In: COLING Proceedings of the 23rd International Conference on Computational Linguistics Proceedings of the Conference, Beijing, China, pp. 295–303. Tsinghua University Press, 23–27 Aug 2010
14. Yardi, S., Boyd, D.: Tweeting from the town square: measuring geographic local networks. In: ICWSM'10 (2010)
15. Kwak, H., Lee, C., Park, H., Moon, S.: What is Twitter, a social network or a news media? In: WWW'10 (2010)
16. Ritter, A., Clark, S., Mausam., Etzioni, O.: Named entity recognition in tweets: an experimental study. In: Proceedings of EMNLP (2011)
17. Faiz, R.: Identifying relevant sentences in news articles for event information extraction. Int. J. Comput. Process. Orient. Lang. (IJCPOL) 19(1), 1–19 (2006)
18. Becker, H., Naaman, M., Gravano, L.: Learning similarity metrics for event identification in social media. In: WSDM'10 (2010)
19. Metzler, D., Cai, C., Hovy, E.: Structured event retrieval over microblog archives. In: Proceedings of HLT-NAACL (2012)
20. Sankaranarayanan, J., Samet, H., Teitler, B.E., Lieberman, M.D., Sperling, J.: Twitterstand: news in tweets. In: GIS'09 (2009)
21. Petrović, S., Osborne, M., Lavrenko, V.: Streaming first story detection with application to Twitter. In: NAACL'10 (2010)
22. Chakrabarti, D., Punera, K.: Event summarization using Tweets. In: ICWSM (2011)
23. Robertson, S., Walker, S., Hancock-Beaulieu, M.: Okapi at TREC-7: automatic ad hoc, filtering, VLC and interactive. In: Text REtrieval Conference TREC, pp. 199–210 (1998)
24. Cherichi, S., Faiz, R.: Analyzing the behavior and text posted by users to extract knowledge. In: Proceedings of the International Conference on Computational Collective Intelligence Technologies and Applications ICCCI 2014, Seoul, Korea, ACM 2014 Lecture Notes in Artificial Intelligence of Springer (2014)
25. Cherichi, S., Faiz, R.: New metric measure for the improvement of search results in microblogs. In: Proceedings of the International Conference on Web Intelligence, Mining and Semantics (WIMS 2013), New York, NY, USA. ACM (2013)
26. Cherichi, S., Faiz, R.: Relevant information discovery in microblogs: new metric measure for the improvement of search results in microblogs. In: Proceedings of INSTICC International Conference on Knowledge Discovery and Information Retrieval (KDIR 2013), Vilamoura, Portugal, ©SciTePress, 19–22 Sept 2013

Part V
Image, Video, Motion Analysis and Recognition

Building 3D Object Reconstruction System Based on Color-Depth Image Sequences

Ngoc Quoc Ly, Tai Cong Tan Vu and Thanh Quoc Tac

Abstract In this paper, we present a 3D object reconstruction system based on color and depth image sequences. The main intention is to build a sufficient system that can digitize the shape and color texture of real objects. Our contributions are the combination of individual processes into one unified procedure and some improvements on these methods. The system consists of four main phases. First, we improve the separation of object from background by performing a saliency map on depth image. Next, we represent the surface of object by a global point cloud and use Poisson reconstruction method to reconstruct the surface mesh. Finally, we propose neighborhood interpolation mapping technique for assigning color of reconstructed model. The experiments conducted on practical datasets have shown that our system is able to scan, reconstruct and simulate real objects completely.

Keywords Surface reconstruction · Object detection · Saliency map · Oriented normal · Texture mapping

1 Introduction

The modeling, recognition, and analysis of the real world are principal goals in Computer Vision & Computer Graphics. The most important problem of these objectives is obtaining a digital representation of real objects [1]. In traditional way, we mostly collect and store the information of objects in the form of 2D data which remains many deficiencies with this data type because of the incomplete details

N.Q. Ly · T.C.T. Vu · T.Q. Tac (✉)
Faculty of Information Technology, University of Science, VNU-HCMC,
Ho Chi Minh City, Vietnam
e-mail: tcqucthanh@gmail.com

N.Q. Ly
e-mail: lqngoc@fit.hcmus.edu.vn

T.C.T. Vu
e-mail: vcttai@gmail.com

© Springer International Publishing Switzerland 2016   323
D. Król et al. (eds.), *Recent Developments in Intelligent Information
and Database Systems*, Studies in Computational Intelligence 642,
DOI 10.1007/978-3-319-31277-4_28

when performing real objects. To improve this, people nowadays use the 3D data having more effectiveness and advantages than 2D data by the fact that it contains much more information.

Recently introduced RGB-D cameras (e.g. Kinect) are able to capture synchronized color and depth images. With its advanced sensing techniques, this technology opened up an opportunity to develop an economical 3D scanner with acceptable quality. There have been many studies introduced to deal with this problems, Kinect Fusion (Microsoft) and 123D Catch (Autodesk) is the typical ones. Kinect Fusion require a RGB-D sensor to capture data while 123D Catch only need a RGB camera to do it, both of them after that can provide a detailed 3D model of the scene. However, both the main object and the surrounding environment are reconstructed, it is hard to model only the expected target. The challenge is to build a system that can separate object automatically and reconstruct its entire appearance. That led to problems need to be solved: detect object, reconstruct surface of object, map color texture to the surface.

In specific, our 3D object reconstruction system needs color-depth image sequences as input. In the beginning, we present a new method that can automatically separate the main object from the environment by estimating a saliency map on the depth images. Subsequently, we perform the registration to organize the collected data to obtain a global point cloud that represents entire surface of object. After that, we reconstruct the surface mesh by performing a Poisson explicit function [2] with oriented normal field as an input along with the point cloud. Finally, the texture of surface mesh will be mapped by our neighborhood interpolation technique. The purpose of this system is to reconstruct a model that can represent both the structure and texture of real objects for recognition, printing or analysis. Furthermore, this procedure can be applied to mobile applications. We use phone's camera to capture a real object, then present it to the screen as 3D data. User can interact with objects in the real-world more naturally.

This paper is organized as follows: Sect. 2 outlines the related works in object segmentation and surface reconstruction. Section 3 describes our approach in object detection and Sect. 4 provide the registration for constructing the global point cloud. Section 5 presents steps to reconstruct the surface and color mapping will be detailed in Sect. 6. Section 7 shows the experimental results of our reconstruction system. Finally, Sect. 8 draws the conclusion and the future works.

2 Related Works and Our Approach

2.1 Related Works

In object detection field, the common methods using color information like K-mean, Mean Shift can cluster pixels having similar color value into one group. However, the difference of color value in one object is very large. Colorful object can be divided into different clusters. On the other hand, using depth information can determine object's cluster more exactly because object's pixels have similar depth value. Research [3] separated object by applying edge detection and connected components algorithm to segment depth image, authors also proposed a depth value defined by user in order to discard background. In case of distance between the camera and objects change frequently, threshold depth value in order to discard background in all cases is a challenge.

In surface reconstruction field, it is a wide range area and there have been many approaches depend on specific contexts. In case of reconstructing a determined shape class (e.g. CAD model), some prior knowledge such as collections of simpler geometric primitives will be very helpful for the reconstruction [4]. Although this method has some advantages such as the accuracy of reconstructed model is relatively high, it is only suitable for those objects that have repetitive patterns. Another approach proposed in [5] reconstructs the model by using the scanner visibility to merge individual range scans. It relies on the information provided by the camera to determine the space regions are empty or unseen, then to extract geometry among those regions. Although this method can incrementally build up a watertight model, it also has some drawbacks such as: poor ability to handle noise and missing data, requiring scanner information.

The surface smoothness methods mentioned in [2, 6, 7] require a surface point cloud and normal field as input can reconstruct entire surface of objects. The approach in [6] computed the signed distance field to extract the surface, while [2] modeled objects by building up a global indicator function. The method proposed in [7] reproduces a linear combination of radically symmetric bias functions to perform surface of objects. While [6] belongs to local surface smoothness group, [2, 7] are in global one. By imposing the smoothness prior, these methods can be effective in producing a watertight surface. However, due to the localism, [6] is very sensitive with noise and non-uniform sampling; [7] also have to face a great difficulties to handle outlier and missing data. For those problems, the data preparation including building the surface point cloud and estimating normal field has also challenges that need to overcome (Fig. 1).

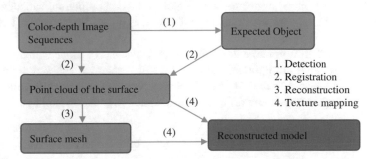

Fig. 1 The framework of our reconstruction system

2.2 Our Approach

Our reconstruction system concentrate on solving the following 4 main tasks: Object detection, registration for building point cloud of surface, reconstructing triangular surface mesh and mapping color texture to the reconstructed surface.

- **Object detection**: We segment depth images base on saliency map, then threshold the segmented image by using two thresholds. Our method decreases the difference of pixels which are in same cluster and increase distance of different clusters.
- **Registration and reconstruction**: We use Poisson implicit function [2] to reconstruct entire surface of object. We focus on building the global point cloud that represent the surface of object from discrete point clouds base on RANSAC and ICP algorithm [7]. For estimating the normal field of the point cloud for the input of reconstruction, we present the direction spreading technique to accomplish this step.
- **Texture mapping**: Due to the similarity of the coordinates between the color surface point cloud and reconstructed surface mesh, we perform neighborhood interpolation mapping technique for assigning texture to achieve finished reconstructed model.

3   Object Detection Method

Image segmentation based on saliency map is an effective method to determine objects have "salient" colors. The main idea is using depth and contrast feature to create a saliency map. "Salient" colors means their value have great contrast with the remaining colors. We note that the depth values of object also have a huge contrast with background, applying saliency map to segment depth image is very suitable. We call this method is object detection based on saliency map. Our procedure has three phases: normalizing depth image, computing saliency map, thresholding. The input is depth image and output is binary image which 1 is pixel of object, rest is 0.

First, we need to normalize depth image to grayscale in order to reduce depth value space, remove unimportant values. Next, we compute saliency map. Each color of image has a "saliency value". Saliency value is calculate follow two factors: the frequency and the difference in distance [8]. The formula is:

$$S(c_l) = \sum_{j=1}^{n} f_j * D(c_l, c_j) \tag{1}$$

where c_l is value of lth color; $S(c_l)$ is saliency value of lth color; n is total color space; f_j is the frequency of jth color; $D(c_l, c_j)$ is the differences between lth color and jth color.

(a)   **(b)**   **(c)**   **(d)**

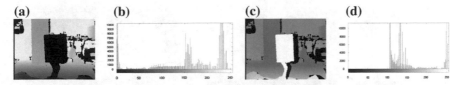

Fig. 2 Example of saliency map. After computing the saliency map, the expected object is more "salient". We can easily separate it from the background. **a** Original depth image. **b** Histogram of original depth image. **c** Saliency image. **d** Histogram of saliency image

Follow the formula, color is "salient" when it has high contrast and low frequency. It is corresponding to human visual, we often notice objects having different colors and were not mistaken when comparing with the remaining surroundings. The complexity of this formula when applied on depth image is only 256^2 (Fig. 2).

Finally, we extract region of object by setting two thresholds for saliency image by the formula as follows:

$$y = \begin{cases} 255, & k_1 < x < k_2 \\ 0, & \text{otherwise} \end{cases} \qquad (2)$$

where y is value after thresholding; x is saliency map value; k_1, k_2 are threshold values $(k_1 < k_2)$. Two threshold values we used is $k_1 = 150, k_2 = 256$, obtained from experiments.

Furthermore, we also apply other supplementary techniques such as background subtraction, extract connected components… to support and enhance the accuracy of this detection procedure.

4 Registration Method

For each color-depth-mask set of image, we create a point cloud to illustrate a part of surface, we call it a local point cloud. Corresponding to each view of data acquisition, the obtained point cloud will be associated with a separated coordinate. We need to figure out the transformations to transform them to one unique coordinate system.

First, we extracted all the feature points of the image by the SURF feature detector [9]. Then we matched that feature points to find out the corresponding pairs. Base on those corresponding points, we mapped them to real coordinates by using depth image. After that, we performed the RANSAC algorithm to estimate the transformation matrix. RANSAC is an iterative method, to estimate the parameters of a transformation given a dataset [10]. For each iteration, a few samples are randomly selected to define a matrix (the minimum number of sample is 3). Then evaluate the model for the whole dataset. This is repeated several times by choosing new samples for each iteration and keeping the best transformation found. All out-model points are also be removed. Since RANSAC defined the

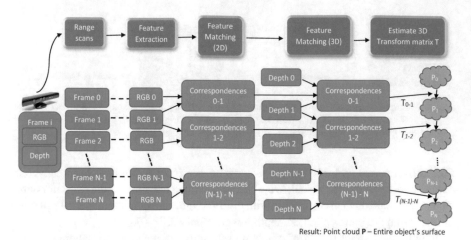

Fig. 3 The overview of our registration process

transformation matrix from only 3 points in model, the founded matrix has not been the best. To improve it, we further use ICP algorithm [11]. This method use entire points to define the transformation. That means, with p_i as source points and q_i as target points, it figures out the Rotation R and translation t by minimizing the squared distance between each corresponding pairs:

$$\min \sum_i \| (Rp_i + t) - q_i \|^2 \tag{3}$$

The overview of our registration process can be summarized as follows (Fig. 3).

5 Surface Reconstruction from Point Clouds

5.1 *Estimating Normal Field*

Let X denote for the point cloud, the tangent plane $T_p(x_i)$ corresponds to the point $x_i \in X$, the normal vector at that point is $\hat{n}_i$. The normal vector of $T_p(x_i)$ is defined by selecting a set of neighboring points of x_i, called $Nbhd(x_i)$—the "k-neighbor" set (k is a parameter). The tangent plane $T_p(x_i)$ is chosen so that it can approximates $Nbhd(x_i)$ best. We then applied PCA method to figure out the unoriented normal vector $\hat{n}_i$. A covariance matrix of $Nbhd(x_i)$ is formed, which is a symmetric 3×3 positive define matrix:

$$C_i = \sum_{y \in Nbhd(x_i)} (y - x_i)(y - x_i)^T \tag{4}$$

If we denote $\lambda_i^1 \geq \lambda_i^2 \geq \lambda_i^3$ as the eigenvalues of C_i which associated with eigenvectors $\hat{v}_i^1, \hat{v}_i^2, \hat{v}_i^3$ respectively, $\hat{n}_i$ is chosen to be either $\hat{v}_i^3$ or $-\hat{v}_i^3$. For the explanation, the eigenvectors $\hat{v}_i^1, \hat{v}_i^2$ express the most variation of 2 dimensions of data. That means it is the tangent plane $T_p(x_i)$. The remaining eigenvector $\hat{v}_i^3$ orthogonal to two eigenvectors above, so that it is obviously the normal vector.

Now, we adjust the direction of normal field. Let consider the point cloud X, supposing that two data points $x_i, x_j \in X$ is are geometrically close. Ideally, when the data is dense and the surface is smooth, the two tangent planes $T_p(x_i)$ and $T_p(x_j)$ are nearly parallel, thus $\hat{n}_i \cdot \hat{n}_j \approx \pm 1$. If the direction of all normal vectors are consistent, then $\hat{n}_i \cdot \hat{n}_j \approx 1$; otherwise, we need to flip either $\hat{n}_i$ or $\hat{n}_j$ [6]. We first start from a single point containing an initial orientation. Then we propagate this orientation to nearby points whose unoriented normal vectors are nearly parallel. Performing this iterative process until all vectors are oriented. Our approach is defined procedurally as:

Input: Point cloud which represents surface of object
Output: Oriented normal field.

```
1. Estimate unoriented normals for all points.
2. Initialize seed point xₖ with normal n̂ₖ
   Initialize a queue q_Directed, add the seed point.
   While(q_Directed.isEmpty() == false)
      CurrentPoint = q_Directed.pop()
      List_Neighbor[] = findNeighbor(CurrentPoint)
      For all points in List_Neighbor[]:
         if(List_Neighbor[i] is undirected)
            direct(CurrentPoint, List_Neighbor[i])
            q_Directed.push(List_Neighbor[i])
         end if
      end For
   end While
```

5.2 Surface Reconstruction

We apply the Poisson implicit function to reconstruct the surface mesh. In details, this method performs a 3D indicator function χ. This function returns value of 1 for the points locating inside the shape and value of 0 otherwise. It is assumed that the gradient of the indicator function is as close as possible to the point cloud with oriented normal. Thus, the oriented point samples can be viewed as the gradient of the indicator function. The approach to solve this problem is proposed in [2].

The problem of computing the indicator function χ reduces to inverting the gradient operator. That means we must find the scalar function χ whose gradient best approximates the normal field $\vec{V}$ defined by the samples. It can be described mathematically as an optimization problem: $\min_\chi \|\nabla \chi - \vec{V}\|$. By applying the divergence operator, it transforms into a standard Poisson problem: compute the scalar function χ whose Laplacian (divergence of gradient) equals the divergence of the vector field:

$$\nabla \cdot \nabla \chi = \Delta \chi = \nabla \cdot N \tag{5}$$

Once solved, the surface is found via function χ with a suitable isovalue that indicates the surface of object. Then we use that information to construct the triangular surface mesh.

6 Color Mapping

The triangular surface mesh includes: a vertex set P, a triangular set T that connects all vertices together. We have to assign color texture to set P. We interpolated the color of each point in P from the original color point cloud PC. For details, with each point p in P, we found N neighborhood points in the point cloud PC, called set np, then mapped color of np to p. Color of p depends on distance from p to each neighborhood point in np and the dependence is represented by a weighted value W. The weighted value of each point in np is calculated as:

$$W_i = 1 - \frac{d_{np_i}}{d_{max}}, \quad i = 1, \ldots, N \tag{6}$$

where W_i is weighted value of ith point in np; d_{np_i} is distance from p to the ith point in np; d_{max} is maximum distance from p to a point in np. The color of point p is interpolated:

$$C_p = \frac{1}{\sum_{i=1}^{N} W_i} \sum_{j=1}^{N} W_j C_j \tag{7}$$

where C_p is the color of point p that need to be set color; W_i, W_j are weighted value of ith, jth point in np; C_j is color of jth point in np.

7 Experimental Results

7.1 Object Detection

Our dataset for object detection is obtained from real objects: Box and Teddy Bear. We use Mean Square Error (MSE) to estimate the difference of our result and ground truth. We used Microsoft Kinect 360 device to collect data (Table 1).

Table 1 Results of object detection

Object	Depth map	Ground truth	Method	
			Our approach	Mean Shift
1				
2				
3				

Table 2 Mean square error results

	Object	1	2	3
MSE	Mean Shift	617.447	12,311.8	6281.3
	Proposed method	229.239	245.325	784.66

Our method rearranged depth value so that we could easily set thresholds to detect object and eliminate completely background. By the same threshold k_1 and k_2, our method detected object region with lower error. With Mean Shift, threshold values must be changed when we changed distance from camera to object, that led to MSE of object 2 and 3 were higher than object 1 (Table 2).

7.2 Point Cloud Registration and Surface Reconstruction

From segmented images, we have built up the global point cloud to represent the surface of object. Then, we estimated the oriented normal field and used both data as input for the surface reconstruction. At final, we mapped color from the point cloud to the reconstructed mesh to obtain the final color mesh of reconstruction (Fig. 4).

By all these results, it proves that all the proposed methods work well for the reconstruction system. We can consolidate completely global point clouds to represent the shape of objects, reconstruct the triangular surface mesh of entire objects and map color for the final models. The reconstructed models can express both the information of shape and color texture of the objects (Fig. 5).

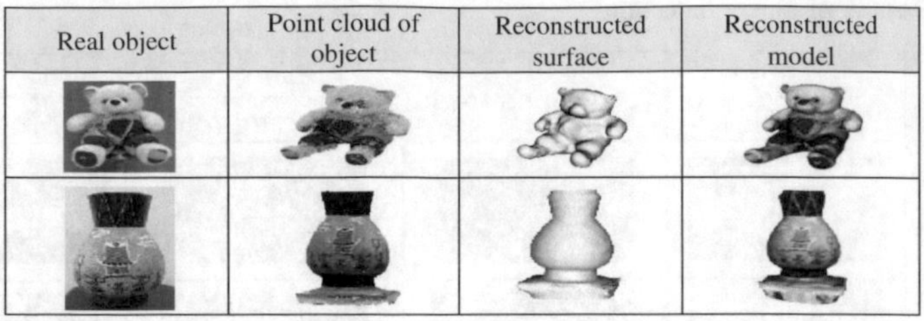

Real object	Point cloud of object	Reconstructed surface	Reconstructed model

Fig. 4 Real objects and the experimental results of registration, reconstruction and texture mapping process

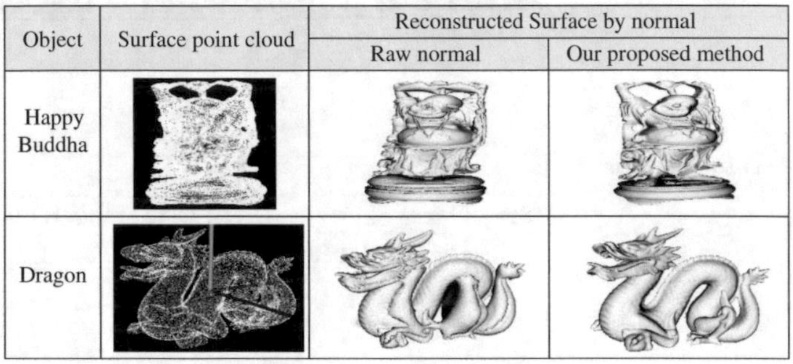

Object	Surface point cloud	Reconstructed Surface by normal	
		Raw normal	Our proposed method
Happy Buddha			
Dragon			

Fig. 5 Improvements in our proposed method for oriented normal estimation

7.3 *Estimating Oriented Normal Field*

We used point clouds dataset published by Stanford University to test our proposed method. It contains only point clouds that represent real objects. As you can see, the incorrect orientation can lead to many unwanted surface components. After performing our method for orienting directions, the reconstructed model is more accurate.

8 Conclusion and Future Works

This paper presented the general procedure to develop a system used as a low-cost 3D scanner for reconstruction and simulation real objects in 3D virtual data. We have also contributed some new approaches in these issues: image segmentation, registration and surface reconstruction. The results can be used as input data for many further applications such as: recognition, printing, analysis…

We continue to improve the segmentation module to separate objects more accurately. Besides, we will also enhance the resolution of images sequence acquired by the technique of Video Super Resolution, the quality of 3D Object reconstruction would achieve a very high quality. We are also working hard to combine this module with our developing Visual SLAM system to achieve more real-life applications in the future.

Acknowledgment This research is funded by Vietnam National University, Ho Chi Minh City (VNU-HCM) under grant number *B2014-18-02*.

References

1. Berger, M., Tagliasacchi, A., Seversky, L., Alliez, P., Levine, J., Sharf, A.: State of the art in surface reconstruction from point clouds. In: Proceedings of Eurographics—Eurographics star reports, vol. 1, pp. 161–185 (2014)
2. Kazhdan, M., Bolitho, M., Hoppe, H.: Poisson surface reconstruction. In: Proceedings of Eurographics Symposium on Geometry Processing, vol. 7 (2006)
3. Triantafyllidis, G., Dimitriou, M., Kounalakis, T, Vidakis, N.: Detection and classification of multiple objects using an RGB-D sensor and linear spatial pyramid matching. J. Electr. Lett. Comput. Vis. Image Anal. **12** (2013)
4. Schnabel, R., Degener, P., Klein, R.: Completion and reconstruction with primitive shapes. In: Proceedings of the Annual Eurographics Conference, Computer Graphics Forum, vol. 28, no. 2, pp. 503–512 (2009)
5. Hoppe, H., DeRose, T.: Surface reconstruction from unorganized points. In: Proceedings of 19th Annual Conference on Computer Graphics and Interactive Techniques, vol. 26, no. 2. ACM, New York (1992)
6. Carr, J.C., Beatson, R.K.: Reconstruction and representation of 3D objects with radial basis functions. In: Proceedings of 28th Annual Conference on Computer Graphics and Interactive Techniques, pp. 67–76. ACM, New York (2001)
7. Wyawahare, M.V., Pradeep, P.M., Abhyankar, H.M.: Image registration techniques: An overview. Int. J. Signal Process. Image Process. Pattern Recogn. **2**, 11–28 (2009)
8. Dat, N.D., Binh, N.T.: Enhancing object quality based on saliency map and derivatives on color distances. In: IEEE RIVF International Conference on Computing Communication Technologies, IEEE, vol. 110, pp. 106–111 (2015)
9. Bay, H., Ess, A., Van Gool, L.: Speeded-up robust features (SURF). J. Comput. Vision Image Underst. **110**, 346–359 (2008)
10. Fischler, M.A., Bolles, R.C.: Random sample consensus: A paradigm for model fitting with applications to image analysis and automated cartography. J. Commun. ACM **24**, 381–395 (1981)
11. Zhang, Zhengyou: Iterative point matching for registration of free-form curves. Int. J. Comput. Vision **13**(2), 119–152 (1992)

Key Frames Extraction from Human Motion Capture Data Based on Hybrid Particle Swarm Optimization Algorithm

Xiaojing Chang, Pengfei Yi and Qiang Zhang

Abstract Extracting key frames from human motion capture data is a hot issue of computer animation in recent years. Though the reconstruction error of the key frames by current methods is small, the number of key frames still needs to be reduced. In order to produce results with less key frames and small reconstruction error, we propose a method employing hybrid particle swarm optimization algorithm to extract key frames. By introducing evolution strategy of Genetic Algorithm (GA) to hybrid particle swarm optimization algorithm, the method can get key frames with optimal compression ratio and small reconstruction error. Experimental results show the effectiveness of our method.

Keywords Key frames · Hybrid particle swarm optimization algorithm · Compression ratio · Reconstruction error

1 Introduction

Human Motion capture has been widely used for animations and Hollywood blockbusters. However, because of the complexity of human body and the variability of human motion, the motion data which is captured is substantial [1]. It is hard to deal with the multidimensional motion data. The storage of the multidimensional human motion data is also a challenge [2]. In order to reduce the storage space of these motion capture data, researchers proposed several methods to extract the key frames from human motion capture data. The main methods can be divided into the following categories:

X. Chang · P. Yi · Q. Zhang (✉)
Key Laboratory of Advanced Design and Intelligent Computing (Dalian University),
Ministry of Education, 116622 Dalian, China
e-mail: zhangq26@126.com

© Springer International Publishing Switzerland 2016   335
D. Król et al. (eds.), *Recent Developments in Intelligent Information
and Database Systems*, Studies in Computational Intelligence 642,
DOI 10.1007/978-3-319-31277-4_29

1. Curve simplification methods. According to the movement of the joint, Bulut and Capin [3] draws motion curves. Then they base Gaussian filtering to extract the characteristic points on the curves. Halit and Capin [4] reduce the dimensions of motion capture data through the method of PCA. Then they use multi-scale Gaussian filtering to gain the key frames, and cluster to determine the final key frames. Peng et al. [5] extract key frames based on the central distance features from the center joint root to limbs. The central distance features are the two-dimensional distance, so they use the method of PCA to extract the one dimensional information.

2. Extraction methods based on clustering. Liu et al. [6] propose that the motion data is divided into N clusters. They take the first frame of each cluster as the key frame. Arikan [7] cluster movement data first. Then they compress data using CPCA. Park and Shin [8] first turn the motion data into quaternion, then using PCA and k—means clustering method processes data to get the key frames.

3. Frame decimation methods. Li et al. [9] extract key frames basing on frame spacing. The frame spacing is calculated by the quaternion. Then set the threshold and calculate the distance of current frame and the next frame. If the frame spacing is less than the threshold, the front frame is eliminated. The frames which are not eliminated are as the final key frames.

4. Matrix factorization method. Gong and Liu [10] expresses the movement sequence with matrix, using the singular value decomposition (SVD) to pick up the key matrix to get the key frames. Cooper and Foote [11] using nonnegative matrix decomposition method to extract the key part of the video. The complex of this kind of methods is high.

5. Intelligent method. Liu et al. [12], present the method which combines the Genetic Algorithm (GA) and the simplex method. GA is applied to calculate the complex nonlinear problem. It is good for global searching and weak for local searching. The simplex method has the strong ability of local searching.

Curve simplification methods may lose the detail information of motion sequences, which impacts on the key frames. Clustering and frame decimation methods need to set threshold. The threshold depends on experience value too much. These methods need multiple tests to get the threshold, which is a waste of time and resources. The method based on the GA and the simplex use two complex frameworks which are difficult to implement.

To overcome the disadvantages mentioned above, we propose an approach using hybrid particle swarm optimization algorithm extracts key frames. The method uses the evolution method that the offspring particles instead of the parent particles of GA, which is easy to realize and needn't to set threshold. Meanwhile, the method adopts the strategy of the compression ratio and reconstruction error as the fitness function to produce results with lower compression ratio and little reconstruction error.

2 Key Frames Extracted Based on Hybrid Particle Swarm Optimization Algorithm

Particle Swarm Optimization algorithm (PSO) was originally proposed in 1995, aiming at the development of computational intelligence [13]. The algorithm looks particles as the birds. Each bird searching food likes a particle finding the optimal value [14]. Comparing with general optimization algorithms, the most prominent characteristic of PSO is the fast convergence speed [15]. Hybrid particle swarm optimization [16, 17] is based on the PSO, adding the hybrid theory of GA. Hybrid particle swarm optimization algorithm is utilized to extract key frames and compress movement data. We must test this algorithm before it is applied. This paper uses the variance test method. Basic steps are as follows:

Step1 Initialize the position and velocity of the particles;

Step2 Calculate the fitness value of each particle. The particle's fitness value is Pi and the optimal fitness value of all particles is Pg;

Step3 The fitness value of Pg is stored in the array V. Take the current 100 Pg to obtain the variance according to the variance formula;

$$s^2 = \frac{(M - x_1)^2 + (M - x_2)^2 + (M - x_3)^2 + \cdots + (M - x_n)^2}{n} \quad (1)$$

$$M = \frac{x_1 + x_2 + x_3 + \ldots + x_n}{n} \quad (2)$$

where n is the number of data, and $x_1, x_2, x_3 \ldots x_n$ represent values of data, and M is the average of the data.

Step4 If the variance is smaller than the set value, the iteration will stop. Then output the key frames, otherwise returns Step3

In our paper, the fitness function is a combination of two constraints [18]. The target of extract the key frames is the minimum compression ratio and error rate. The two goals is a contradiction, so setting the fitness function is:

$$fitness = 0.50 * E_{keyframe} + 0.50 * E_{error} \quad (3)$$

$E_{keyframe}$ is the compression ratio. That is, the ratio of the key frames and the total number of frames. E_{error} is the error rate, namely the ratio of reconstruction error of key frames and maximum reconstruction error.

This paper adopts the method of spherical interpolation to reconstruct the motion sequences. Frame spacing of original motion and the motion of reconstruction:

$$E = \frac{1}{n} \sum_{i=1}^{n} (M_1(i) - M_2(i))^2 \quad (4)$$

where $M_1(i)$ is the original motion data, $M_2(i)$ is the reconstructed motion data, n is the total number of the motion frames.

3 Results

In this paper, we use the motions of the motion capture database [19] of Carnegie Mellon University (CMU) to test our method. The sampling frequency for the human motion capture data is 120 frames/s. There are 2605 trials in 6 categories and 23 subcategories. In the lab tests, we select four kinds sports: walking (315 frames), jumping (439 frames), kicking a ball (801 frames), walk-jump-walk (1199 frames). The format of the data is BVH. The parameter settings of hybrid particle swarm optimization algorithm in the experiments are as follows: The number of particles is 30. Acceleration constant $c1$ is 2.5. Acceleration constant $c2$ is 2.0. The maximum inertia weight w_{max} is 1.2. The minimum inertia weight w_{min} is 0.8. The hybrid probability Pc is 0.9. The proportion of hybrid pool size Sp is 0.2.

Experiment 1: Using the hybrid particle swarm optimization algorithm to extract key frames.

Figure 1 show using our method deals with four kinds of typical motion data and the process of extracting key frames. The horizontal axis shows the number of iteration. The vertical axis represents the best fitness value of iteration. The fitness value decreases with the increase of the number of iteration. When the program gets the stop condition, it output the optimal fitness value and gets the extracted key frames sequence.

Experiment 2: The Saliency method [4] is a relatively new method and has representativeness. The method has been compared with the other two methods and is superior to other methods. So using our method and the method of [4] deal with motion sequence to get the key frames. Calculate compression ratio and reconstruction error.

Seen from Table 1: for the simple walk, the reconstruction error of [4] is much higher than our method. In addition, the compression ratio of [4] is 14.3 %. The compression ratio (5.1 %) has a significant reduction in our method, which significantly reduce the number of key frames. Jump is an action of feet off the ground. For the movement of jumping, the compression ratio of [4] more than 5.2 % of our method. And using our method, the reconstruction error is smaller than [4], so the reconstructed movement is close to the capture motion. For kicking a ball, the compression ratio of the method [4] is 11.1 %, 4.2 % more than our method. And the reconstruction error of the article is also smaller than the method [4]. It is concluded the number of the key frames which is extracted by our method is few, and the key frames can accurately representative of the original movement. For more complex walk-jump-walk which consists of some simple movements, the compression ratio which uses hybrid particle swarm optimization algorithm is 2.6 % less than that of the method of [4]. But the reconstructed motion sequence is a

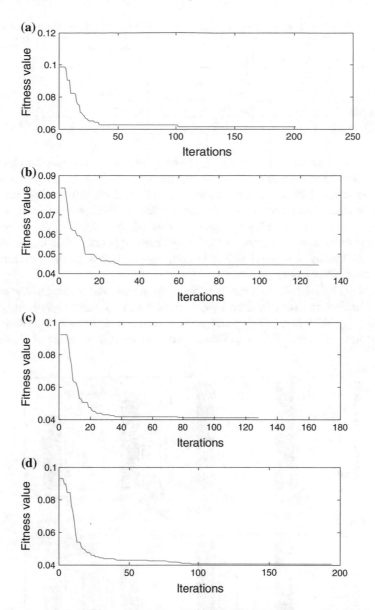

Fig. 1 The fitness value of four sports varies with the number of iteration. **a** The fitness value of walking varies with the number of iteration. **b** The fitness value of jumping varies with the number of iteration. **c** The fitness value of kicking a ball varies with the number of iteration. **d** The fitness value of walk-jump-walk varies with the number of iteration

little poor, namely that the reconstruction error is larger. It shows that for motion capture data that contains a variety of simple movement forms, the adaptability of our method needs to be improved.

Table 1 Compression ratio and reconstruction error using our method and the method of [4]

Motion type	Total frames	Number of the key frame	Compression ratio (%)	Reconstruction error
Walking	315	45(29)	14.3(9.2)	44.71(32.59)
Jumping	439	58(35)	13.2(8.0)	45.65(30.60)
Kicking a ball	801	89(55)	11.1(6.9)	35.50(31.76)
Walk-jump-walk	1199	112(80)	9.3(6.7)	25.83(33.11)

In summary, for the gently walking, the basic movement jumping and kicking a ball of a little more complex movement, the reconstruction error gotten by our method is lower than the method Saliency of [4], and the effect of reconstructed motion is also better. For more complex motion walk-jump-walk, the effect of reconstructed motion is slightly inferior. But whether for simple or complex movement the compression ratio gotten by our method is lower than the Saliency of [4]. So our method has good ability to compress motion data and can more efficiently extract the key frames.

According to Table 1, compare the compression ratio and reconstruction error of the motion capture data extracted by our method and the method Saliency of [4].

Figure 2a shows that using hybrid particle swarm optimization algorithm to get the compression ratio of four different types movements is about 7 %, and the

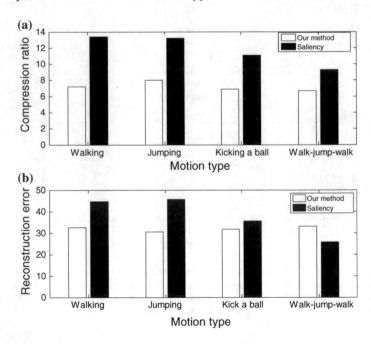

Fig. 2 **a** The compression ratio of the key frames extracted by our method and the method Saliency of [4]. **b** The reconstruction error of the key frames extracted by our method and the method Saliency of [4]

compression ratio gotten by the method Saliency of [4] is more than 10 %. By the above know our method compresses movement data better and make it easier to store and application. From the Fig. 2b, we can realize using the hybrid particle swarm optimization algorithm to extract the key frames from most sports, not only gets low compression ratio and reconstruction error, but also makes the key frames from the movement data more accurately.

4 Discussion

In this paper, we propose using hybrid particle swarm optimization algorithm to extract key frames. The method uses the evolution method of the offspring particles instead of the parent particles of the GA, which is easy to realize. Our method adopts the compression ratio and reconstruction error as the fitness function. The method can guarantee the integrity of the motion sequences information and needn't to set threshold. For most motions, our method is conducive to the compression and storage of data.

For some simple movements like walking, jumping, kicking a ball, our method can extract less number of key frames with lower reconstruction error. For complex movement which is consisted of a variety of simple motions like walk-jump-walk, although the reconstruction error received by our method may not be minimal, the number of the key frames is still less. Therefore, our method can be apply to the tasks which require minimal storage space for human motion capture data.

Acknowledgement This work is supported by the National Natural Science Foundation of China (No.61370141, 61300015), the Program for Liaoning Innovative Research Team in University (Nos. LT2015002), the Program for Science and Technology Research in New Jinzhou District (No. KJCX-ZTPY-2014-0012).

References

1. Bregler, C.: Motion capture technology for entertainment. Signal Process. Mag. **24**, 158–160 (2007)
2. Wang, P.J., Lau, R.W., Zhang, M.M., Wang, J., Song, H.Y., Pan, Z.G.: A real-time database architecture for motion capture data. In: 19th ACM International Conference on Multimedia, pp. 1337–1340. ACM Press, New York (2011)
3. Bulut, E., Capin, T.: Key frame extraction from motion capture data by curve saliency. In: Computer Animation and Social Agents, pp. 119–123. IEEE Press (2007)
4. Halit, C., Capin, T.: Multiscale motion saliency for keyframe extraction from motion capture sequences. Comput. Anim. Virtual Worlds **22**, 3–14 (2011)
5. Peng, S.J.: Key frame extraction using central distance feature for human motion data. J. Syst. Simul. **24**, 565–569 (2012)
6. Liu, F., Zhuang, Y.T., Wu, F., Pan, Y.H.: 3D motion retrieval with motion index tree. Comput. Vis. Image Underst. **92**, 265–284 (2003)

7. Arikan, O.: Compression of motion capture databases. ACM Trans. Graph. (TOG) **25**, 890–897 (2006)
8. Park, M.J., Shin, S.Y.: Example-based motion cloning. Comput. Anim Virtual Worlds **15**, 245–257 (2004)
9. Li, S.Y., Hou, J., Gan, L.Y.: Extraction of motion key-frame Based on inter-frame Pitch. Comput. Eng. **41**, 242–247 (2015)
10. Gong, Y.H., Liu, X.: Video summarization using singular value decomposition. In: IEEE Conference on Computer Vision and Pattern Recognition, pp. 174–180. IEEE Press, Hilton Head Island (2000)
11. Cooper, M., Foote, J.: Summarizing video using non-negative similarity matrix factorization. In: IEEE Workshop on Multimedia Signal Processing, pp. 25–28. IEEE Press (2002)
12. Liu, X.M., Hao, A.M., Zhao, D.: Optimization-based key frame extraction for motion capture animation. In: Nadia Magnenat-Thalmann. The visual computer, vol. 29, pp. 85–95. Springer, Berlin (2012)
13. Engelbrecht, A.: Particle swarm optimization. In: the Genetic and Evolutionary Computation Conference, pp. 65–91. ACM Press, New York (2015)
14. Anantathanavit, M., Munlin, M. A.: Radius particle swarm optimization. In: International Computer Science and Engineering Conference (ICSEC), pp. 126–130. IEEE Press, Nakorn Pathom (2013)
15. Khosravi, S., Akbarzadeh-T, M. R.: Real-parameter compact supervision for the particle swarm optimization (RCSPSO). In: Intelligent Systems (ICIS), pp. 1–6. IEEE Press, Iranian (2014)
16. Sarkar, S., Das, S.: A hybrid particle swarm with differential evolution operator approach (DEPSO) for linear array synthesis. In: Panigrahi, B.K., Das, S., Suganthan, P.N., Dash, S.S. (eds.) Swarm, Evolutionary, and Memetic Computing. LNCS, vol. 6466, pp. 416–423. Springer, Berlin (2010)
17. Lu, H.Y., Sriyanyong, P., Song, Y.H., Dillon, T.: Experimental study of a new hybrid PSO with mutation for economic dispatch with non-smooth cost function. Int. J. Electr. Power Energy Syst. **32**, 921–935 (2010)
18. Lu, G.L., Kudo, M., Toyama, J.: Robust human pose estimation from corrupted images with partial occlusions and noise pollutions. In: the IEEE International Conference on Granular Computing (GrC), pp. 433–438. IEEE Press, Kaohsiung (2011)
19. CMU Graphics Lab Motion Capture Database. Available online: http://mocap.cs.cmu.edu/

Living Labs for Human Motion Analysis and Synthesis in Shareconomy Model

Marek Kulbacki, Kamil Wereszczyński, Jakub Segen, Artur Bąk,
Marzena Wojciechowska and Jerzy Paweł Nowacki

Abstract The article presents the Living Labs for Human Motion Analysis and Synthesis (LivMASS), which is a shareconomy resource. It is intended to integrate local and geographically distributed stakeholders. The current resources and potential of LivMASS are described, including its laboratories, research activities, specific projects, collaborations and the collected datasets. A telepresence infrastructure is used to support research activities across Poland, within the Pionier Network and in Europe within the GEANT Network. The funding resources and planned funding strategies include targeted programs for research based on a LivMASS network, a communication campaign, and programs to foster networking and the inclusion of SMEs, universities, medical institutions and other prospective stakeholders.

Keywords Shareconomy · Living Labs · Motion Labs

1 Introduction

The Polish-Japanese Academy of Information Technology (PJAIT) conducts research and development in information and communications technology (ICT) specializing in Artificial Intelligence, Bioinformatics, Social Informatics, Computer Graphics, Image and Video Processing and Human Motion Analysis and Synthesis. During the last five years, in the Research and Development Center of PJAIT in Bytom, Poland concept evolved of motion analysis and synthesis labs in shareconomy model. In this period four research laboratories: Human Motion Lab (HML), Human Microexpression Lab (HMX), Human Seeing Lab (HSL) focused

M. Kulbacki (✉) · K. Wereszczyński · J. Segen · A. Bąk · M. Wojciechowska · J.P. Nowacki
Polish-Japanese Academy of Information Technology, Koszykowa 86,
02-008 Warszawa, Poland
e-mail: mk@pja.edu.pl

K. Wereszczyński
Silesian University of Technology, Institute of Informatics, Akademicka 16,
44-100 Gliwice, Poland

© Springer International Publishing Switzerland 2016
D. Król et al. (eds.), *Recent Developments in Intelligent Information
and Database Systems*, Studies in Computational Intelligence 642,
DOI 10.1007/978-3-319-31277-4_30

on video and image analysis, Human Facial Modeling Lab (HFML) have been created with the help of using EU and national grants. Also, two more are being formed: Human Dynamics and Multimodal Interaction Lab (HDMI), and Wearable Technology Lab (WTL) to study performing movement analysis in real-time [1, 2], with immediate feedback to both expert and patient.

This article provides a description of the PJAIT R&D Center and conducted there research activities. It begins, in Sect. 2 with an overview of the R&D Center's research areas, laboratories and resources and explains the shareconomy model of activity. Shareconomy is a peer-to-peer model including non-profits, government, corporations and individuals, with shared access to resources and services which optimizes their utilization. Section 3 continues with a more detailed description of selected labs, Sect. 4 lists collected datasets and Sect. 5 addresses specialized projects.

2 R&D Overview

Authors have been conducting research on motion analysis and synthesis since 1990 in US and started to develop Laboratories for Human Motion Analysis and Synthesis in 2008 in Poland. Currently laboratories are used for research carried out by interdisciplinary research teams including the experts in various domains of science, mostly in medicine, with special interests in usage of advanced computer measurement instruments supporting motor system therapy. The building of this advanced infrastructure fulfills a huge social demand especially the catastrophic statistical data on the health status of children and the elderly.

Only in the last years, 70 % of examined population in the age of 10–12 have suffered from posture deviation; every fourth person in the age of 70 and older is not able to walk without help through the distance longer than 500 m. The trouble with unassisted getting up or walking down the stairs (onto the first floor) is the problem concerning every twentieth person generally but every twelfth one in the age of over 60. About 45 % of disabilities are caused by injuries and diseases of the motoric system. The LivMASS lab supports research on nature and causes of disorders of human motor system and on diagnostic methods in collaboration within Active and Assisted Living programme.

2.1 R&D Center Resources

The resources of R&D Center (Fig. 1) contain: IT infrastructure and physical, human organizational and technological resources. The connection of these two compounds avails remote and efficient use in research and development process in shareconomy model. The R&D Center provides computation and storage servers with a wide-range of applicable software able to process data from laboratory

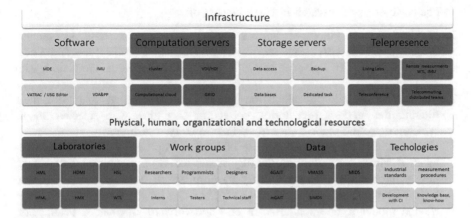

Fig. 1 Infrastructure of R&D center

equipment using mastered technologies, with cooperation and support of researchers, programmers, designers and technical staff. The use of telepresence makes possible virtual collaboration with external contributors. It creates wide and complete environment for:

1. Development of existing concepts and methods.
2. New ideas.
3. The increase of an existing technology level of the readiness.

2.2 Research Areas

Currently there are more than 10 large research projects in range of national and international research consortiums (more than 100 involved entities in total) including following areas of interest:

1. Biomechanics in medicine: diagnosis of disease entities and description of new symptoms, cooperation with medical universities and clinical hospitals.
2. Medicine: image processing and recognition for improved patient care—cooperation with clinical hospitals.
3. Public Security: research leading to creation of detection algorithms based on motion analysis, such as detection of situations preceding illegal and dangerous activities, illegal signals and intentions, detection of specific activities, person re-identification based on the gait as well as behavioral biometrics—cooperation with the police and municipal services.
4. Defense: increasing the local awareness of a soldier in unsafe areas using a range of solutions in pattern recognition and motion analysis domain—cooperation with the army.

The basic areas of research in HML consist of:

1. Multimodal acquisition of human motion in Vicon system: kinematics, dynamics, EMG, GRF, four streams High Definition (HD) video and Local Field Potential (LFP) planned after obtaining an access to the simulators with BMI.
2. Motion representation: recalculation between different formats of motion files, filtering the markers trajectories, correction of frames with selected bones, skeleton detection algorithms and skeleton quality measurement.
3. Human motion acquisition with use of Kinect: implementation and improvement of Microsoft algorithm, the algorithms based on various particle filters, pose sequence measurement quality by comparison with reference sequence from Vicon system.
4. Motion modeling and synthesis: reverse and forward kinematics, Featherston algorithms, the implementation and its extension. Motion analysis with use of motion descriptors: motion segmentation, similarity criteria for different motion representation like Dynamic Time Warping (DTW) and its alternatives, dimension reduction, discovering manifolds for various kinds of motion (dimensionality, mapping), clustering and classification of motion. Study of motion as an entity feature: extraction of entity descriptors.

2.3 Shareconomy Model of Activity

The Shareconomy is a socio-economic system created around the sharing of human and physical resources. It includes the shared creation, production, distribution, trade and consumption of goods and services by different people and organizations. The following are some of the common business models adapted in shareconomy [3]: Service Fee model, Freemium model, Re-Cycle and- Sell. The slogan "Shareconomy" describes an important social trend: changing attitudes of the desire to have for the need to share. Shareconomy has a decisive influence on the course of individual processes in the enterprise—even the increasingly widespread use of social networks by companies. The network becomes the main place of work and fulfillment of tasks by project groups—both within the organization and beyond. Business partners, consultants, suppliers and customers are engaged in various elements of our business becoming a part of a larger process. The boundaries between the different parts of the organization and between the company and its surroundings are blurring. Therefore, both managers and employees must change the philosophy of thinking and acting, and learn to share knowledge, contacts and resources. In order to reach a successful market leaders they have already deployed at each other modern tools that allows for fast and efficient sharing of knowledge. Blogs, Wikipedia, collaboration systems, software for surveying and other such applications will change significantly our working environment in the next coming years. The transformation of the entire system will communicate, how to make decisions, the role of management in the company's life, and even the expectations

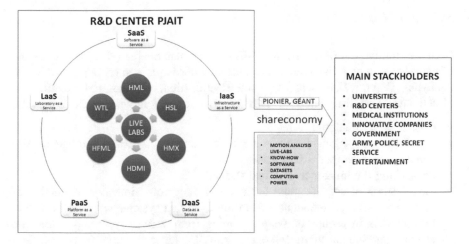

Fig. 2 Shareconomy model of activity

of candidates to potential employers. This philosophy of action is visible even in the fact that more and more industries establish close cooperation. Companies from different sectors are working together on finding new solutions and development of innovation [4]. Development and application providers of ICT services increasingly depend on the growth of industries that are major consumers of ICT solutions: the automotive industry and the energy and the health sector [5] (Fig. 2).

3 Laboratories

3.1 HML

The motion laboratory cooperates with medical institutions in the range of measurements and analysis of human motion in following aspects:

- a possibility of early diagnosis of diseases (and their degree of advancement) that are manifested by motility disorders (e.g. Parkinson's disease);
- an assessment of the process and efficiency of advanced treatments based on the comparative analysis of different methods;
- an assessment of rehabilitation procedures for diseases and injuries of the muscle-skeletal system;
- an assessment of training methods for sport institutions, also for the research related to the physical culture;
- a research investigating the occurrence level of various disorders of the muscle-skeletal system in scope of examined population, e.g. in order to estimate the risk of occurrence of selected disorders in examined population (e.g. children, miners etc.).

3.2 HMX and HSL

Both the Human Microexpression (HMX) and Human Seeing (HSL) labs carry out research in computer vision area. The current work is focused on Intelligent Video Analytics (IVA), which includes person identification, tracking, and identification of activities. The HMX and HSL labs are involved in following research aspects:

- multi-spectral imaging;
- algorithms for segmentation and classification with use of semi-supervised learning and orthogonal projection;
- mosaicking the images from video stream;
- identification of activities and detection of dangerous situations, prediction of intentions based on the signals from multi-camera systems with use of behavioral models in groups of people coming from social psychology, biometric technics and human figure inference from 3D video;
- understanding the semantics of video information with use of 3D pose estimation from video streams;
- reconstruction of 3D scene on the base of video stream;
- computational optimization of selected algorithms of OpenCV library with use of multi-core processors in Compute Unified Device Architecture (CUDA).

The HSL already contains the data from public area collected during more than 500 days which gives more than 5000 h of recordings of full-HD quality in Motion JPEG format with the frame rate from 19 to 23 frame per second (fpr). Such video dataset, which is still extending, is the powerful basis for the research based on Computer Vision. Some part of the dataset was labeled manually for around 3000 of distinguished activities. Taking a rough estimates into consideration, the unlabeled part of dataset contains around 3.5 millions of activities that can be potentially usable for training the IVA systems. Currently there are provided works on the automatic and semi-automatic labeling process with use of the software implemented by Lab's resources. The semi-automatic process leads to creation of the training datasets for particular activities (e.g. Walk, run, conversation) that can be used by the activity detection system. The learned activities are then detected and labeled by the fully automatic system. As a final result, this process leads to creation of large versatile activity dataset that is extendible in sense of size and content versatility. There are some additional reference information attached to the video data including calibration data and data related to different phase of detection. The dataset is powerful tool for the research in Computer Vision domain in both the raw representation (only the raw video data) as well as in the processed form. Due to its big size the dataset is unique worldwide as the currently exposed datasets have a size of maximally a few Gigabytes of low resolution video data containing recorded motion and camera calibration data. A the moment the dataset of HSL has a size of more than 20 TB of processed and labeled video data where it is increasing by another 150 GB everyday. Currently the integration of HSL with Motion Data

Framework (MDF) is being introduced in order to facilitate the access to dataset for other national and international institutions.

The HMX laboratory is also involved in the interdisciplinary research by linking the computer science and psychology domains. It concerns possibility of Computer Vision application in emotion detection area on the base of microexpression of face. Two aspects are taken into consideration: (1) detection of microexpression regarding the revealing of the emotion of person and (2) potential possibility of an emotion detection.

4 Datasets Collected in Bytom

In the range of realized research projects, the PJATK Academy department localized in Bytom created the following datasets as a result of projects and various experiments:

1. Application of video surveillance systems for activities detection and persons identification as well as detection of dangerous situations with use of biometric techniques and human figures inference in 3D from video streams:

   - video dataset [6] plus multilayered index describing the motion on different levels of abstraction that has the annotation layer which is simultaneously the dataset indexation in sense of an activity pattern recognition;
   - microexpression dataset.

2. System with modules library for advanced and interactive synthesis of human motion:

   - referential multimodal dataset from measurement of movement, walk, exercises of healthy people and the patients with classification according to disease entities (motion capture, 4 video streams, emg, grf);
   - dataset of realistic animation assets dedicated for computer games (motion capture and rigged meshes).

3. Designing of a quantitative motion measurements, rationalizing, on the base of multimodal motion measurement, subjective criteria UPDRS for improvement the diagnostics before and after DBS implantation for patients suffering from Parkinson's disease:

   - multimodal dataset consisting of selected motion tasks of patients with Parkinson's disease.

4. Automated Assessment of Joint Synovitis Activity from Medical Ultrasound and Power Doppler Examinations using Image Processing and Machine Learning Methods (Polish-Norwegian project) [7, 8]:

   - dataset of segmented and annotated ultrasound images of hand including the Power Doppler information.

5. New technologies for high-resolution acquisition and animation of face mimics (Innotech, Intech).
6. Human motion suit for human motion acquisition based on sensor—IMU with software for collection, visualization and analysis of motion data [9]:

   - human motion dataset in different technologies: referential motion capture, referential xsense, kinematics and dynamics.

Some of datasets have been collected and organized in the cloud based Human Motion Database [10]. A client application software Motion Data Editor (MDE) [11], enables visualization, and processing of data from any number of multimodal measurements simultaneously and synchronously supporting dozens of industrial formats used for medical data storage. The application has modular structure, that enables a simple extensions of its capabilities with a dedicated plugin system. MDE uses an intuitive data flow approach for data processing [12]. The developed system provides centralized storage for medical data and associated descriptions, allowing users to process data in the cloud model.

5 Specialized Projects

The unique attribute of the measurement systems at the Bytom center is their precision. Example are the high image resolutions including Full HD, and large number of sensors included in motion analysis, where even 500 modalities can be used at the same time. It gives the unique possibility of non-invasive diagnostics e.g. diagnosis of disease entities that cannot be precisely determined by regular observation without specialized measurements (e.g. joint implants).

The current infrastructure supports also the projects involving computer science and medicine domains like:

(a) system for optical imaging of tissues and supporting the diagnostics in selected cases of cancer. It is the research project where the goal is to create the information system for diagnosis and prognosis of selected cancer cases including the finding of optimal configuration and parametrization of image processing modules which allows for early detection of precancerous and cancer states, more precise determination of the border between healthy and pathological tissues, shortening the time of decision on treatment, assessment of treatment efficiency, non-invasive control and observation of patients after the treatment and early localization of cancer recurrence;

(b) testing and development of classification algorithms on the base of hyperspectral images in photodynamic diagnostics as well as diagnostics of fundus. The main research area is the analysis, implementation, numerical tests and development of algorithms for segmentation and classification of multispectral images dedicated to specialized application like endoscopy including photodynamic diagnostics, ophthalmology and fundus diagnostics. The final effect will be the algorithms of segmentation and classification of multispectral and hypespectral images whose usability will be confirmed by clinical examinations.

6 Practical Perspectives of HSL Laboratory

The Human Seeing Laboratory (HSL) is a unique world resource for data research and analysis in computer vision and artificial intelligence. Research and development in intelligent video surveillance at the HSL lab cover the following aspects:

- automatic identification of persons basing on gait;
- assessment of emotional state on the base of face mimics and gestures;
- identification of activities on the base of automatically learned patterns of group motion;
- automatic detection of untypical and dangerous activities of persona or group of people;
- identification of activities on the base of gestures and non-verbal communication;
- detection of situations preceding the illegal and dangerous activities;
- signaling the illegal intentions on the base of analysis of trajectories representing people;
- assessment of forces strength between people;
- automatic detection and tracking the human figures in the image sequences from multi-camera systems in scope of supervised areas;
- searching the persons and specific behaviors in the video archives on the base of learned pattern of activities.

The context of the deployment of research results is also significant. Already, public order agencies are interested in application of developed technologies, namely municipal police and regular police. For this reason the ongoing works consider following applications:

- creation of evidences on the base of biometric analysis;
- registering the motion in crime scene;
- searching lost children in areas of intensive movements and mass events;
- searching the persons registered materials with use of individual characteristic of motion;
- identification of incidents and automatic photos delivery presenting the their participants to the police patrols located near the place of incident;
- support in video monitoring the persons moving in the field of view of many cameras by automatic sliding the window in the virtual picture collected from all cameras;
- alerting the surveillance service on detection of critical situations attaching the selected frames and video sequences for verification;

- automatic selection of information about persons being monitored and sending the following information to the intervening staff: clear images, video from cameras, localization and trajectories for displaying on the mobile screen;
- *Seeing through walls* is the application of technology where the picture of monitored persons and their background look like it would be seen from the perspective of watcher after removing all visual obstacles like walls.

7 Conclusions

The Living Labs for Human Motion Analysis and Synthesis is a shareconomy model integrating local and geographically distributed stakeholders, enabling collaborative research and innovations in human motion analysis and synthesis. All resources from Research and Development Center PJAIT in Bytom, Poland are elements of an ecosystem capable of supporting advanced research activities remotely and efficient utilization of research and development resources in shareconomy model. For telepresence we use communication system with unified communication and voice-over-IP solutions providing an optimal user experience, regardless of location or device and reducing interworking complexity.

Acknowledgments This work has been supported by the National Centre for Research and Development (project POIG.02.03.00-24-149/13 "Interdisciplinary laboratories for human motion analysis and synthesis in shareconomy model").

References

1. Geijtenbeek, T., Steenbrink, F., Otten, B., Even-Zohar, O.: D-flow: immersive virtual reality and real-time feedback for rehabilitation. In: Proceedings of the 10th International Conference on Virtual Reality Continuum and Its Applications in Industry (VRCAI 11). ACM, New York, pp. 201–208 (2011)
2. van den Bogert, A.J., Geijtenbeek, T., Even-Zohar, O., Steenbrink, F., Hardin, E.C.: A real-time system for biomechanical analysis of human movement and muscle function. Med. Biol. Eng. Comput. **51**(10), 1069–1077 (2013). (Springer, Berlin, Heidelberg)
3. http://www.entrepreneurial-insights.com/introduction-to-sharing-economy/
4. Hamari, J., Sjöklint, M., Ukkonen, A.: The sharing economy: why people participate in collaborative consumption. J. Assoc. Inf. Sci. Technol. (2015)
5. Schiederig, T., Herstatt, C.: Shareconomy—Performance-oriented Systems as a Strategy. Schiederig/Herstatt. Working Paper No. 77, pp. 1–13 (2014)
6. Kulbacki, M., Segen, J., Wereszczynski, K., Gudys, A.: VMASS: massive dataset of multi-camera video for learning, classification and recognition of human actions. ACIIDS **2014**, 565–574 (2014)
7. Automated Assessment of Joint Synovitis Activity from Medical Ultrasound and Power Doppler Examinations using Image Processing and Machine Learning Methods http://eeagrants.org/project-portal/project/PL12-0015

8. Segen, J., Kulbacki, M., Wereszczynski, K.: Registration of ultrasound images for automated assessment of synovitis activity. ACIIDS **2015**, 307–316 (2015)
9. Kulbacki, M., Koteras, R., Szczesna, A., Daniec, K., Bieda, R., Slupik, J., Segen, J., Nawrat, A., Polanski, A., Wojciechowski, K.: Scalable, Wearable, Unobtrusive Sensor Network for Multimodal Human Monitoring with Distributed Control. In: Lackovic, I., Vasic, D. (eds.) IFMBE Proceedings, pp. 914–917 (2015)
10. Filipowicz, W., Habela, P., Kaczmarski, K., Kulbacki, M.: A generic approach to design and querying of multi-purpose human motion database. ICCVG **2010**, 105–113 (2010)
11. Kulbacki, M., Janiak, M., Kniec, W.: Motion data editor software architecture oriented on efficient and general purpose data analysis. ACIIDS **2014**, 545–554 (2014)
12. Janiak, M., Kulbacki, M., Kniec, W., Nowacki, J.P., Drabik, A.: Data flow processing framework for multimodal data environment software. New Trends Intell. Inf. Database Syst. **2015**, 353–362 (2015)

Data-Driven Complex Motion Design for Humanoid Robots

Xiaojie Zheng, Pengfei Yi and Qiang Zhang

Abstract Humanoid robot motion plan based on the similarity of human motion is a hot topic in recent years. For outer space or some other disaster scene, which is not safe and workable for humans, how to autonomously, quickly and accurately complete the task is extremely important for the humanoid robot. This paper presents an analysis method based on data-driven for complex choreography of humanoid robots. Firstly we convert the BVH motion capture data to joint angle trajectory of the humanoid robot. Secondly we optimize process to ensure the balance, so that robots can reproduce human motion. Then we can get a motion diagram of several motion sequences, through Tarjan algorithm by using the similarity between frames to reduce data redundancy. Finally the shortest path between the frames is obtained by Floyd algorithm, namely a sequence between arbitrary frames, driving robot to realize different trajectories rapidly. Experiment verified the feasibility.

Keywords Data-driven · BVH · Motion synthesis · Humanoid robots

1 Introduction

Due to its proportion of joint structure is similar to humans, having strong ability to adapt to the environment, and being flexible and closed to human characteristics, therefore, through real-time interaction to generate the stable, harmonious and natural movement, coordinating or replacing human work has become the research focus in a people-oriented working environment today [1]. However, for some complex motion, solving trajectory by conventional analytical motion equation is difficult to achieve. Its motion equation is complicated and difficult to establish, or even has no solution and operability poorly. Accordingly, scholars have put forward

X. Zheng · P. Yi · Q. Zhang (✉)
Key Laboratory of Advanced Design and Intelligent Computing (Dalian University),
Ministry of Education, 116622 Dalian, China
e-mail: zhangq26@126.com

© Springer International Publishing Switzerland 2016
D. Król et al. (eds.), *Recent Developments in Intelligent Information
and Database Systems*, Studies in Computational Intelligence 642,
DOI 10.1007/978-3-319-31277-4_31

the humanoid robot motion plan thought based on the similarity of human motion [2], making humanoid robot reproduce human motion visually and effectively.

The motion capture technology based on sensor after decades of development, has been widely applied to animation, medical, virtual reality, ergonomics and many other high-tech fields. In the field of ergonomics, motion capture technology is used for robot motion planning, real-time remote control, the implementation and control of complex action, reflecting its high application value [3]. So far, most of the work for data-driven humanoid robot motion is done by using optical-based motion capture system to achieve [4–6]. Yamane et al. [7] make the robot to imitate dance motion by obtaining the marker position of tracking system, that they extend it to the size of the human form, and solve the joint angles trajectory by inverse kinematics. Teachasrisaksakul et al. [8] proposed a framework based BSN, which optimize joint angle trajectory of data conversion to meet the robot's mechanical constraints and balance, and reproduce lifelike human motion. Seekircher et al. [9] use Microsoft's Kinect to collect data, and separately use three modern optimization algorithms (CMA-ES, xnes, PSO) to make movement balance, and map to the humanoid robot joint angle trajectory to complete a stable movement for comparing their advantages and disadvantages. Koo et al. [10] also use Kinect to collect motion capture data, then map joint angle data of arm to achieve arms control based on gestures robots intuitively and visually. Ho et al. [11] proposed a spatial relationship-based approach to the synthesize motion of control humanoid robot. They will refer the motion and highly constrained environment as an input to produce a stable movement, so that humanoid robot can automatically adapt to the environment changes. Luo et al. [12] make the original trajectory and the trajectory of the manipulator to minimize differences in time and geometry between the motion of writing "hello", by decoupling the geometry and time optimization method to select good initial trajectory. Gärtner et al. [13] propose the method of imitation gestures. Considering the physical limit of the robot and the end executor's biggest position, they utilize nonlinear optimization to maximize human and robot joint trajectory similarity.

Like Hawk, Eagle and Vicon [14] and other motion capture equipment influences by space and the environment etc. and when hundreds of human movement joints data is mapped to only dozens of joint robot joint angle data, it will lead to slip, foot strike through ground and posture deformation because of the huge amount and redundancy for motion capture data [15]. So whether it is based on motion capture system or various optimization algorithms for humanoid robot motion design, which basically focus on how to correctly match in the dynamic and kinematic, few considers the preliminary analysis and processing work for motion capture data.

To solve above problems, we propose that experiment will use the existing human action data instead of these devices to improve their adaptability to the environment. We also use motion synthesis method to increase the diversity of action to adapt to complex environments. The correctness of the method is verified by experiment.

2 Methodology

This article uses the BVH file at CMU database which records the completed skeleton model information, sampling frequency and the detailed joint action information of each frame. The file data is a total of two parts. The first part describes the joints relationship between the father and son and the offset of joint nodes which form a joint node tree, as well as the way of Euler rotation. The other part describes the sampling frames, frame frequency and the Euler rotation angles.

NAO H25 has 25 degrees of freedom. Its movement based on generalized inverse kinematics can handle Descartes Coordinates, joint control, balance, redundancy and task priority [16]. And the Roll, Pitch, Yaw angles describe every robot posture.

2.1 Data Conversion

Because the humanoid robot with rigid structure is similar to that of human beings, and the NAO robot inputs each joint angle value, so we only consider the rotation angle of each joint point and we can ignore the offset. Then compared with the robot model and the simplified human skeleton model, human skeleton has three degrees of freedom for each joint, and the robot is only 1 to 2, meantime the BVH joint angle of each frame is "child node" as opposed to "parent node" local coordinate system of the Euler rotation angles, but joint angles used in the robot motion control are in the Cartesian coordinate system which turn under its own coordinate. So the direct mapping is obviously not possible. Therefore, we convert the BVH data by the coordinate system to the robot coordinate joint node, and then we solve the Euler angles are rotated around its own coordinate system by inverse kinematics. Then we drive the robot corresponding articulation motion. Given the limitations of robot joints physical structure, the robot is not at the upper end that the human body joint can reach in this experiment. If the range is beyond the scope of robot joint angle, we set the joint angle as maximum.

We can get the offset between the joints from the BVH data, which is in the zero state. It is a vector that is in the local coordinate system of their "parent" node. According to the representation method of Euler Angle, we can get the equation of the rotation matrix:

$$R = R_z R_x R_y \tag{1}$$

where R_z is rotation around the Z axis, R_x is rotation around the X axis, R_y is rotation around the Y axis, α, β, γ are the Euler rotation angle that nodes turn around its own coordinate system X, Y, Z in the Cartesian coordinates.

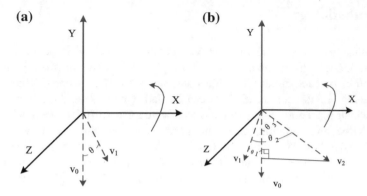

Fig. 1 Joint coordinate system. **a** One DOF. **b** Two DOFs

Through the rotation matrix, we can get a coordinate in a frame, which is "child" under the "parent" local coordinate system. Then we can get the coordinate of humanoid robot in the same frame through coordinate system transformation.

Since the robot joints only have 1 to 2 DOFs, and the coordinate system will be accompanied by its own rotation in rotation around an axis, so the order of rotation should be considered when the robot joint has two degrees of freedom. Here we fall into two cases on Fig. 1. We set zero state unit vector of joint nodes is $(0, -1, 0)$ under its "parent" local coordinate system (since the shoulder is $(0, 0, 1)$. According to the right-hand Cartesian coordinate system, we simply make a compensation angle $\pm\pi/2$ on the ShoulderPitch). For the case of one DOF, $\theta = \mathrm{atan}(z/y)$, for the case of two DOFs, $\theta_1 = \theta_3 = \mathrm{acos}\left(y \big/ \sqrt{(x^2+y^2+z^2)}\right), \quad \theta_2 = \mathrm{atan}(x/z).$

2.2 Motion Synthesis

Because of the robot joint motor mechanical limit and the redundancy of motion capture date, we process it inefficiently. So we optimize the motion capture data, and synthesize some motion for driving robot.

Since the Euclidean distance is easy to achieve and it has a high efficiency on processing date, it became the most common method for calculating the distance. We also use Euclidean distance to calculate the distance between frames. The smaller value indicates that the higher similarity and the larger value indicates that the lower similarity. When the value is beyond a certain threshold, we consider that they are not similar. Any similarity value of two frames may constitute a motion graph, which is a non-connected graph.

Tarjan algorithm is a search algorithm for graph that is based on the depth-first. Each strongly connected component of the search tree is a subtree. We calculated the graph using its maximum fully connected subgraph. In this case, the redundant

Fig. 2 Flow chart of our method

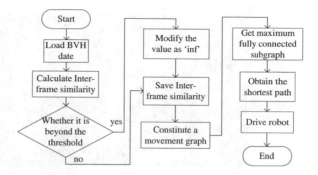

data have been further optimized, and this graph is connected, that is, any inter-frame can compose a set of actions.

Floyd algorithm is suitable for APSP (All Pairs Shortest Paths), that is a dynamic programming algorithm. Since we use the distance indicates the similarity of the frames, so we use Floyd algorithm solving the shortest path of any frames in a connected graph. These paths are stored in the robot memory, then we give a state value of the beginning and end of motion. In this way, the robot will find the most similar frame, and then find the corresponding motion sequence to complete the motion. The process is shown on Fig. 2.

3 Results

Experimental hardware platform is a NAOqi H25 robot that has 25 degrees of freedom. The robot is in communication with a Linux desktop. It is based on NAOqi SDK, using Python programming language and being data-driven.

Walk is a complex action, and has distinct characteristics, which is a challenging task. We adopt 07_01.bvh, 107_01.bvh, 107_03.bvh in the CMU database that have three different ways of forward motion sequence as an input, and calculate the degree of similarity between any frames. Then we can get the motion graph of motion through our method in 2.393 s. By our method we find the shortest path between any frames. We give a state value of the beginning and end of motion. Then we find out the similar key-frame, and search the shortest path to drive robot in 0.053 s. It is not hard to discover that actions of humanoid robot and human are similar in Fig. 3, which is a certain stance contrast figure of the synthetic action of the human body skeleton model correspond to the robot motion.

Figure 4 shows the joint angle trajectory contrast figures of the human body skeleton model and the humanoid robot on a certain degree of freedom of the limbs, and the in-out error of the limbs. The horizontal axis represents the number of frames, and the vertical axis represents the joint angle. So we can find a similar joint angle trajectory in Fig. 4a, b. Due to joint motor mechanical limit, the value of the joint angle is slightly different at a certain frame. There is some minor angle in-out

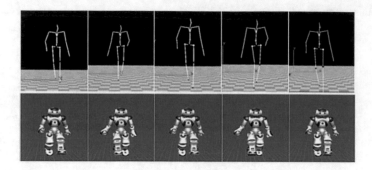

Fig. 3 The chart accordingly of human BVH data key posture and the robot posture

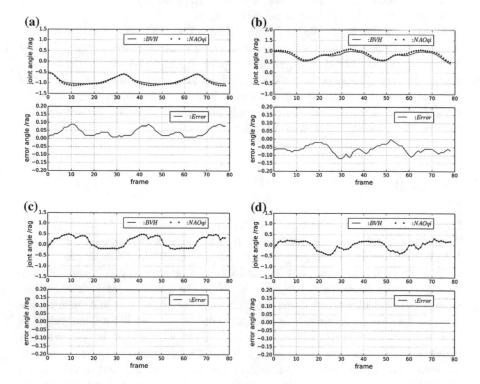

Fig. 4 Joint angle trajectory contrast figure of limbs and the corresponding in-out error. **a** The DOF of LElbowYaw. **b** The DOF of RElbowYaw. **c** The DOF of LHipRoll. **d** The DOF of RHipRoll

error as shown on Fig. 4a, b. Since the lower limb plays in preference to the upper limb, joint angle trajectory of the lower limbs remains the same to ensure its consistency in Fig. 4c, d. We can find that the error of lower limbs is almost negligible. From Fig. 4, we see that similar joint angle trajectory can explain a high similarity on the synthesis motion and human motion. We verify the feasibility of

Table 1 The comparison of different methods

Type of method	Device dependence	Difficulty of applying to whole body	Fidelity of motion	Environmental adaptability	Speed performance of method
Similarity-based [5, 7–10]	High	Low	High	Low	Low
Optimizing-based [11, 12]	Low	High	Low	High	Low
Our method	Low	Low	High	High	High

converting human data to the robot model and effectiveness of our proposed method.

Compared with the method and theory that literature mentioned, our advantages are presented in Table 1.

1. The motion design methods based on the similarity of action (such as literature [5, 7–10]) used Kinect motion capture device obtain the data source. The use of these devices has a higher request for site conditions, and it does not apply to outdoor complex environment or outer space. We directly use the data of motion capture database as an input to drive the robot. It can overcome the limitations of environment. And motion of the data-driven directly has a high fidelity and a fast response.

2. The motion design methods based on optimizing of motion synthesis (such as literature [11, 12]) optimized joint angle trajectory based on space or decoupling geometric methods. They made synthesis movements of certain limb, while optimizing the joint angle trajectory is slow and their applicability is not strong. And the motion has a low fidelity about human. We use the similarity between frames synthesize a new action. It can be applied to the body movement, and the synthesis of motion has a strong adaptability.

4 Discussion

This paper discusses the use of BVH data based on CMU database, and performs man–machine mapping by inverse kinematics to achieve the conversion between human data and machine joint angle data. We verify its feasibility. At this time, we propose a new motion design method based on the similarity of human motion. By this method, several BVH motion sequences constitute a motion graph, for solving a variety of different paths. These paths are stored in memory to drive the robot, which operate quickly and accurately relative to the wearable device. It is also able to perform some tasks such as outer space or high-risk missions.

However, the limitations of this approach is that motion must be already in the motion database, and you need to process data in advance, in other words, the scope of task is narrow. If the data is large, it will has a greater long computation time and a higher price for memory. In future, our experiments will further consider the balance of the synthesis motion, as well as the environmental adaptability. Simultaneously we will also try some complicated motion designs.

Acknowledgement This work is supported by the National Natural Science Foundation of China (No.61370141, 61300015), the Program for Liaoning Innovative Research Team in University (Nos. LT2015002), the Program for Science and Technology Research in New Jinzhou District (No. KJCX-ZTPY-2014-0012).

References

1. Dalibard, S., Nakhaei, A., Lamiraux, F., Laumond, J.P.: Whole-body task planning for a humanoid robot: a way to integrate collision avoidance. In: Humanoids 9th IEEE-RAS International Conference on Humanoid Robots, pp. 355–360. IEEE Press, New York (2009)
2. Kofinas, N., Orfanoudakis, E., Lagoudakis, M.G.: Complete analytical inverse kinematics for NAO. In: Autonomous Robot Systems 13th International Conference on Autonomous Robot Systems, pp. 1–6. IEEE Press, New York (2013)
3. Kofinas, N., Orfanoudakis, E., Lagoudakis, M.G.: Complete analytical forward and inverse kinematics for the NAO humanoid robot. J. Intell. Rob. Syst. **77**, 251–264 (2015)
4. Nakaoka, S., Nakazawa, A., Yokoi, K., Ikeuchi, K.: Leg motion primitives for a dancing humanoid robot. In: IEEE International Conference on Robotics and Automation, pp. 610–615. IEEE Press, New York (2004)
5. Kim, S., Kim, C., You, B., Oh, S.: Stable whole-body motion generation for humanoid robots to imitate human motions. In: IEEE/RSJ International Conference on Intelligent Robots and Systems, pp. 2518–2524. IEEE Press, New York (2009)
6. Boutin, L., Eon, A., Zeghloul, S., Lacouture, P.: From human motion capture to humanoid locomotion imitation application to the robots HRP-2 and HOAP-3. Robotica **29**(02), 325–334 (2011)
7. Yamane, K., Hodgins, J.: Simultaneous tracking and balancing of humanoid robots for imitating human motion capture data. In: IEEE/RSJ International Conference on Intelligent Robots and Systems, pp. 2510–2517. IEEE Press, New York (2009)
8. Teachasrisaksakul, K., Zhang, Z., Yang, G.Z.: The use of BSN for whole body motion training for a humanoid robot. In: 11th International Conference on Wearable and Implantable Body Sensor Networks (BSN), pp. 45–51. IEEE Press, New York (2014)
9. Seekircher, A., Stoecker, J., Abeyruwan, S., Visser, U.: Motion capture and contemporary optimization algorithms for robust and stable motions on simulated biped robots. In: Chen, X. P., Stone, P., Sucar, L.E. (eds.) RoboCup 2012. LNCS, vol. 7500, pp. 213–224. Springer, Heidelberg (2013)
10. Koo, K.M., Kim, Y.J., Seo, Y.H.: Motion capture based dual arm control of a humanoid robot using kinect. In: Park, J., Barolli, L., Xhafa, f. (eds.) Information Technology Convergence 2013. LNEE, vol. 253, pp. 913–921. Springer, Netherlands (2013)
11. Ho, E.S., Shum, H.P.: Motion adaptation for humanoid robots in constrained environments. In: IEEE International Conference on Robotics and Automation (ICRA), pp. 3813–3818. IEEE Press, New York (2013)
12. Luo, J., Hauser, K.: Interactive generation of dynamically feasible robot trajectories from sketches using temporal mimicking. In: IEEE International Conference on Robotics and Automation (ICRA), pp. 3665–3670. IEEE Press, New York (2012)

13. Gärtner, S., Do, M., Asfour, T., Dillmann, R., Simonidis, C., Seemann, W.: Generation of human-like motion for humanoid robots based on marker-based motion capture data. In: 41st International Symposium on and 2010 6th German Conference on Robotics (ROBOTIK), pp. 1–8. VDE Press (2010)
14. VICON, http://www.vicon.com/
15. Almetwally, I., Mallem, M.: Real-time tele-operation and tele-walking of humanoid robot Nao using Kinect depth camera. In: 10th IEEE International Conference on Networking, Sensing and Control (ICNSC), pp. 463–466. IEEE Press, New York (2013)
16. Nao documentation, http://www.aldebaran-robotics.com/documentation

Hand Detection and Gesture Recognition Using Symmetric Patterns

Hassan Mashad Nemati, Yuantao Fan
and Fernando Alonso-Fernandez

Abstract Hand detection and gesture recognition is one of the challenging issues in human-robot interaction. In this paper we proposed a novel method to detect human hands and recognize gestures from video stream by utilizing a family of symmetric patterns: log-spiral codes. In this case, several log-family spirals mounted on a hand glove were extracted and utilized for positioning the palm and fingers. The proposed method can be applied in real time and even on a low quality camera stream. The experiments are implemented in different conditions to evaluate the illumination, scale, and rotation invariance of the proposed method. The results show that using the proposed technique we can have a precise and reliable detection and tracking of the hand and fingers with accuracy about 98 %.

Keywords Hand detection · Gesture recognition · Symmetric patterns · Log-spiral codes · Human-robot interaction

1 Introduction

Hand detection and gesture recognition is a promising application for human-machine interaction purposes. In this paper we proposed a method using a family of symmetric patterns as features for detecting human hand as well as recognizing several gestures. Utilizing features for identifying position of the object is well-established and widely used paradigm in computer vision and pattern recognition. A number of popular features has been proposed and developed in previous

H.M. Nemati (✉) · Y. Fan · F. Alonso-Fernandez
Center for Applied Intelligent Systems Research (CAISR), Halmstad University,
Halmstad, Sweden
e-mail: Hassan.Nemati@hh.se

Y. Fan
e-mail: Yuantao.Fan@hh.se

F. Alonso-Fernandez
e-mail: feralo@hh.se

© Springer International Publishing Switzerland 2016
D. Król et al. (eds.), *Recent Developments in Intelligent Information
and Database Systems*, Studies in Computational Intelligence 642,
DOI 10.1007/978-3-319-31277-4_32

decades for general purpose tasks such as Harris detector [1] and local scale-invariant features [2], there can be used for hand detection and feature recognition as well. Several other research projects e.g. [3, 4] propose to use human skin as visual features for hand detection.

A hand gesture detector can assist human interaction with machines under various circumstances. For example: (a) operators in factory that has constraints in using other ways of interfacing the machine; (b) disabled or elderly people can interact with machines using natural and universal patterns of motion; (c) for authentication purposes, a person just need to pose the hand in front of a camera and make a series of gestures; (d) provide various ways for interfacing with embedded devices with only camera sensors etc.

Some of the applications such as industrial machine operating require precision and high reliability in tracking the hand. Therefore, unique, robust features must be introduced for this purpose. The work [5] suggests the use of symmetric patterns and corresponding filters for pattern recognition which is later explored by Karlsson and Bigun in [6], where it is described a method to use log-spiral codes as feature for visual positioning. Log-spirals are invariant to rotation, scale and drastic changes in intensity. Therefore they are suitable for the type of application that need precision and high reliability. In this paper, we proposed to attach spirals on hand gloves and exploit them as features for both hand detection and gesture recognition.

The main contributions of the paper are the use of Log-spiral codes for hand detection and simple gesture recognition, the use of a spiral pair to make the method more adaptive and robust to changes, the implementation of a way to reject false detection of spirals as well as the analysis of testing the performance of the method under different conditions.

2 Related Work

State-of-art approaches for hand detection essentially propose segmentation methods based on features such as skin color, texture, shape and edge detection, e.g. [3] utilize features including shape, context and skin color as multiple proposals for detecting hand and approach described in [4] proposed a way of using color based method to detect hand and fingers for user interaction purpose. Several color-based methods employ artificial markers as features, e.g. [7] utilize a multi-colored glove for hand tracking and method described in [8] uses several types of color markers, attached to fingers, for hand gesture interaction with tabletop. Some other method [9] also take information of background into account for detection. The work in [10] described a method based on Viola-Jones [11] object detection frame work, utilize a hierarchical detection for estimating the positions.

Gesture recognition method were based on the feature extracted from the detection method. One common approach is to use machine learning method.

Works [12–14] utilize Hidden Markov Model to exploit temporal information to recognize hand gesture. Works [15, 16] use Multilayer Perceptron and [17, 18] employ Support Vector Machine to recognize gestures. Those method are conducted in supervised manner but unsupervised methods can be exploited in some cases.

3 Method

This section describes the algorithm employed for hand detection and gesture recognition. We use symmetry features for such tasks. Symmetry features enable the description of symmetric patterns such as lines, circles, parabolas, and so on (Fig. 1). These features are extracted via symmetry filters, Eq. 1, which output how much of a certain symmetry exist in a local image neighborhood [19, 20]. Concretely, we employ the log-spiral family (first family of Fig. 1). The exact patterns employed are shown in Fig. 2, which are attached to a glove as depicted in Fig. 3.

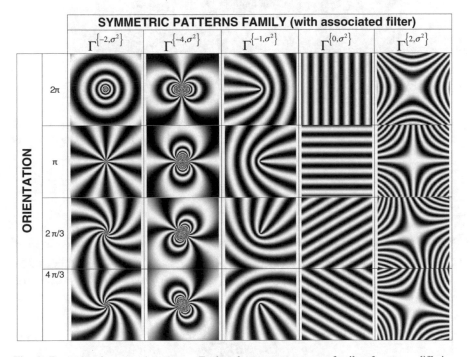

Fig. 1 Example of symmetric patterns. Each *column* represents one family of patterns differing only by their orientation (given in column 2). The associated filter suitable to detect each family (Eq. 2) is also indicated in *row 2*

Fig. 2 Spirals employed in our experiments

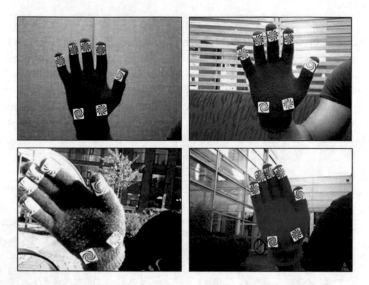

Fig. 3 Example of different test conditions: indoor with simple background (*top left*), indoor with complex background (*top right*), in/outdoor with high illumination (*bottom left*), and outdoor (*bottom right*)

3.1 Symmetry Filters

Symmetry filters are a family of filters computed from symmetry derivatives of Gaussians. The nth symmetry derivative of a Gaussian, $\Gamma^{\{n,\sigma^2\}}$, is obtained by applying the partial derivative operator $D_x + iD_y = (\partial/\partial x) + i(\partial/\partial y)$, called 1st symmetry derivative, to a Gaussian:

$$\Gamma^{\{n,\sigma^2\}} = \begin{cases} \left(D_x + iD_y\right)^n g(x,y) & (n \geq 0) \\ \left(D_x - iD_y\right)^{|n|} g(x,y) & (n < 0) \end{cases} \tag{1}$$

Since $D_x + iD_y$ and $\left(-\frac{1}{\sigma^2}\right)(x+iy)$ behave identically when acting on a Gaussian [19, 20], Eq. 1 can be rewritten as

$$\Gamma^{\{n,\sigma^2\}} = \begin{cases} \left(-\frac{1}{\sigma^2}\right)^n (x+iy)^n g(x,y) & (n \geq 0) \\ \left(-\frac{1}{\sigma^2}\right)^{|n|} (x-iy)^{|n|} g(x,y) & (n < 0) \end{cases} \tag{2}$$

The interest is that these symmetry derivatives of Gaussians are able to detect patterns as those of Fig. 1 through the computation of the second order complex moment of the power spectrum via $I_{20} = \left\langle \Gamma^{\{n,\sigma_2^2\}}, h \right\rangle$, where h is the complex-valued orientation tensor field given by $h = \left\langle \Gamma^{\{1,\sigma_1^2\}}, f \right\rangle^2$ and f is the image under analysis [20]. Parameter σ_1 defines the size of the derivation filters used in the computation of h, whereas σ_2, used in the computation of I_{20}, defines the size extension of the sought pattern.

For each family of symmetric patterns, there is a symmetry filter $\Gamma^{\{n,\sigma^2\}}$ (indexed by n) suitable to detect the whole family [21]. Figure 1 indicates the filters that are used to detect each family. The local maxima in $|I_{20}|$ gives the location, whereas the argument of I_{20} at maxima locations gives the group orientation of the detected pattern (except for the first family in Fig. 1, $n = -2$, where the 'orientation' represents the chirality of the curves). Therefore, I_{20} encodes how much of a certain type of symmetry exists in a local neighborhood of the image f. In addition, a single symmetry filter $\Gamma^{\{n,\sigma^2\}}$ is used for the recognition of the entire family of patterns, regardless of their orientation (or chirality). Symmetry filters have been successfully applied to a wide range of detection tasks such as cross-markers in vehicle crash tests [22], core-points and minutiae in fingerprints [23, 24], or iris boundaries [25]. The beauty of this method is even more emphasized by the fact that I_{20} is computed by filtering in Cartesian coordinates without the need of transformation to curvilinear coordinates (which is implicitly encoded in the filter).

3.2 Hand Detection Using Log-Spirals

In this work, we employ the set of spirals of order $n = -2$. These are called log-spirals since they are generated by a $\log(z)$ coordinate transformation of 2D-sinusoids $\exp\left(i\left(\omega_x x + \omega_y y\right)\right)$ [6]. In this case, as introduced above, 'orientation' describes the twist-angle or chirality of the spirals (left/right handedness). Due to the separability property of 2D Gaussian, the detection filter can be re-written as:

$$\Gamma^{\{-2,\sigma^2\}} = \left(-\frac{1}{\sigma^2}\right)^2 (x-iy)^2 g(x)g(y) \tag{3}$$

so the 2D convolutions can be computed by 1D convolutions. Moreover, in computing h, 1D convolutions can be used as well due to the same property. This results a considerable higher speed, allowing real-time detection. Another advantage of the proposed system is that it does not need training.

In our experiments, we employ 4 different spirals with twist angles $\frac{\pi}{5}, \frac{3\pi}{5}, \frac{7\pi}{5}, \frac{9\pi}{5}$ as shown in Fig. 2. They are generated according to the description given in [6], with $L_0 = 12$ legs. These spirals are mounted at the following positions on a glove, as shown in Fig. 3:

- One unique spiral on bottom left of the palm
- One unique spiral on bottom right of the palm
- One unique spiral on thumb
- Four same spirals on point finger, middle finger, ring finger, and small finger

Therefore the glove contains 7 spirals which define the hand border. The relative position of these points, the number of detected or not detected points, and the angle of the two spiral at the bottom of the palm are used for hand detection and gesture recognition.

4   Experimental Results

In this work we use an ordinary video webcam with resolution of 640 * 480 and maximum frame rate of 24. The tests are performed in an offline situation, where 17 different videos containing 22,168 frames are recorded in different scenarios and then the proposed method is applied on these recorded videos. The calculation time for each frame is about 0.056 s in a Macbook that runs OSX 10.9.5 and has an Intel core i7 2.3 Ghz processor, which means the method can be applied in real time videos. The distance between the hand and the camera at the initial frame is between 10 and 20 cm; this distance may change after the first frame.

4.1   Hand Detection

The performance of the proposed method is evaluated in four different conditions: indoor with simple background, indoor with complex background, indoor or outdoor with high illumination, and outdoor. An example of these conditions are shown in Fig. 3. In addition to the environmental condition and effects of illumination, we test the detection technique while the hand is rotating into left or right and also moving forward or backward to the camera (change of scale). These experiments are implemented to prove that the proposed technique is illumination, scale, and rotation invariant.

In order to have a scale invariant detection we need to adjust the size of the filter based on the distance between hand and camera. In this case, when the hand become closer to the camera the size of the filter should increase and when it becomes farther the filter size should decrease. This can be done by defining the size of the filter not as a fixed value but as a function of distance between hand and camera. To be able to detect this distance, the position of the two spirals at the bottom palm of the hand is considered as the ground truth. Therefore, when in a video frame this length become smaller than a certain value means the hand become farther from the camera and when the length is greater than a certain value means the hand become closer. Using this analogy we are able to modify and update the size of filter depending on the distance between hand and camera and improve the spiral detection algorithm.

In order to evaluate the performance of the proposed method, we used two ways to measure error. In the first way, we count the number of undetected spirals as false negative and then sum the number of these false negatives in all the frames. The expected number of spiral at these experiments is 7, i.e. all the spirals in all the frames are visible. Consequently, from these 7 spiral if we can detect any of them it will be considered as true positives. The expected number of true positives can be calculated by multiplying the number of frames by 7 (the number of spirals). Therefore, the ratio of the sum of the false negatives over the expected number of true positives gives us the false negative ratio error in the proposed method.

In the second way, we count the number of imperfect frames. In this case, the imperfect frame is defined as the frame that there is exist at least one spiral which is not detected by the method. The ratio of the sum of the imperfect frames over the total number of frames is used as the second way to estimate the performance of the proposed method.

The result of the performance evaluation is shown in Table 1. As can be seen in this table, the ratio of error in detecting the spirals in all the tests is less than 2 %, i.e. the proposed technique can detect the spirals (the border of the hand) with accuracy about 98 %. Furthermore, in more than 92 % of all the test frames the proposed techniques can detect all the spirals without any imperfect frame.

Note that in the high illumination test, when we had direct sun light over the spirals (see Fig. 4), the method can not detect all the spirals correctly. In fact, strong sunlight on printed spirals causes reflection with high intensity and consequently increases the imperfect frame ratio error up to 0.082 (see Table 1).

A sample scan of the detected spirals from different experiments are shown in Fig. 5. In these figures the frame number, filter size, angle of the hand, estimated

Table 1 The result of the performance evaluation—the ratio of error in detecting the spirals in all the tests is less than 2 %, the imperfect frame ratio error in all the tests is less than 8.2 %

Test environment	Number of frames	False negative		Imperfect frame	
		Ratio	Percent (%)	Ratio	Percent (%)
Indoor simple background	10,801	0.0096	~1	0.0542	~5.5
Indoor complex background	3917	0.0151	~1.5	0.0638	~6.5
In/outdoor high illumination	2432	0.0117	~1.2	0.0818	~8.2
Outdoor	5018	0.0060	~1	0.0296	~3

Fig. 4 Effect of direct sun light over the spirals

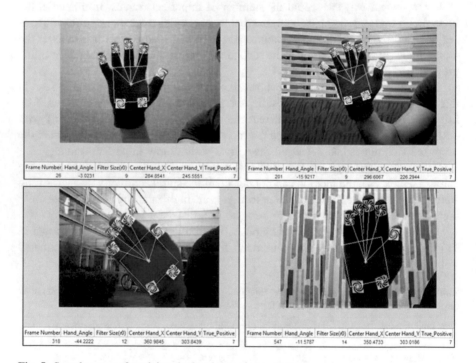

Fig. 5 Sample scans of applying the hand detection technique

center of the hand, and the number of true positives in the current frame are specified. Note that the center and angle of the hand are estimated based on the relative position between the two spiral at the bottom palm.

4.2 Gesture Recognition

In this section we describe the result of applying our gesture recognition technique. In this case we have defined eight simple gestures as the commands to be sent to an intelligent robot. These commands are defined as "Move to Right", "Move to Left", "Stop", "Turn", "Do Action 1", "Do Action 2", "Do Action 3", and "Do Action 4". These commands and the corresponding results from the gesture recognition technique are shown in Fig. 6.

To recognize a gesture we take advantage of the relative location between each unique spiral. Indeed, each individual spiral specifies an specific part of the hand i.e. fingers, thumb, bottom palm left, and bottom palm right. For example, the angle of the hand can be computed by using the two spirals in the bottom palm. This can be used for recognizing commands such as "Move to Right" and "Move to Left". Combining the information from other spirals and the angle of hand will give us the possibility to recognize other gestures.

At each frame, based on the number and the position of the spirals, a command is generated. In order to avoid falsely recognize random movement of the hand as a command or a gesture from an imperfect frame, the proposed technique waits to receive the same command for at least 12 consecutive frames (half a second). When it became certain about the captured command it will demonstrate the commands on the screen. Using this will avoid sending unexpected commands during changes in the hand gestures. Therefore, the proposed technique will be less sensitive to fast changes and receiving hand position from an imperfect frame. Using the proposed hand detection and gesture recognition technique, we are able to detect all the gestures and corresponding commands correctly i.e. all the hand gestures and corresponding commands in all the recorded videos are correctly detected and classified.

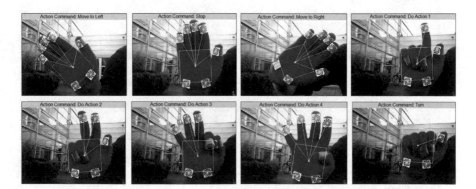

Fig. 6 Simple commands used for hand gesture detection

5   Conclusion and Discussion

In this paper we presented a novel technique for human hand detection using symmetric log-spiral patterns. Four different types of log-spirals are generated and mounted on a hand glove. The relative location of these spirals are used for positioning of the palm, fingers, and the angle of the hand. By taking advantage of the location of each spiral we are able to recognize simple commands from the hand gestures. The experimental results showed that the proposed technique can provide a reliable hand detection and gesture recognition with accuracy about 98 %.

To improve the hand detection technique we can use tracking methods such as Kalman Filters and Extended Kalman Filters. In this case, by estimating the position of the fingers in the next frame we can decrease the effects of imperfect frames.

Furthermore, we can create a precise dictionary of language signs using a person records the hand gestures with the glove. This can be used for example to reproduce the messages by an avatar, or to create new messages (not recorded) by combining individual signs. With this, we do not need to store videos, but sequences of spiral points which takes less space for storage and for transmission.

Acknowledgments F.A.-F. thanks the Swedish Research Council for funding this research. Authors also acknowledge the CAISR program of the Swedish Knowledge Foundation.

References

1. Harris, C., Stephens, M.: A combined corner and edge detector. In: Alvey Vision Conference, vol. 15, Citeseer, p. 50, (1988)
2. Lowe, D.G.: Object recognition from local scale-invariant features. In: Computer Vision, 1999. The Proceedings of the Seventh IEEE International Conference on, vol. 2, IEEE, pp. 1150–1157, (1999)
3. Mittal, A., Zisserman, A., Torr, P.H.: Hand detection using multiple proposals. In: BMVC, Citeseer, pp. 1–11, (2011)
4. Kang, S.K., Nam, M.Y., Rhee, P.K.: Color based hand and finger detection technology for user interaction. In: Convergence and Hybrid Information Technology, 2008. ICHIT'08. International Conference on, IEEE, pp. 229–236, (2008)
5. Teferi, D., Bigun, J.: Multi-view and multi-scale recognition of symmetric patterns. In: Image Analysis. Springer, Berlin, pp. 657–666, (2009)
6. Karlsson, S., Bigun, J.: Synthesis and detection of log-spiral codes. In: SSBA Symposium i bildanalys, 17–18 mars 2011, Linköping, p. 4, (2011)
7. Wang, R.Y., Popović, J.: Real-time hand-tracking with a color glove. ACM Trans. Graph. (TOG) **28**(3), 63, (2009)
8. Bellarbi, A., Benbelkacem, S., Zenati-Henda, N., Belhocine, M.: Hand gesture interaction using color-based method for tabletop interfaces. In: Intelligent Signal Processing (WISP), 7th International Symposium on IEEE, pp. 1–6, (2011)
9. Zhu, X., Yang, J., Waibel, A.: Segmenting hands of arbitrary color. In: Proceedings of Fourth IEEE International Conference on Automatic Face and Gesture Recognition, IEEE, pp. 446–453, (2000)

10. Stenger, B., Thayananthan, A., Torr, P.H., Cipolla, R.: Hand pose estimation using hierarchical detection. In: Computer Vision in Human-Computer Interaction. Springer, Berlin, pp. 105–116, (2004)
11. Viola, P., Jones, M.J., Snow, D.: Detecting pedestrians using patterns of motion and appearance. Int. J. Comput. Vis. **63**(2), 153–161 (2005)
12. Yang, Z., Li, Y., Chen, W., Zheng, Y.: Dynamic hand gesture recognition using hidden markov models. In: 7th International Conference on Computer Science and Education (ICCSE), IEEE, pp. 360–365, (2012)
13. Kohn, B., Belbachir, A.N., Nowakowska, A.: Real-time gesture recognition using bio inspired 3d vision sensor. In: IEEE Computer Society Conference on Computer Vision and Pattern Recognition Workshops (CVPRW), IEEE, 2012, pp. 37–42
14. Chen, F.-S., Fu, C.-M., Huang, C.-L.: Hand gesture recognition using a real-time tracking method and hidden markov models. Image Vis. Comput. **21**(8), 745–758 (2003)
15. Vatavu, R.-D., Pentiuc, Ş.-G., Chaillou, C., Grisoni, L., Degrande, S.: Visual recognition of hand postures for interacting with virtual environments. Adv. Electr. Comput. Eng. **6**(13), 55–58, (2006)
16. Wysoski, S.G.: A rotation invariant static hand gesture recognition system using boundary information and neural networks. Ph.D. thesis, NAGOYA INSTITUTE OF TECHNOLOGY (2003)
17. Sultana, A., Rajapuspha, T.: Vision based gesture recognition for alphabetical hand gestures using the svm classifier. Int. J. Comput. Sci. Eng. Technol. **3**(7)
18. Pradhan, A., Ghose, M., Pradhan, M., Qazi, S., Moors, T., EL-Arab, I.M.E., El-Din, H.S., Mohamed, H.A., Syed, U., Memon, A., et al.: A hand gesture recognition using feature extraction. Int. J. Curr. Eng. Technol. **2**(4), 323–327 (2012)
19. Bigun, J., Bigun, T., Nilsson, K.: Recognition by symmetry derivatives and the generalized structure tensor. IEEE Trans. Pattern Anal. Mach. Intell. **26**(12), 1590–1605 (2004)
20. Bigun, J.: Vision with Direction. Springer, Berlin (2006)
21. Bigün, J.: Pattern recognition in images by symmetries and coordinate transformations. Comput. Vis. Image Underst. **68**(3), 290–307 (1997)
22. Bigün, J., Granlund, G.H., Wiklund, J.: Multidimensional orientation estimation with applications to texture analysis and optical flow. IEEE Trans. Pattern Anal. Mach. Intell. **8**, 775–790 (1991)
23. Nilsson, K., Bigun, J.: Localization of corresponding points in fingerprints by complex filtering. Pattern Recogn. Lett. **24**(13), 2135–2144 (2003)
24. Fronthaler, H., Kollreider, K., Bigun, J., Fierrez, J., Alonso-Fernandez, F., Ortega-Garcia, J., Gonzalez-Rodriguez, J.: Fingerprint image-quality estimation and its application to multialgorithm verification. IEEE Trans. Inf. Forensics Secur. **3**(2), 331–338 (2008)
25. Alonso-Fernandez, F., Bigun, J.: Iris boundaries segmentation using the generalized structure tensor. A study on the effects of image degradation. In: IEEE Fifth International Conference on Biometrics: Theory, Applications and Systems (BTAS), IEEE, pp. 426–431, (2012)

Non Verbal Approach for Emotion Detection

Bharati Dixit and Arun Gaikwad

Abstract Non verbal approaches of emotion detection plays vital role in various applications like E-learning, automatic pain monitoring, driver alert system, cognitive assessment etc. which are developed to enhance quality of human life. Facial expressions based approach is one of the very effective approaches which is used widely on standalone basis or combined with other approaches known as multimodal techniques of emotion detection. The paper discusses facial expressions based emotion detection which uses patch based face features and SVM Classifier. The experimentation carried out on JAFFE database provides 90.65 % average accuracy for emotion detection for basic emotions happy, anger, sad, surprise, disgust, fear and neutral. The other performance parameters through experimentation are obtained as Average True positive rate is 90.5942 %, average false positive rate is 9.3671 % and average false negative rate is 9.4 %. The average feature extraction time is 18.49 s and emotion detection time through these extracted features is 1.1 s.

Keywords Facial expressions · Emotion detection · SVM · E-learning

1 Introduction

Many researchers have done studies of human emotions and its relation with human behavior and impact on human personality. Emotion detection has found its own space in some important application like E-learning, automatic pain monitoring, driver alert system, cognitive assessment etc. Efficient and accurate emotion detection is the need for all such applications. Scientific studies reveal that Rational intelligence which is related to memory, decision making etc. and social intelligence which is related with adaption of changes in surroundings, communication

B. Dixit (✉)
Sinhgad College of Engineering, Pune, India
e-mail: dixit.bharati@gmail.com

A. Gaikwad
Zeal College of Engineering, Pune, India

© Springer International Publishing Switzerland 2016   377
D. Król et al. (eds.), *Recent Developments in Intelligent Information and Database Systems*, Studies in Computational Intelligence 642,
DOI 10.1007/978-3-319-31277-4_33

etc. are interlinked with emotions which are very crucial for learning abilities and behavioral patterns of any human being.

Biometric techniques based emotion detection falls majorly under two categories verbal and non verbal. Verbal approach of emotion detection is based on voice as input. Non verbal approaches can use Brain signals, Facial Expressions, Cues, Body Posture, Gesture and Actions as inputs. All these approaches can be used on standalone basis or can be combined as multimodal approaches to enhance the performance of emotion detection system at the cost of increased computations.

Verbal approach is based on voice as input which can be analyzed over more than 200 features but Non verbal approaches are also equally challenging. Some insights of major non verbal approaches is discussed right here. CNS—Central nervous system is the origin of all the emotions and the inputs for analysis of CNS are available in the form of EEG, MEG, PET etc. Fundamental issue of reliability of voice signals and facial expression recognitions are overcome in this approach however the availability and collection of data is challenging task [1]. Cues based non verbal approach uses signals received through movement/positioning of individual part of the body or the group of parts of body in relation to each other. Facial expressions, body posture and gestures can involve more than one cues to interpret emotional state of the person [2]. Body posture comprises of various body parts like torso, arms and legs. An Example is Clenching of fist and raising it up appear like the subject is trying to attack someone and emotion is interpreted as anger. Jumping up and down with high frequency is interpreted as emotion of happiness.

Facial expressions based approaches can take into account the fiducial points as well as texture, shape and color of the skin to interpret the emotional state of the person. This study focuses on emotion detection through facial expressions.

The theme of the paper is further developed under different sections. Overview of the work done so far is discussed in Sect. 2. Details of the methodology is discussed in Sects. 3 and 4. Section 5 describes the experimental work. Section 6 highlights performance analysis. Conclusion and directions for future work are part of Sect. 7.

2 Literature Survey

This section summarizes the relevant work done by researchers and scientists across the world and published in renowned Journals.

Zhou et al. [3] discusses that High performance for face recognition systems occurs in controlled environments and degrades with variations in illumination, facial expression, and pose. Efforts have been made to explore alternate face modalities such as infrared (IR) and 3-D for face recognition. Studies also demonstrate that fusion of multiple face modalities improve performance as compared with single modal face recognition. This paper categorizes these algorithms into single modal and multimodal face recognition and evaluates methods within each category via detailed descriptions of representative work and summarizations.

Ramırez Rivera [4] introduces dynamic-micro-texture descriptor called spatiotemporal directional number transitional graph (DNG), which describes both the spatial structure and motion of each local neighborhood by capturing the direction of natural flow in the temporal domain. Structure of the local neighborhood is used to compute the transition of such directions between frames. Statistics of the direction transitions in a transitional graph, which acts as a signature for a given spatiotemporal region in the dynamic texture. Results validate the robustness of the proposed descriptor in different scenarios for expression recognition and dynamic texture analysis.

Majumdar et al. [5] has proposed an emotion recognition model using system identification. Twenty six dimensional geometric feature vector is extracted using three different algorithms. Classification is done using an intermediate Kohonen self-organizing map layer. A comparative study with Radial basis function, Multi-layer perceptron and Support vector machine is carried out.

Guo-Feng et al. [6] this paper reviews the latest progress of the domestic and international facial expression recognition technology since 2006. Paper focuses on the method of expressional feature extraction of basic expression, mixed expression and micro expression, non-basic expression.

Mower et al. [7] emphasize on complex emotions through emotion profiles and Song [8] highlights the importance of emotion recognition for varied applications. Ligang Zhang in [9] takes the basis as variety in representing features of face as region based appearance features and point based geometric features. These features can be static or dynamic in nature. Study focuses on static images and experimented for distance features. These features are extracted using patch based Gabor Features. The method promises to provide good results.

Heni et al. [10] has experimented for the real time emotion detection to synthesize the behavior while user is playing on Smartphone. Smart devices have the ability to recognize facial expressions but robust recognition of facial expressions in real time remains a challenge due to difficulties in accurately extracting the most pertinent emotional characteristics. Author has worked towards this.

Mourao et al. [11] has experimented for reduction in computation cost for emotion detection through facial expressions by reducing the Action Units under consideration. EMFACS (Emotion Facial Action Coding System) is taxonomy of face muscle movements and positions called Action Units (AU). Author aim at finding a minimal set of AU to represent a given expression and apply reconstruction to compute the deviation from the average face as an additive model of facial micro-expressions (the AUs).

Gao et al. [12] has worked for application to keep the driver alert by monitoring the attentive and emotional status of the driver for the safety and comfort of driving. In this work a real-time non-intrusive monitoring system is developed, which detects the emotional states of the driver by analyzing facial expressions. The system considers two negative basic emotions, anger and disgust, as stress related emotions. Experimental results show that the developed system operates very well on simulated data even with generic models.

As this is an open area for research and even though popular and efficient techniques are available but there is a need to make those techniques more effective as application space is quite wide.

3 Methodology of Implementation

This section discusses the various universally accepted facial expressions, approaches of facial expression analysis and methodology used for this study.

3.1 Basic Facial Expressions

Six basic classes of facial expressions and a neutral class are universally accepted worldwide. These six emotions are fear, anger, sad, surprise, disgust and happy along with neutral class. An example for the same is as shown in Fig. 1. (JAFFE—Japanese Female face Expression database).

3.2 Facial Expression Recognition Approaches

Facial expression analysis can be done for static images as well as for image sequences. The basic building blocks for any facial expression analysis system are face detection/face localization block, facial feature extraction block and classifier/expression detection block. Pre processing and face alignment can be done as per the requirement after face detection and before facial features extraction. Skin color segmentation is widely used technique for face detection. Either local features or global features can be extracted from face depending upon the overall approach followed to construct the emotion detection system. Many efficient and popular classifiers are available to perform last step of expression classification.

There are three main approaches of facial expression analysis which can be used for static images and videos.

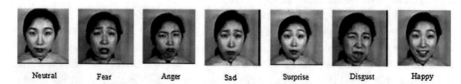

Neutral   Fear   Anger   Sad   Surprise   Disgust   Happy

Fig. 1 JAFFE database showing all emotions [13]

These approaches are:

1. Appearance based method, represented by Eigen faces, Fisher Faces and other methods using machine learning techniques such as neural networks and Support Vector machine
2. Model Based methods, including graph matching, optical flow based method and others
3. Hybrid of Appearance based and model based methods, such as AAM—Active Appearance Model.

Appearance based methods are superior to model based methods in terms of system complexity and performance reproducibility [14].

4 Implementation Methodology

The complete flow of emotion detection from capturing input image to facial expression classification is shown in Fig. 2. Various important steps are distinctly shown in flow diagram of Fig. 2. The critical step is facial feature extraction and the approach followed here uses Gabor filter bank for this purpose.

In this study 2D Gabor filter is used which can be represented mathematically as

$$F(x, y) = \exp(-(X2 + \gamma2Y2)/2\sigma2) * \cos(2\pi X/\lambda) \qquad (1)$$

$$X = x\cos\theta + y\sin\theta, \quad Y = -x\sin\theta + y\cos\theta$$

where θ is orientation, σ is effective width; λ is wavelength and γ is aspect ratio.

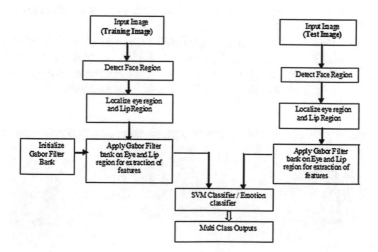

Fig. 2 Flow diagram of emotion detection process

The proposed work uses patch based facial features where lip patch and eye patch is used for feature extraction. Scientific study has revealed that patch based features are robust against variations in scale, position and orientation however the experimentation is not extended to prove this robustness.

The patch for lip region is localized from face which has the size 64 × 64. This patch is operated by Gabor filter bank of 8 scales and 5 orientations. The extracted feature vector is of size 2560. Similarly the patch of eye region is localized and feature vector is obtained for eye region. These features are saved as trained database.

For testing of this system, the test image is chosen from database and its features are extracted in similar fashion. These extracted features are fed to SVM classifier.

SVM classifier takes decision of class of test image based on decision boundary concept. Multiclass outputs are obtained through SVM classifier.

5 Experimental Results

Experimentation is carried out in simulation environment from matrix laboratory with system specification as 3 GB internal RAM and Intel core2duo Processor operating at 2.8 GHz.

Japanese Female Face Database (JAFFE) is used for experimentation. For experimentation of 7 different emotions 214 images of 10 subjects are used. Each emotion contains approximately 30 to 32 images. These images are used as train and test database during training and validation phase of the system respectively. Few images in bit map format and Jpeg format are also used for experimentation. Some snapshots of the implementations are shown in Figs. 3, 4, 5 and 6.

Figures 3 and 4 represents the snapshots of the initialization of Gabor filter bank, face detection, localization of lip and eye region. Gabor filter bank is used for creation of features from lip and eye region. This depicts training phase of the emotion detection system.

The selection of test image from the test database, extraction of Gabor features and detection of emotion are shown in the snapshots in Figs. 5 and 6. Images of different formats like tiff, bit map and jpeg are used as test images for validation of system.

Quantitative analysis of the obtained results is discussed in next section.

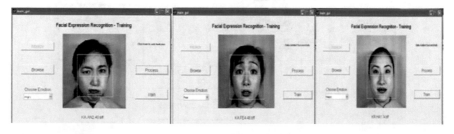

Fig. 3 Face detection, localization of lip and eye region for tiff images

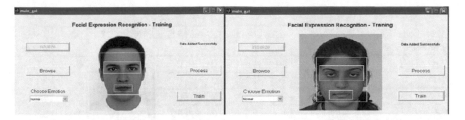

Fig. 4 Face detection, localization of lip and eye region for bmp and jpeg images

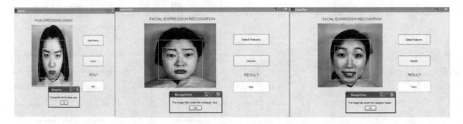

Fig. 5 Selection of test image, feature extraction and detection of emotion for tiff images

Fig. 6 Selection of test image, feature extraction and detection of emotion for bmp and jpeg images

6 Performance Analysis

Experimentation is carried over tiff, bit map and jpeg format images but performance analysis is carried out only for results obtained for tiff images i.e. JAFFE database.

JAFFE database contains total 214 images which are combination of around 30 images of each emotion. Twenty images of each emotion category is used for training and ten images are used for testing purpose. Three fold cross validation is used for validation of result. Obtained results are summarized in the form of

Table 1 Confusion matrix

Actual class (row-wise) \predicted class (column-wise)	HA	SU	SA	FE	AN	DI	NE	Total images
HA	30	0	0	0	1	0	0	31
SU	3	27	0	0	0	0	0	30
SA	0	0	28	0	2	1	0	31
FE	1	0	0	30	0	0	1	32
AN	0	0	0	0	27	0	3	30
DI	0	0	2	0	2	26	0	30
NE	4	0	0	0	0	0	26	30
Total images	38	27	30	30	32	27	30	214

Table 2 Calculation of TPR, FPR and FNR for each emotion

Emotion	TPR	FPR	FNR
HA	96.77	25.80	3.22
SU	90	0.0	10.00
SA	90.32	6.45	9.67
FE	93.75	0.0	6.25
AN	90	16.6666	10.00
DI	86.66	3.3333	13.33
NE	86.66	13.3333	13.33

confusion matrix shown in Table 1. Emotion Happy, Surprise, Sad, Fear, Anger, Disgust and Neutral are abbreviated as HA, SU, SA, FE, AN, DI AND NE respectively.

Various performance parameters like accuracy, True positive rate—TPR, False positive rate—FPR and False negative rate—FNR are calculated for each emotion. The results are summarized in Table 2.

The average accuracy, average FPR and average FNR are obtained through the contents of Table 2. The calculated values of performance parameters are as follows.

1. **Average Accuracy**: $(TP + TN)/TP + FP + FN + TN = 90.65\%$
2. **Average Weighted Accuracy**: $[31(96.77) + 30(90) + 31(90.32) + 32(93.75) + 30(90) + 30(86.66) + 30(86.66)]/214 = 90.6513\%$.
3. **Average TPR**: 90.5942 %, Average FPR: 9.3671 % and Average FNR: 9.4 %

Some of the timing parameters are also obtained for this system. The feature extraction time and Emotion recognition time after extraction of features are summarized in Table 3.

The average feature extraction time is 18.49 s and average emotion detection time is 1.1 s.

Table 3 Feature extraction time and emotion recognition time

Emotion	Feature extraction time (s)	Recognition time (s)
HA	20.85	1.0
SU	18.63	1.0
SA	16.23	1.1
FE	25.23	1.1
AN	20.98	1.1
DI	14.64	1.2
NE	12.87	1.2
Average	18.49	1.1

7 Conclusion and Future Work

Non verbal approach for emotion detection has significant role in many applications. Few important applications are E—learning where trainer and learners are remotely located so the analysis of learning experiences, learning curve, feedback and cognitive assessment of the learners can be done through spontaneous facial expressions. Facial expressions based emotion detection system can be developed to know the fatigue of driver and an alert can be provided to driver to avoid further consequences. The pain of bed ridden patients can be monitored automatically through facial expression. The work is not effectively applicable for certain segment of human beings suffering from specific problems like autism etc.

Facial patch based method of emotion detection which uses Gabor filter bank of 8 scales and 5 orientations for facial feature extraction is experimented for JAFFE database. The experimental results are quite promising and provide average accuracy of 90.65 % and overall emotion detection time is less than 20 s.

The work can be further extended for rotation and scale invariance. The experimentation can be performed in presence of noise as well. Illumination invariance can also be explored depending upon the requirement of application. The work can be extended for other publically available emotion database like Cohn Kanade database or database prepared with local subjects for specific study.

References

1. Taleb, T., Nasser, N.: A novel middleware solution to improve ubiquitous healthcare systems aided by affective information. IEEE Trans. Inf. Technol. Biomed. **14**(2), 335–349 (2010)
2. Butalia, A., Ramani, A.K., Kulkarni, P.: Emotional recognition and towards context based decision. Int. J. Comput. Appl. **9**(3), 42–53 (2010)
3. Zhou, H., et al.: Recent advances on single modal and multimodal face recognition: A survey. IEEE Trans Human-Mach Syst **44**(6), 701–771 (2014)
4. Ramırez Rivera, A.: Spatiotemporal directional number transitional graph for dynamic texture recognition. IEEE Trans. Pattern Anal. Mach. Intell. **37**(10), 2146–2152 (2015)

5. Majumder, A., etal.: Emotion recognition from geometric facial features using self-organizing map. In: Special issue of Handwriting Recognition and Other PR Applications by Elsevier Publications (2013)
6. Guo-Feng, Z., et al.: New research advances in facial expression recognition. In: IEEE International Conference on Control and Decision Conference (CCDC) (2013)
7. Mower, E., Mataric, M.J.: A frame work for automatic human emotion classification using emotion profiles. IEEE Trans. Audio Speech Lang. Process. **19**(5), 1057–1070 (2011)
8. Song, M., Tao, D., Liu, Z., Li, X., Zhou, M.: Image ratio features for facial expression recognition application. IEEE Trans. Syst. Man Cybern. **40**(3), 779–788 (2010)
9. Zhang, L.: Facial expression recognition using facial movement features. IEEE Trans. Affect. Comput. **2**(4), 219–229 (2011)
10. Heni, N., Hamam, H.: Facial emotion detection of Smartphone games users. In: 28th IEEE Canadian Conference on Electrical and Computer Engineering (CCECE), pp. 1243–1247 (2015)
11. Mourao, A., Borges, P., Correia, N., Magalhaes, J.: Sparse reconstruction of facial expressions with localized Gabor moments. In: Proceedings of the 22nd European Signal Processing Conference (EUSIPCO), pp. 1642–1646 (2014)
12. Gao, H., Yuce, A., Thiran, J.-P.: Detecting emotional stress from facial expressions for driving safety. In: IEEE International Conference on Image Processing (ICIP), pp. 5961–5965 (2014)
13. Standard Dataset Available: http://www.kasrl.org/jaffe_download.html
14. Lajevardi, S.M., Hussain, Z.M.: Higher order orthogonal moments for invariant facial expression recognition. In: Elsevier Journal on Digital Signal Processing, available online from March 2010, pp. 1771–1779 (2010)

Part VI
Advanced Computing Applications
and Technologies

Technological Devices for Elderly People with Alzheimer's Disease: A Review Study

Blanka Klimova and Petra Maresova

Abstract Current trends indicate a gradual increase in the number of elderly people in the developed countries, whose life expectancy thanks to better living conditions is prolonging. The higher age means more frequent occurrence of aging diseases such as dementia. In order to maintain quality of their life, more relevant support is needed. This can be done not only by their family members, additional caregivers, but also by different types of technological and health devices satisfying their specific needs. The aim of this article is to specify possibilities of using modern information technologies for patients with Alzheimer's disease (AD). This is explored with respect to the diagnosis of this disease and with respect to the technological devices such as monitoring, assistance, therapeutic, or diagnosis technologies which can enable an improvement of quality of life of these people in individual stages of this disease.

Keywords Technological devices · Alzheimer's disease · Elderly people · Benefits

1 Introduction

At present there is big potential in the use of information and communication technologies in different spheres of human activities. It is especially the IEEE Computer Society that is involved in their use. The IEEE Computer Society associates computer experts and managers in the field of ICT all over the world. These people specify

B. Klimova (✉)
Department of Applied Linguistics, Faculty of Informatics and Management,
University of Hradec Kralove, Hradec Kralove, Czech Republic
e-mail: blanka.klimova@uhk.cz

P. Maresova
Department of Economics, Faculty of Informatics and Management,
University of Hradec Kralove, Hradec Kralove, Czech Republic
e-mail: petra.maresova@uhk.cz

© Springer International Publishing Switzerland 2016
D. Król et al. (eds.), *Recent Developments in Intelligent Information and Database Systems*, Studies in Computational Intelligence 642,
DOI 10.1007/978-3-319-31277-4_34

global trends in this field, which are as follows: Internet of Things, Cybersecurity, Big data and their visualization, cloud computing, connection of the Internet, cloud and mobile technologies, Interactive Public Displays, 3D visualization, patients' mobility with the help of haptic equipment, and multi-core memory sharing. In the area of health care there are the following trends: a wider use of computational, visualization and communication technologies and SW, the so-called *eHealth*, coordination and optimization of care (cost cuts), electronic records of health state and medical records, software (SW) and hardware (HW) equipment for the monitoring and control of *home-care* [1], and higher comfort and safety for patients.

Furthermore, there is a number of significant technologies and emerging models that are making a big splash on the health care information technologies and applications [2–5]. The following trends can be observed:

Health sensing: There has been a sharp incline in the quantity and variety of consumer devices and medical sensors that capture some aspect of physiological, cognitive and physical human health. The implementation of these technologies empowers the end-users (e.g. chronic patients) by providing means to monitor and record the status continually and, if the need arises, seek remote assistance.

Big data analysis in health care: With the increasing digitization of health care, a large amount of health care data has been accumulated and the size is increasing in an unprecedented rate. Discovering the deep knowledge and values from the big health care data is the key to deliver the best evidence-based, patient-centric, and accountable care.

Cloud computing in health care: With health care providers looking at solutions to lower the operating costs, emerging technologies such as cloud computing can provide an ideal platform to achieve highly efficient use of computing resources, simplify management, and improve services in a safe and secure manner. Cloud computing can support the analysis of the big data mentioned above. There is no doubt that the adoption of these innovative technologies in medical fields can create significant opportunities. Nevertheless, many challenges still need to be addressed in order to achieve truly enhanced health care services [6]. These technologies are starting to allow health care practitioners to offer cheaper, faster and more efficient patient care than ever before. The health care industry is slowly but surely becoming more agile, effective and cost-effective for patients looking for care. The biggest innovations in health care technology with far reaching impacts according to [7] are as follows:

- Microchips modelling clinical trials;
- Wearable technology like google glass;
- 3D printed biological materials;
- Optogenetics;
- Hybrid operating rooms;
- Digestible sensor;
- Cloud-based provider relationship management software.

The aim of this article is to specify possibilities of using modern information technologies for patients with Alzheimer's disease (AD). This is explored with respect to the diagnosis of this disease and with respect to the devices for the improvement of quality of life of these people in individual stages of this disease.

2 Methods

For the purpose of this article a method of literature review of available sources describing current modern information technologies and their role in the diagnosing of dementia was applied.

In the database of Web of Science the authors reviewed research studies connected with the topic on the basis of the keywords "Alzheimer disease AND technology". Figure 1 indicates four times higher occurrence of these keywords in the course of 2004–2014.

Many results were too much widely focused and they were connected with the use of modern technologies in health care. Therefore the authors examined only those research studies which were closely connected with the explored topic.

The selection procedure of the final number of studies was done in the following four steps:

- Identification (identification of the key words and consequently, available relevant sources);
- Duplication check;
- Assessment of relevancy (verification on the basis of abstracts whether the selected study corresponds to the set goal); and
- Use of available studies (Fig. 2).

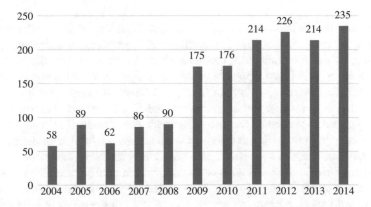

Fig. 1 Occurrence of the keywords Alzheimer disease and technology in the database web of science. *Source* Authors' own processing

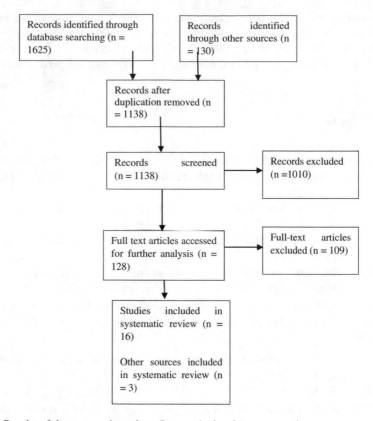

Fig. 2 Results of the systematic review. *Source*: Authors' own processing

3   Technologies for People with AD

AD is an irreversible neurodegenerative disorder characterized by a progressive and significant decline in memory, cognitive and physical functioning [8–11]. According to the International Classification of Diseases (ICD), there are three widely used criteria-based approaches to the diagnosis of Alzheimer's disease. Three common misconceptions regarding Alzheimer's disease—that it is a global disorder, that it is a diagnosis of exclusion and that it can be diagnosed only at autopsy—are all eschewed by the three diagnostic frameworks [12]. This affects not only quality of their own life but also imposes a substantial burden on their families, respectively on their caregivers. Therefore both groups welcome any kind of reducing this load, either physical or mental, which can be partially reduced by implementing suitable technologies. These technologies are particularly useful in the early, mild and moderate, phases of AD or at its diagnosing. In addition, they can slow down the onset of AD symptoms.

Generally, these technologies can be divided into monitoring, assistance, therapeutic, and diagnosis technologies [13]. All these technologies aim to:

- Promote independence and autonomy, both for the person with dementia and those around them;
- Help manage potential risks in and around the home;
- Reduce early entry into care homes and hospitals;
- Facilitate memory and recall; and
- Reduce the stress on caregivers, improving their quality of life, and that of the person with dementia.

Monitoring technologies mainly focus on monitoring well-being of AD patients and avoiding maintenance and safety problems at home. These technologies involve, for example, Care Media or Smart Carpet technology. Care Media technology concentrates on recognizing various behavioral patterns and social interactions such as sitting, walking or standing. It can also eliminate aggressive behavior. Smart carpet technology can measure walking speed and/or gait and therefore it can help in the identification of any deterioration in physical ability [13].

Assistance technologies include, for instance, window-on-the-world consisting of a remote camera with an RF link to a receiver plugged into the TV; conversation prompter; and sequence support. One of the most popular and tested technologies is also a computer-aided telephone system [11, 14], which helps AD patients with a simple miscroswitch response (i.e. it helps the person avoid all the difficulties connected with remembering, retrieving or dialing telephone numbers) make a phone call.

These technologies can be divided into four basic groups according to their purposes [14]: *prevention and engagement* (e.g. home automation technologies which can monitor and ensure home safety features such as fire and smoke alarms, ventilation, sensors for water temperature, or power control); *compensation and assistance* (e.g. robotics which can perform household maintenance such as vacuum or assist with bathing or eating (a butler) or companionship activities; *care support* (e.g. communication such as access to telecare and medical networks and social support networking, e-mails, real-time alarms); and *enhancement and satisfaction* (e.g. sensors for monitoring, initiating alarms and data collection, among which the most common types are environmental, e.g. motion detection, thermostats or water usage, radiofrequency transmitters and computer-vision such as webcam, user recognition or motion analysis).

Therapeutic technologies are, for example, multimedia biographies helping individuals remember their past or SenseCam [13], which is a small portable camera that can be worn throughout the day and which makes about 1000–1500 snapshots of a person's daily life. The AD patients can replay this as a three-minute flash-frame movie at the end of the day. This might help these people remember what they experienced during their day. In addition, their cognitive and memory skills can be enhanced by computer games if they are played on a regular basis [15]. Most recently Lancioni et al. [16] have introduced and tested a computer-aided program aimed at fostering patient's verbal engagement on a number of life experiences/topics previously selected for him or her and presented in the sessions through a friendly female, who appeared on the computer screen. The female asked the patient about the aforementioned experiences/topics, and provided him or her with positive attention, and possibly verbal guidance (i.e., prompts/encouragements).

As research studies [17–19] illustrate, *diagnosis technologies* can reliably and effectively contribute to non-invasive detection of AD in the early stages. The so-called Computer Aided Diagnosis (CAD) tools can offer meaningful comparison between normal and diseased subjects using the analysis of certain features in a functional brain image, and detecting the appearance of the abnormalities. These CAD systems in fact apply machine learning techniques to build classifier system that may help the clinicians to establish a diagnosis on the available, usually imaging data. The diagnosis technologies now very often include devices from the area of virtual reality. The reason is that virtual reality environments have a major advantage for the assessment of spatial navigation and memory formation, as computer-simulated first-person environments can simulate navigation in a large-scale space. Mild cognitive impairment (MCI) is considered as a stage of cognitive, especially mnemonic impairment beyond what is considered normal for age, but not of sufficient magnitude as to warrant the diagnosis of dementia or Alzheimer's disease (AD) [20, 21]. Several studies have demonstrated reduced hippocampus size in individuals with amnesic MCI (aMCI) when compared with healthy controls (consult [22]). Hippocampus size reduction in aMCI (approximately 10–15 %) is, however, less strong than that observed in individuals with clinically probable AD (up to 40 %; consult [23]). In another study it was found that the AD patients' performances were inferior to that of the MCI and even more to that of the healthy aged groups, in line with the progression of hippocampus atrophy reported in the literature [24].

To improve the diagnosis of AD, VR-DOT[1] performance score data were compared with markers of neurodegeneration over different time intervals. Consequently, significant correlations of VR-DOT score with MRI-derived AD signature biomarker and with electrophysiological correlates, Enterprise Resource Planning (ERPs) were revealed. Furthermore, with respect to these markers, VR-DOT was also associated with a lower neuropsychological test performance in the attention and executive domains. Specifically, executive tasks (verbal fluency, trails) requiring high attention resources highly correlated with VR-DOT. This study also confirmed the usefulness of P3b latencies at the parietal site as a tool for assessing declining progression but also as an AD-converter predictor. Thus, using simple computer-based neuropsychological assessment, including VR-DOT combined with functional neurophysiologic markers, could optimize the diagnosis of AD in an early stage of the disease and can provide the greatest benefit in terms of cost and risk compared with other techniques [25].

Although there seem to be a lot of benefits of these technologies for AD patients, there are obvious pitfalls as well. Apart from being quite costly, these technologies

---

[1]virtual reality day-out task (VR-DOT) environment.

may have problems with maintenance, alert or data management. In addition, there might arise hardware issues, e.g., sensors can be damaged, or automatic software updates can cause interference. The methodology, e.g., of prompts, is not so much tested yet. And there are many more such issues.

4 Conclusion

With the increasing number of younger elderly people suffering from Alzheimer´s disease and their ability to exploit technological devices, there is a gradual trend to maintain and improve quality of their life by implementing technologies into their lives in order to delay their cognitive decline and accompanying behavioral and psychological symptoms of AD. Table 1 below summarizes the main technological device groups and their benefits.

Table 1 Summary of technologies for Alzheimer disease

The group of technologies	The type of technology	Modern technologies which make the corner of this support	Benefits	Studies
Diagnosis technologies	CAD systems	Cloud computing	Delay of the onset of the disease at its early diagnosis	[17–25]
		Data analysis in real time		
Monitoring technologies	Care Media or Smart carpet technology	Big data (NoSQL database)	Provision of patients' greater security Time-saving for caregivers	[13]
		Connection of the Internet, cloud and mobile technologies		
Assistance technologies	Robotics, telecare and medical networks and social support networking, e-mails, real-time alarms, initiating alarms and data collection	Multicore memory sharing	Improvement of patients' quality of life Better chance of communication with surrounding, support for social contacts	[11, 14]
		Software-defined anything (open code or modular building of systems) [26, 27]		
Therapeutic technologies	Multimedia biographies, computer-aided program		Support for the prevention of the disease Support for the treatment of the disease, reduction of the symptoms	[13, 15, 16]

Source Authors' own processing

The importance of technologies and their use as assistance tools will be growing with the rise of ageing population. The main trends of their development mainly include: cloud computing, mobile technologies, multicore memory sharing or big data.

References

1. Tesar, V., Zdimera, A.: Globalni trendy ve zdravotnictvi a farmacii Global trends in health care and pharmacy. www.mzv.cz/file/971363/OMEV_globalni_trendy_zdravotnictvi_fin.ppt (2013)
2. Jensen, P.B., Jensen, L.J., Brunak, S.: Mining electronic health records: Towards better research applications and clinical care. Nat. Rev. Genet. 13(6), 395–405 (2012)
3. Khousa, K.A., Mohamcd, N., Jaroodi, J.: E-Health cloud: Opportunities and challenges. Future Internet 4, 621–645 (2012)
4. Murdoch, T.B., Detsky, A.S.: The inevitable application of big data to health care. J. Am. Med. Assoc. 309(13), 1351–1352 (2013)
5. Zheng, Y.L., Ding, X.R.: Yan Poon, B.P., Lai Lo, H.: Unobtrusive sensing and wearable devices for health informatics. IEEE Trans. Biomed. Eng. 61(5), 1538–1554 (2014)
6. Mulder, J., Wang, Y., Shi Chen, S.:Emerging information technologies for enhanced health care. Computers in industry (2015, in press)
7. van Hoof, J., Zwerts-Verhelst, E.L., Nieboer, M.E.: Innovations in multidisciplinary education in healthcare and technology. Perspect. Med. Educ. 4(3), 146–148 (2015)
8. Arkin, S.: Language-enriched exercise plus socialization slows cognitive decline in Alzheimer's disease. Am. J. Alzheimer's Dis. Other Dementias 22, 62–77 (2007)
9. Giovannetti, T., Bettcher, B.M., Libon, D.J.: Environmental adaptations improve everyday action performance in Alzheimer's disease: Empirical support from performance-based assessment. Neuropsychology 21, 448–457 (2007)
10. Gure, T.R., Kabeto, M.U., Plassman, B.L.: Differences in functional impairment across subtypes of dementia. J. Gerontol. (Series A). 65, 434–441 (2010)
11. Perilli, V.: Persons with Alzheimer's disease make phone calls independently using a computer-aided telephone system. Res. Dev. Disabil. 33, 1014–1020 (2012)
12. WHO: ICD classification of mental and behavioural disorders, http://cnsdiseases.com/definitions-of-alzheimers-disease (1993)
13. Carrillo, M.C., Dishman, E., Plowman, T.: Everyday technologies for Alzheimer's disease care: Research findings, directions, and challenges. Alzheimer's Dement. 5, 479–488 (2009)
14. Perilli, V.: A computer-aided telephone system to enable five persons with Alzheimer's disease to make phone calls independently. Res. Dev. Disabil. 34, 1991–1997 (2013)
15. Jimison, H.B., Pavel, M., Bissell, P.: A framework for cognitive monitoring using computer game interactions. Stud. Health. Technol. Inform. 129, 1073–1077 (2007)
16. Lancioni, G.E.: A computer-aided program for helping patients with moderate Alzheimer's disease engage in verbal reminiscence. Res. Dev. Disabil. 35, 3026–3033 (2014)
17. Martinez-Murcia, F.J., Gorriz, J.M., Ramirez, J.: Computer aided diagnosis tool for Alzheimer's disease based on mann-whitney-wilcoxon u-test. Expert Syst. Appl. 39, 9676–9685 (2012)
18. Papakostas, G.A., Savio, A., Grana, M.: A lattice computing approach to Alzheimer's disease computer assisted diagnosis based on mri data. Neurocomputing 150, 37–42 (2015)
19. Savio, A., Grana, M.: Deformation based feature selection for computer aided diagnosis of Alzheimer's disease. Expert Syst. Appl. 40, 1619–1628 (2013)
20. Petersen, R.C., Negash, S.: Mild cognitive impairment: An overview. CNS Spectr. 13, 45–53 (2008)

21. Petersen, R.C., Smith, G.E., Waring, S.C.: Mild cognitive impairment: Clinical characterization and outcome. Arch. Neurol. **56**(3), 303–308 (1999)
22. Wolf, H., Jekic, V., Gertz, H.J.: A critical discussion of the role of neuroimaging in mild cognitive impairment. Acta Neurol. Scand. **107**(Suppl. 179), 52–76 (2003)
23. Barnes, J., Bartlett, J.W., van de Pol, L.A.: A meta-analysis of hippocampal atrophy rates in Alzheimer's disease. Neurobiol. Aging **30**, 1711–1723 (2009)
24. Plancher, G., Tirard, A., Gyselinck, V., Nicolas, S., Piolino, P.: Using virtual reality to characterize episodic memory profiles in amnestic mild cognitive impairment and Alzheimer's disease: Influence of active and passive encoding. Neuropsychologia **50**, 592–602 (2012)
25. Tarnanas, I., Tsolaki, M., Nef, T., Mosimann, U.P.: Can a novel computerized cognitive screening test provide additional information for early detection of Alzheimer's disease? Alzheimers Dement. **10**(6), 790–798 (2014)
26. Zhang, S., Zhang, S., Chen, X., Huo, X.: Cloud computing research and development trend. 2010 Second international conference on future networks, pp. 93–97 (2010)
27. Syputa, R.: Current trends in ICT, http://www.4gtrends.com/articles/29487/35-current-trends-in-ict/ (2011)

Integrated System Supporting Research on Environment Related Cancers

Wojciech Bensz, Damian Borys, Krzysztof Fujarewicz, Kinga Herok,
Roman Jaksik, Marcin Krasucki, Agata Kurczyk, Kamil Matusik,
Dariusz Mrozek, Magdalena Ochab, Marcin Pacholczyk,
Justyna Pieter, Krzysztof Puszynski, Krzysztof Psiuk-Maksymowicz,
Sebastian Student, Andrzej Swierniak and Jaroslaw Smieja

Abstract There are many impediments to progress in cancer research. Insufficient
or low quality data and computational tools that are dispersed among various sites
are one of them. In this paper we present an integrated system that combines all
stages of cancer studies, from gathering of clinical data, through elaborate patient
questionnaires and bioinformatics tools, to data warehousing and preparation of
analysis reports.

Keywords Cancer research · Integrated systems · Data warehouses

1 Introduction

Lack of easy-to-use software and hardware infrastructure that would support all
stages of epidemiological cancer studies, as well as poor availability of data from
clinical and experimental groups is one of the main reasons of slower, than technology and state of biological and clinical knowledge would indicate, advances in
cancer research. Though there exist many bioinformatics databases, supported with
computational tools to analyze data stored there, these tools are often very specific
to these databases and cannot be easily applied anywhere else. Additionally,
research in epidemiology may be very specific, which is why it is difficult to
compare and analyze data between different research centers.

In search for tools that would facilitate biomedical research, we have created an
integrated system that encompasses all its stages. Though it has been developed with
studies into environmental-associated cancers, its architecture and computational

W. Bensz · D. Borys · K. Fujarewicz · K. Herok · R. Jaksik · M. Krasucki · A. Kurczyk
K. Matusik · D. Mrozek · M. Ochab · M. Pacholczyk · J. Pieter · K. Puszynski
K. Psiuk-Maksymowicz · S. Student · A. Swierniak · J. Smieja (✉)
Faculty of Automatic Control, Electronics and Computer Science, Silesian University
of Technology, Akademicka 16, Gliwice, Poland
e-mail: Jaroslaw.Smieja@polsl.pl

© Springer International Publishing Switzerland 2016 399
D. Król et al. (eds.), *Recent Developments in Intelligent Information
and Database Systems*, Studies in Computational Intelligence 642,
DOI 10.1007/978-3-319-31277-4_35

tools that have been implemented may support virtually any clinical and biological research. In the paper, we describe the architecture of the system and provide insights into some of its tools.

2 Features of Epidemiological Studies

The epidemiological studies, not only in the field of cancer, are characterized by two distinct features. First, the patients are chosen from existing databases, and asked to participate in the project. Second, among various data gathered in these projects, a huge part comes from personal questionnaires that are tailored to a specific study and usually are unique, though some of the questions or question types may reappear in different studies.

A simplified flowchart of an epidemiological study is presented in Fig. 1. Following selection of a pool of prospective patients, they are contacted by clinical staff to obtain written consent to participate in the study. Those who agree to participate, enter the project. The first step then is filling in detailed surveys, which so far has been done in paper, followed by entering data to spreadsheets. Afterwards, the clinical path is defined, i.e. the list of necessary physical and/or physiological examinations is compiled, complete with time intervals between subsequent health checks. After the samples are collected from the patients (or, more specifically, project participants), they are analyzed in laboratories and the results are stored in the database. Usually the research ends at that point with conclusions drawn for future studies.

There are three problems to be addressed at this point. First, data is not shared among different research centers, not necessarily because of lack of will to share. Second, keeping track of the position of each patient in the study path is difficult

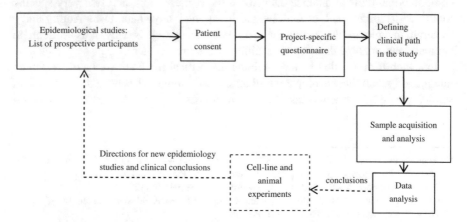

Fig. 1 Simplified flowchart of epidemiological studies. In addition supporting standard actions, the system facilitates additional interactions shown by the *dashed lines*

and it leads to flawed or missing data. Finally, despite the fact that quite often the research teams comprise specialists from various fields, they cannot fully utilize data that has been acquired due to the differences in IT systems. We propose to integrate all steps using an integrated system that would facilitate creation of an additional feedback employing also experimentalists, as depicted in the Fig. 1.

3 Overview of the System Architecture

The system is composed of three interconnected types of subsystems (Fig. 2). In the first line of research there are local databases, managed by so called LIMS. They can be proprietary, under assumption that they provide some kind of an interface allowing data export. However, to take full advantage of the system properties, we developed our own LIMS-like system that is briefly presented in the subsequent section. These locally maintained databases communicate with a central data warehouse.

Data warehouse systems are complex information systems designed to perform advanced reporting, on-line multidimensional analytical analysis (OLAP), trend and what-if analysis, results prediction and advanced data mining [1–4]. The results generated by data warehouse systems are often an essential element of enterprise level strategic management process, and that is why data warehouse systems must provide exact, reliable and up-to-date information. Multidimensional analysis is also becoming increasingly popular in the area of Life sciences, in which multiple views of the same data allow to get insights about real living organisms. This concept was successfully utilized in various existing projects such as BioWarehouse [5], Atlas [6], and BioDWH [7].

The central data warehouse developed in the SysCancer project is used to support multidimensional analysis from local databases after data earmarked for public access has been exported from them. Additionally, it is used as a gateway for

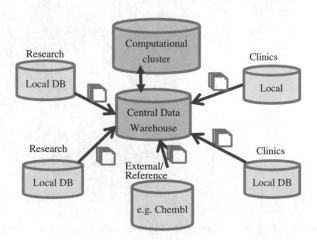

Fig. 2 Architecture of the integrated system supporting biomedical research

the computational cluster, responsible for performing complex analyses of data using advanced algorithms that are accessible through a self-explanatory simple interface.

4 Local Databases (LIMS-like Systems)

Having epidemiological research oriented on environment related cancer in mind, we have developed a system for managing all stages of such research. It comprises subsystems for patient management complete with tools for defining clinical paths and supporting acquisition of medical samples results, survey subsystems as well as subsystems for animal and molecular biology laboratories (Fig. 3). Data stored in the local database may be earmarked for export to the central data warehouse, following necessary anonimization procedures.

The most significant advantage of our system over other approaches like Prolab3 or LabCollector is the ability to integrate knowledge from other, remote databases through the use of web-services (e.g. NCBI gene data, Web of Science scientific literature information) and the ability to use the data stored in local databases as an input into various custom data analysis applications (e.g. PCR primer sequences to

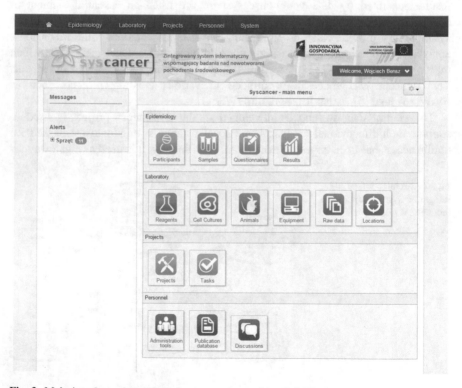

Fig. 3 Main interface of the LIMS for management of local databases

determine possible amplicons). The system has a modular architecture which can be adjusted to the current requirements of the laboratory, and despite numerous features it very easy to use.

4.1 Typical Laboratory Management Operations

Users belonging to different laboratories are authorized to access their respective resources and obtain tools specific to their work characteristics. For example, molecular biologists that run experiments on cell cultures, have access to the subsystems managing reagents and cell cultures, with separate interfaces for entering data from experiments. Similarly, employees involved in animal experiments have access to the respective animal subsystem. In addition to interactive forms facilitating data entering, there are also tools for uploading raw results in the form of files specific to the equipment used in experimental work.

Additionally, there is a separate subsystem for equipment management with two main features. It may be used for keeping track of equipment maintenance, with alerts generated when periodic servicing is required. Additionally, stand-alone mobile applications may be used to request access to a specific equipment in the desired time slots.

4.2 Questionnaires Subsystem

Arguably the subsystem that should bring the greatest improvements into the research at the local database level is the one that facilitates creation and filling in the questionnaires in epidemiological projects (Fig. 4). The creator for project-

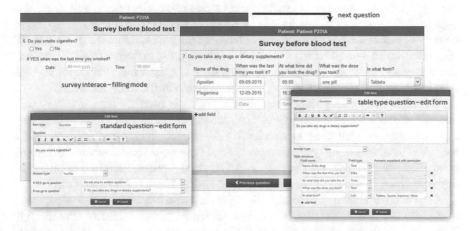

Fig. 4 Sample view of the questionnaires subsystem

specific surveys allows for various types of questions to be defined by the user. Once the questionnaire is ready, it can be made available for mobile devices of the volunteers or staff who are responsible to gather data. In order to provide safety of the system, the filled-in questionnaires can be saved to sandboxes and only from there can be imported by the research center staff to local databases.

4.3 Subsystem for Patient Data Acquisition, Management of Samples Collection and Storage of the Results

The subsystem for patient data acquisition provides standard functions. However, it also facilitates the use of pre-designed barcodes for easier sample and patient identification as well as tools for multiple use of larger tissue or blood samples. It also facilitates export of anonimized data to the central data warehouse.

5 Computational Tools

The computational cluster connected to the central data warehouse provides an interface to virtually any type of analysis. We have implemented a series of tools that can be divided into three basic categories:

- Preprocessing of data form biological experiments, necessary if various experimental platforms are used;
- Data analysis, including clustering and classification algorithms;
- Selected algorithms for biomedical image processing and analysis.

The user interface is the same for all algorithms. The user is requested to upload the input data file and choose one of the implemented algorithms, following the choice of the type of the task that is to be performed (Fig. 5). The pop-up list is context-dependent, i.e. only those algorithms that are appropriate for a given task. If the algorithm requires a setup of parameter values, these also appear on the screen. In most cases, the results are stored in the indicated output file, though for several simple analyses there are graphical reports available in the main interface.

The first group of algorithms consists of methods used for data filtering, normalization and automatic correction. They are mainly dedicated to microarray and PCR experiments, allowing to standardize the measurements and unify the data format. This allows to integrate results obtained in various studies which often were conducted using distinct platforms (e.g. Affymetrix, Agilent, Illumina) or based on different types of genetic material (e.g. DNA, mRNA, microRNA, proteins).

As we have shown previously [8–10] data standardization has a significant impact on the post-processing of the data, including identification of features that differentiate selected groups of samples. Since most of the sample preparation

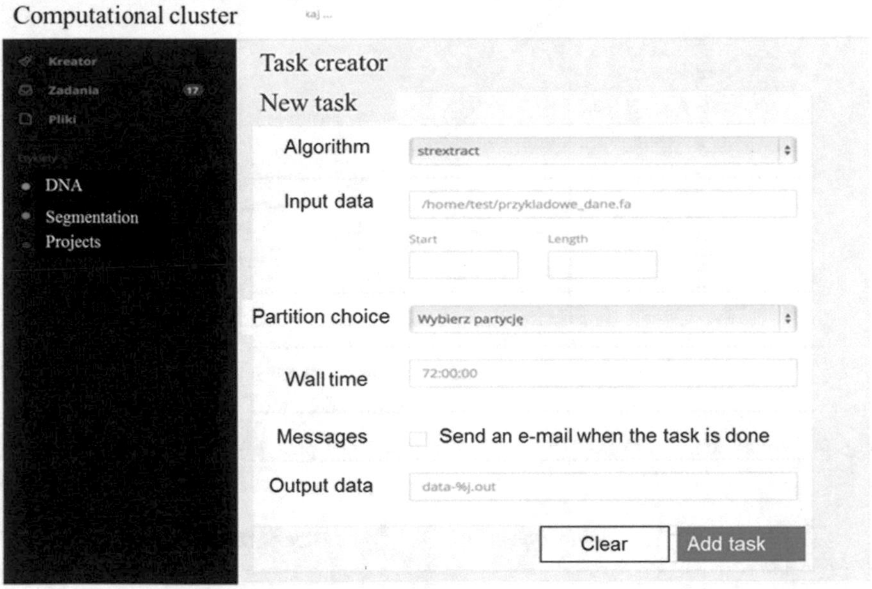

Computational cluster   taj ...

Kreator	**Task creator**
Zadania (17)	**New task**
Pliki	
	Algorithm strextract ⬍
DNA	
Segmentation	Input data /home/test/przykladowe_dane.fa
Projects	
	Start Length
	Partition choice Wybierz partycję ⬍
	Wall time 72:00:00
	Messages ☐ Send an e-mail when the task is done
	Output data data-%j.out
	Clear Add task

Fig. 5 Interface to the computational cluster

methods are shared between various measurement platforms, like DNA/RNA isolation and amplification, methods which are designed to reduce bias introduced by their use can be shared among distinct studies. One of such methods implemented in our computational environment introduces signal correction based on the gene or microarray probe GC content, enhancing the specificity and sensitivity of algorithms used to detect differentially expressed genes, comparing to typical data analysis workflows [8]. It is needed for reliable analysis of microarray experiment data, as nucleotide composition affects signal intensity obtained in gene expression studies using various methods and platforms [9]. Nucleotide sequence dependent bias can significantly affect the expression estimates when using standard normalization procedures that do not take into account the nucleotide composition of genes. High or low GC content genes are mainly affected but, as shown in Fig. 6, even genes with 54 and 56 % GC can show significant differences post-normalization, when two compared microarrays differ in the total signal intensity. Figure 6 was created based on the results of over 10 thousand microarrays downloaded from the ArrayExpress repository (LOESS smoothing was applied). It shows the relation between total un-processed signal intensity of two microarrays and expression intensity difference, after normalization, of selected housekeeping genes GAPDH and ACTB (which are expected to have a stable expression level). The relation shows that standard normalization approaches can be inefficient resulting in artificial differences between certain genes. The dashed line marks the

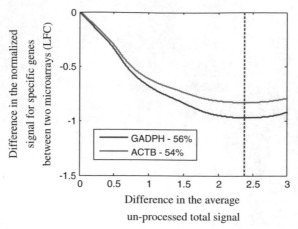

Fig. 6 Comparison of data from two different experimental platforms

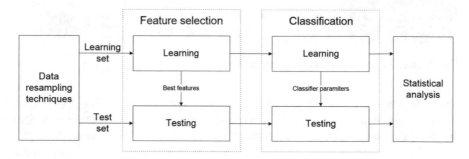

Fig. 7 Classification system scheme

average difference between samples which is typical for the experiments studied, indicating that this effect can be commonly observed in real data.

Once data have undergone appropriate preprocessing, it can be fed into computational-cluster-run analysis. This may, for example, consist in high-throughput sequencing data analysis, using newly developed approaches to find functionally relevant conclusions, independent of original database that is the source of data [11] or classification of multiclass microarray data [12].

Data analyzed in the classification system come out from various experiments and from different biomedical techniques. Most of them generate different types and volumes of data. For that reason there is hard to find the universal classification model for all these data. This is the reason why we implement the set of different classification and feature selection methods. Our classifier is based on the classification system showed in Fig. 7. Data preprocessing is a separate part of the system, choice of proper classification and feature selection methods.

Various types of classification have been implemented, including Diagonal Linear Discriminant Analysis classifier (DLDA) and Support Vector Machines

(SVM). For feature selection the GS method, BSS/WSS-statistics, and PLS method can be chosen from.

Validation methodology is an important part of the classification system. Here, we have used a bootstrap-based technique as a resampling technique and for the classification accuracy estimation method. We have implemented different bootstrap based selection and classification indicators, described in detail in [12]. These model validation techniques used can help the user to select the proper implemented method mainly for feature selection and classification [13].

Medical and microscopic imaging requires a separate set of tools to support analysis of data coming from these techniques. They include resizing of the images, segmentation [10], morphological white top-hat transformation in case of Time of Flight images, noise reduction and gamma transformation [14].

6 Data Warehouse

The central data warehouse is designed to collect, aggregate and facilitate analysis of data collected from various research centers that participate in a large project. It is assumed that those willing to utilize the system will export their data to the central data warehouse (Fig. 8). Since for each type of analysis a separate ETL is required, only predefined data types and analysis goals are currently implemented. Three examples have been created to illustrate feasibility of the entire integrated system:

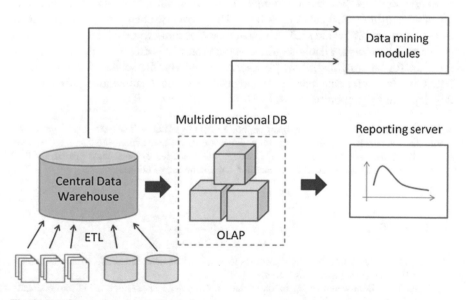

Fig. 8 Interface to the computational cluster

- Analysis of blood test results (the annotation for blood tests is relatively uniform and it can be assumed that even proprietary LIMS systems that would connect to the central warehouse would provide data in the ready-to-use format)
- Analysis of questionnaires data (assuming that the same questionnaire subsystem and the same type of predefined questionnaire has been used)
- Analysis of data from external/reference database such as ChemBL (such databases usually are equipped with the tools for analysis of data stored there; however, these tools are often very limited in scope).

Additionally, the data warehouse is used to aggregate data concerning the use of computational cluster resources to facilitate further reporting.

7 Conclusions

The solution presented in this paper is the first system that integrates all stages of the research on environmental-related cancer in one place. Its architecture allows connection of new partner institutions and further development of analytical and computational tools for biomedical research. While all system components have been designed for a specific project, they can support many other biological or medical research projects or management in biomedical institutions. Modularity of local database subsystems allows to choose components needed. All its modules were implemented in Nofer Institute of Occupational Medicine in Lodz, Poland and another, individualized version in Center of Oncology- Maria Sklodowska-Curie Memorial Institute, Branch in Gliwice, Poland, where they underwent thorough testing procedures. The computational cluster does not impose any constraints on types of algorithms that can by run on it. Therefore it is open for future extensions. However, it should be stressed that each new algorithm to be added to the pool of computational tools available requires preparation of a dedicated context-dependent interface for entering algorithm parameters. Similarly, the scope of analysis facilitated by the central data warehouse cannot be expanded automatically but require development of supporting tools.

Acknowledgments The work has been supported by the POIG.02.03.01-00-040/13 grant. We would like to thank our collaborators from Nofer Institute of Occupational Medicine, led by prof. Konrad Rydzyński and from Institute of Oncology, Gliwice, led by prof. Piotr Widłak for their valuable comments and suggestions concerning local database solutions.

References

1. Kimball, R., Reeves, L., Margy, R., Thornthwaite, W.: The data warehouse. Lifecycle Toolkit, 3rd edn. John Wiley and Sons (2013)
2. Ponniah, P.: Data warehousing fundamentals. A Comprehensive Guide for IT Professionals, John Wiley and Sons (2001)

3. Inmon, W.H., Linstedt, D.: Data Architecture: A Primer for the Data Scientist: Big Data, Data Warehouse and Data Vault. 1st edn. Morgan Kaufmann (2014)
4. Malysiak-Mrozek, B., Mrozek, D., Kozielski, S.: Processing of crisp and fuzzy measures in the fuzzy data warehouse for global natural resources. In: García-Pedrajas, N. (ed.) IEA/AIE 2010, Part III, LNAI 6098, pp. 616–625 (2010)
5. Lee, T.J.: BioWarehouse: A bioinformatics database warehouse toolkit. BMC Bioinform. **7**, 170 (2006)
6. Shah, S.P.: Atlas—A data warehouse for integrative bioinformatics. BMC Bioinform. **6**, 34 (2005)
7. Topel, T., Kormeier, B., Klassen, A., Hofestädt, R.: BioDWH: A data warehouse kit for life science data integration. J. Integr. Bioinform. **5**, 93 (2008)
8. Jaksik, R., Bensz, W., Smieja, J.: Nucleotide composition-based measurement bias in high throughput gene expression studies. Proceedings of the international conference on man-machine interactions, ICCMI (2015)
9. Jaksik, R., Iwanaszko, M., Rzeszowska-Wolny, J., Kimmel, M.: Microarray experiments and factors which affect their reliability. Biol Direct **10**(1), 46 (2015)
10. Student, S., Danch-Wierzchowska, M., Gorczewski, K., Borys, D.: Automatic segmentation system of emission tomography data based on classfication system. LNBI Bioinf. Biomed. Eng. **9043**, 274–281 (2015)
11. Stokowy, T., Eszlinger, M., Swierniak, M., Fujarewicz, K., Jarzab, B., Paschke, R., Krohn, K.: Analysis options for high-throughput sequencing in miRNA expression profiling. BMC Res. Notes **7**, 144 (2014)
12. Student, S., Fujarewicz, K.: Stable feature selection and classification algorithms for multiclass microarray data. Biol. Direct **7**, 33 (2012)
13. Jaksik, R., Marczyk, M., Polanska, J., Rzeszowska-Wolny, J.: Sources of high variance between probe signals in affymetrix short oligonucleotide microarrays. Sensors **14**(1), 532–548 (2014)
14. Psiuk-Maksymowicz, K., Borys, D., Smieja, J.: Magnetic resonance angiogram processing and modelling of the cerebral vascular network. Proceedings of the sixth international conference bioinformatics, biocomputational systems and biotechnologies BIOTECHNO 2014, pp. 59–62 (2014)

An Efficient Tree-based Rule Mining Algorithm for Sequential Rules with Explicit Timestamps in Temporal Educational Databases

Hoang Thi Hong Van, Vo Thi Ngoc Chau and Nguyen Hua Phung

Abstract In educational data mining, frequent patterns and association rules are popular to help us get insights into the characteristics of the students and their study. Nonetheless, frequent patterns and rules discovered in the existing works are simple with no temporal information along the student's study paths. Indeed, many sequential pattern and rule mining techniques just considered a sequence of ordered events with no explicit time. In order to achieve sequential rules with explicit timestamps in temporal educational databases that contain timestamp-extended sequences, our work defines a tree-based rule mining algorithm from the frequent sequences generated and organized in a prefix tree enhanced with explicit timestamps. Experimental results on real educational datasets have shown that the proposed algorithm can provide more informative sequential rules with explicit timestamps. Besides, it is more efficient than the brute-force list-based algorithm by optimizing the manipulations on the prefix tree for sequential rules with explicit timestamps.

1 Introduction

Nowadays, educational data mining is known as an application of data mining in the education domain to gain useful knowledge for educational decision making support. Among the mining tasks, many existing works such as [2, 6, 9, 11] have derived frequent patterns and association rules in educational databases. However, only non-temporal association rules in [6, 11] or only frequent sequential/temporal

H.T.H. Van (✉)
Ton Duc Thang University, Ho Chi Minh City, Vietnam
e-mail: hoangthihongvan@tdt.edu.vn

V.T.N. Chau · N.H. Phung
Ho Chi Minh City University of Technology, Ho Chi Minh City, Vietnam
e-mail: chauvtn@cse.hcmut.edu.vn

N.H. Phung
e-mail: phung@cse.hcmut.edu.vn

© Springer International Publishing Switzerland 2016   411
D. Król et al. (eds.), *Recent Developments in Intelligent Information
and Database Systems*, Studies in Computational Intelligence 642,
DOI 10.1007/978-3-319-31277-4_36

patterns in [2, 9] have been discovered. Therefore, in order to provide educational decision makers with more informative actionable knowledge, our work is dedicated to sequential rules with explicit timestamps in detail from temporal educational databases in an academic credit system (Fig. 1).

As for pattern mining in general in the data mining area, we are aware of many related works such as [3, 8, 17, 18] which unfortunately did not aim at sequential rules with explicit temporal information from the resulting frequent temporal patterns discovered in a temporal educational database as recently defined in [9]. Regarding the existing works on rule mining, [4, 10], some of the most recent rule mining works, have not yet considered the temporal aspect of the resulting rules. [5, 12–14] considered only occurrence ordering of the itemsets in the resulting sequential rules. [16] provided more temporal relations in the resulting temporal association rules. [7, 15] just augmented each resulting rule with explicit temporal information in its entirety, not for each single component of the rule.

In this paper, we propose a tree-based rule mining algorithm from the frequent sequences generated and organized in a prefix tree enhanced with explicit timestamps. In comparison with the aforementioned related works, our resulting sequential rules with explicit timestamps are more informative and detailed to capture the temporal aspects of knowledge hidden in the temporal educational databases. Regarding the efficiency, we have conducted several experiments on real educational datasets and found that the tree-based algorithm outperforms the brute-force list-based algorithm by optimizing the manipulations on the prefix tree. In short, our work has the merit of efficiently discovering sequential rules with explicit timestamps from frequent timestamp extended sequences in a temporal educational database. This task has not yet been supported by any existing rule mining approaches (Fig. 2).

Algorithm 1: Rule_Mining_From_ a_Prefix_Tree_of_Frequent_Sequences
Input: a prefix tree of all frequent sequences, a minimum confidence *min_conf*
Output: a collection R of resulting sequential rules
Method:
1. *Mining rules from a list FS₁ of frequent 1-sequences derived from the prefix tree*
   R_1 := **Rule_Mining_ From_ List_ of_1-Sequences**
2. *Mining rules from frequent k-sequences for k>1*
   R_2 := **Rule_Mining_From_Tree_of_2-Sequences**
3. *Return a collection R of resulting sequential rules*
   R := R_1 U R_2

Fig. 1 Algorithm 1: Rule mining from a prefix tree of frequent sequences

```
Algorithm 2: Rule_Mining_ From_ List_ of_1-Sequences
Input: a list FS₁ of frequent 1-sequences, a minimum
confidence min_conf
Output: a collection R₁ of resulting sequential rules
Method:
1. Sort all sequences of FS₁ in ascending order of
   sequence length
2. For each sequence Sₗ in FS₁ in the form of <(E,0)>
3.     TS := all sequences in FS₁ following Sₗ in order
4.     For each sequence Sᵣ in TS in the form of <(E',0)>
5.         If E⊂E' then
6.             Generate a rule r: E→E'-E
7.             Calculate support and confidence of r:
                   support(r) := support(Sᵣ)
                   confidence(r) := support(Sᵣ)/support(Sₗ)
8.   If confidence(r) ≥ min_conf then R₁ ← r
```

Fig. 2 Algorithm 2: Rule mining from list of 1-sequences

2 Towards an Efficient Tree-based Approach to Mining Sequential Rules with Explicit Timestamps

In our work, we consider the frequent sequences discovered from a timestamp-extended sequence database and organized in a prefix tree enhanced with explicit timestamps as defined in [9]. For mining sequential rules with explicit timestamps from such sequences, we determine one efficient tree-based approach to speed up the sequential rule mining process in the two following cases:

Case1:   In this case, frequent 1-sequences are considered in the form of $<(E,0)>$ where E is a set of events taking place at the same moment. For each S_L of such frequent 1-sequences, a sequential rule r is generated step by step as follows:

Step 1:   Determine each frequent 1-sequence S_R in the form of $<(E',0)>$ where E' is a proper superset of E.

Step 2:   Determine the right hand side of r as the difference between E' and E and form r: E→E'−E.

Step 3:   Calculate support and confidence of r:

$$\text{support } (r) = \text{support } (S_R) \tag{1}$$

$$\text{confidence } (r) = \text{support } (S_R)/\text{support } (S_L) \tag{2}$$

Step 4:   Return r as a resulting rule if confidence(r) $\geq$ *min_conf* where *min_conf* is a given minimum confidence.

Case2: In this case, we take into consideration other frequent k-sequences for k≥2. Each S of such frequent k-sequences is divided into two parts: prefix *pre* and postfix *post* of S. A sequential rule r is generated as: *pre→post*. support and confidence values of r are calculated as follows:

$$\text{support } (r) = \text{support } (pre + + post) = \text{support } (S) \qquad (3)$$

$$\begin{aligned} \text{confidence } (r) &= \text{support } (pre + + post)/\text{support } (pre) \\ &= \text{support } (S)/\text{support } (pre) \end{aligned} \qquad (4)$$

Due to the Apriori property in the frequent sequence mining task, as S is frequent, so are its prefix and postfix, *pre* and *post*, respectively (Fig. 3).

In particular, we define a tree-based algorithm to discover sequential rules with explicit timestamps from a set of frequent sequences organized in a prefix tree as previously mentioned. Discussed in [9], our prefix tree is good enough to capture necessary information about each frequent sequence and expected to efficiently help speeding up the sequential rule process in comparison with the traditional brute-force list-based algorithm which directly forms sequential rules from a list of the resulting frequent sequences. Indeed, each node at level k in the prefix tree corresponds to a frequent sequence with the length of k. Different from prefix trees in [14], that node represents only the k-th item and its timestamp of the corresponding frequent k-sequence. In order to obtain the corresponding frequent k-sequence, we simply traverse the tree from the root to that node at level k. While traversing the tree from the root, we build up such a frequent k-sequence from a frequent (k-1)-sequence in the two ways as follows:

i. *Sequence extension*: an item along with its timestamp at the node at level k is added into the frequent (k-1)-sequence corresponding to its parent at level (k-1). In addition, such a node at level k is specified as an s node.
ii. *Itemset extension*: an item at level k is added into the latest itemset of the current sequence and its timestamp is also the timestamp of this latest itemset. In this way, such a node at level k is specified as an i node.

Based on these two ways of extending a (k-1)-sequence, a sequence α which is composed of (k-1) items is a prefix of all sequences extended from α if its

---

Algorithm 3: Rule_Mining_From_Tree_of_2-Sequences

```
Input: a prefix tree Prefix_Tree of frequent sequences
with the root node root, a minimum confidence min_conf
Output: a collection R₂ of resulting sequential rules
Method:
1. F1 = all child nodes of root at level 1 of Prefix_Tree
2. For each node p in F1
3.        Call Mining_from_tree (null, p) to update R₂
```

Fig. 3 Algorithm 3: Rule mining from tree of 2-sequences

corresponding node at level (k-1) in the prefix tree is an s node. Otherwise, α is an incomplete prefix of all sequences extended from α.

Given a prefix tree of all frequent sequences and a minimum confidence *min_conf*, our tree-based algorithm is sketched in Algorithm 1 with two main parts: the first part in Algorithm 2 for mining sequential rules from frequent 1-sequences as handled in Case 1 and the second one in Algorithm 3 for mining sequential rules from frequent k-sequences for k>1 as described in Case 2.

Different from the list-based algorithm, the tree-based algorithm deals with case 2 while the prefix tree is traversed instead of completely generating a frequent k-sequence, dividing it into a prefix and postfix, and then forming an appropriate sequential rule. Particularly in Mining_from_tree procedure of Algorithm 3 in Fig. 4, let us consider any node p of the prefix tree different from its root.

At the beginning, a sequence S_p corresponding to node p is obtained in either sequence extension or itemset extension manner depending on the type of node p which is either an s node or an i node, respectively.

If a child node of node p, called pchild, is an s node, a sequence S_p is a prefix of the sequence corresponding to pchild. In addition, all sequences S_c in the subtree whose root is pchild are postfixes of S_p. As a result, we generate sequential rules in the form of $S_p \rightarrow S_c$.

If pchild is an i node, a sequence S_p will be extended in an itemset extension manner via another invocation of Mining_from_tree procedure. After that, all child nodes of pchild will be next considered to generate sequential rules.

For example, we illustrate how sequential rules hidden in a timestamp-extended sequence database are generated efficiently from the prefix tree in Fig. 5a.

```
Procedure: Mining_from_tree(Sequence Prefix, Node p)
Input: Prefix which is a frequent sequence at the parent
  node of node p, node p which will be appended to Prefix
Output: updating R2
Method:
      //Generate a sequence Sp corresponding to node p
1. If p.type = s then Sp = Prefix.AddAppendSequence(p)
2. Else  Sp = Prefix.AddAppendItem(p) // p.type = i
3. For each child node pchild of node p
4.     If pchild.type = s then
5.         PList := a collection of all sequences from the
           subtree whose root is pchild
6.         For each sequence Sc in Plist
7.             Generate a rule r: Sp → Sc
8.             Calculate support and confidence of r:
                  support (r) := support (Sc)
                  confidence (r):= support (Sc)/support (Sp)
9.             If confidence (r)≥min_conf then R2←r
10.        Mining_from_tree (Sp, pchild)
11.    Else Mining_from_tree (Sp , pchild) //pchild.type=i
```

Fig. 4 Procedure: Mining from tree

(a)   **(b)**

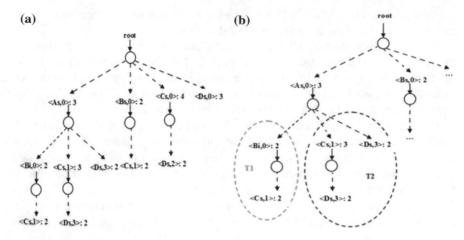

Fig. 5 **a** A prefix tree T of all frequent sequences; **b** Examining node <A,0> of T

Firstly for case 1, we generate all sequential rules from a set of frequent sequences of size 1 which are: <(A,0)>, <(B,0)><(C,0)>, <(D,0)>, and<(AB,0)>. Among these sequences, only <(AB,0)> is processed to obtain the two following sequential rules:

$$R_1 : \; <(A,0)> \; \to \; <(B,0)> \; (0.5; \; 0.67)$$
$$R_2 : \; <(B,0)> \; \to \; <(A,0)> \; (0.5; \; 1)$$

Where:

- R_1 implies that there is 50 % of the students who failed both course A and course B in the same semester and among those students, 67 % of the students who failed course A also failed course B in that semester.
- Similarly, R_2 implies that 50 % of the students who failed both course A and course B in the same semester and among those students, the students who failed course A also certainly failed course B in that semester.

Secondly for case 2, we generate sequential rules by traversing the prefix tree node by node. Let us examine node <A,0> at level 1 of the tree as shown in Fig. 5b.

Based on the type of node <A,0> and its child nodes, a sequence <(A,0)> is an incomplete prefix of all sequences in the subtree T1 whose root is node <B,0> and a prefix of all sequences in the subtrees T2 whose roots are nodes <C,1> and <D,3>.

Let us check the nodes in T1. As a child node <B,0> of node <A,0> is an i node, the sequence <(A,0)> is extended to be a sequence <(AB,0)>. In turn, the sequence <(AB,0)> is a prefix of a sequence corresponding to node <C,1> and its child nodes. Therefore, we next traverse the subtree whose root is node <C,1> to obtain a postfix, a sequence <(C,1)>. As a result, we generate a sequential rule with the prefix <(AB,0)> and the postfix <(C,1)> which is R_3 in Table 1.

Let us check the nodes in T2. As a child node <C,1> of node <A,0> is an s node, the sequence <(A,0)> is a prefix of the sequence corresponding to node <C,1> and

Table 1 Sequential rules discovered at node <A,0>

Prefix	List of Post	Rule
<(AB,0)>	<(C,1)>	R_3: <(AB,0)> → <(C,1)> (0.5; 1)
<(A,0)>	<(C,1)>, <(C,1), (D,3)>	R_4: <(A,0)> → <(C,1)> (0.75; 1)
		R_5: <(A,0)> → <(C,1), (D,3)>(0.5; 0.67)
<(A,0), (C,1)>	<(D,3)>	R_6: <(A,0), (C,1)> → <(D,3)>(0.5; 0.67)
<(A,0)>	<(D,3)>	R_7: <(A,0)> → <(D, 3)>(0.5; 0.67)

all its child nodes. Traversing the subtree rooted by node <C,1>, we collect two postfixes which are<(C,1)> and <(C,1), (D,3)>. Therefore, two corresponding sequential rules R_4 and R_5 are created as shown in Table 1. After that, the sequence <(A,0)> is extended to be a sequence <(A,0), (C,1)> corresponding to node <C,1>. In turn, the sequence <(A,0), (C,1)> is a prefix of all sequences corresponding to node <D,3> and all its child nodes because node <D,3> is an s node. In the same way, traversing the subtree whose root is node <D,3>, we obtain a postfix which is <(D,3)>. At that moment, we generate a sequential rule R_6 with the prefix <(A,0), (C,1)> and the postfix <(D,3)> as displayed in Table 1. Similarly for node <D,3> which is another child node of node <A,0>, we have a prefix <(A,0)> of all sequences at node <D,3> and its child nodes as node <D,3> is an s node. Traversing the subtree rooted by node <D,3>, we get one postfix <(D,3)> and form a sequential rule R_7 with the prefix <(A,0)> and the postfix <(D,3)>.

As a result, a set of sequential rules discovered at node <A,0> is presented in Table 1. Regarding the semantics of the resulting rules, we describe the meaning of R_3 for instance. R_3 is a sequential rule with explicit timestamps implying that there is 50 % of the students who failed both course A and course B in one semester and then failed course C in the following semester; among those students, if they failed both course A and course B in the same semester, certainly they failed course C one semester later.

3 An Evaluation of the Proposed Algorithm

From the theoretical point of view, we evaluate our algorithm as follows. As soon as sequential rule mining is performed on the prefix tree, we cut the cost of checking a valid prefix of a sequence. This is because if a node has a child node whose type is an s node, its corresponding sequence is always a prefix of all sequences derived from the subtree whose root is that child node. Such a feature leads to a higher efficiency of the tree-based algorithm as compared to the brute-force list-based algorithm which needs to check prefixes and postfixes of each sequence by manually dividing each sequence into a prefix and a postfix time and time again. In comparison with an existing prefix tree in mining sequential rules with no timestamp in [14], our prefix tree is different in the following points. In our prefix tree, each node except the root contains only one item in a sequence while one in the prefix tree in [14] captures an entire sequence. Thus, the storage space can be better utilized for our prefix tree. Besides, it is natural for us to generate a sequence at a node while the tree is traversed in the two aforementioned sequence and itemset extension ways. At the same time, at each node different from the root, it is simpler for us to determine the postfixes of a sequence which is now a prefix corresponding to that node by examining the type of a child node and if it is an s node, we collect all postfixes from the subtree rooted by that child node. In contrast, [14] had to remove a prefix corresponding to a node p from all sequences at the child nodes of p to obtain all appropriate postfixes. Especially, our tree captures an explicit timestamp at each node so that sequential rules with explicit timestamps can be derived from the prefix tree while there is no timestamp in [14]. Hence, our tree can be regarded as a timestamp-extended prefix tree which is a temporal version of the existing prefix trees in sequential rule mining.

For an evaluation from the empirical point of view, several experiments and discussions about their results are given in the following part. We carried out the experiments on a 2.53 GHz Intel Core i3 PC with 4.00 GB RAM, Windows 7 Home Premium, a 64-bit operating system.

Our experiments used educational data which are unsuccessful study results in the three semesters 4, 5, and 6 of the undergraduate students. They enrolled in 2005–2008 following the program in Computer Science at Faculty of Computer Science and Engineering, Ho Chi Minh City University of Technology, Vietnam, [1]. For the first semester, we used the study results of 151 students among 361 students enrolled in 2005 who have failed at least one course. Similarly, the study results of 1334 students enrolled in 2005–2008 have been examined and used.

In addition, in the following tables, with the students enrolled in 2008, 567_x for x in {4, 5, 6} has 5678_x as an incremental version; in turn, with the students enrolled in 2007, 56_x has 567_x as an incremental version; with the students enrolled in 2006, 5_x has 56_x as an incremental version; and 5_x is an original data set with the unsuccessful study results of the students enrolled in 2005. More details about our data sets can be found in [9].

In our experiments, we examine the proposed algorithm in two cases: (1). Keeping the minimum support *min_sup* unchanged at 0.03 and the minimum confidence *min_conf* at 0.4, each algorithm is performed for various data sets in six semesters; (2). With a single data set 5678_6 in semester 6, the minimum support *min_sup* is varied from 0.02 to 0.1 with a gap of 0.01 and the minimum confidence *min_conf* is fixed at 0.6. In addition, we compare our proposed tree-based algorithm with the brute-force list-based algorithm. For each algorithm in each case, we recorded the number of frequent sequences Sequence# and rules Rule# generated and how long the algorithm is executed for sequential rules with explicit times-tamps in milliseconds. Tables 2 and 3 present our experimental results where LTime stands for the processing time of the brute-force list-based algorithm, TTime for the processing time of the tree-based algorithm, and LTime/TTime is the ratio of LTime to TTime.

In the first case, it is realized that the processing time of the tree-based algorithm is always much less than one of the list-based algorithm. It is also worth noting that

Table 2 Experimental results with *min_sup* = 0.03, *min_conf* = 0.4, and different datasets

Dataset	Sequence#	Rule#	LTime	TTime	LTime/TTime
5_4	134	61	2.02	≈ 0	–
56_4	72	25	≈ 0	0.16	–
567_4	193	119	3.60	≈ 0	–
5678_4	1403	1377	282.04	4.39	*64.25*
5_5	163	92	3.14	0.15	*20.93*
56_5	84	48	1.42	≈ 0	–
567_5	310	251	12.93	0.63	*20.52*
5678_5	3073	3920	1887.92	10.90	*173.20*
5_6	197	131	4.51	0.93	*4.85*
56_6	113	97	1.71	≈ 0	–
567_6	545	611	48.69	2.18	*22.33*
5678_6	7176	12629	14299.60	32.45	*440.67*

Table 3 Results with 5678_6 dataset, Varying *min_sup* in [0.02, 0.1], and *min_conf* = 0.6

min_sup	Sequence#	Rule#	LTime	TTime	LTime/TTime
0.02	137236	233662	6529098	881.69	*7405.2*
0.03	7176	9395	14068.38	29.93	*470.04*
0.04	1262	1106	283.15	3.28	*86.33*
0.05	319	214	14.48	0.47	*30.81*
0.06	167	117	4.47	0.30	*14.90*
0.07	95	61	1.54	≈ 0	–
0.08	52	21	0.62	≈ 0	–
0.09	36	11	≈ 0	0.32	–
0.1	26	8	0.93	0.16	*5.81*

as the number of frequent sequences discovered from the database is small, the two algorithms are equivalent to generate corresponding sequential rules. However, as the size of the database increases and the number of frequent sequences gets larger and larger, the tree-based algorithm consistently outperforms the list-based algorithm more and more. In general, the processing time of the tree-based algorithm is a few times to a few hundred times better than one of the list-based algorithm and thus, the tree-based algorithm is more efficient than the list-based algorithm.

Similar results have been obtained in the second case. There has been a consistent difference in the processing time of the tree-based algorithm from that of the list-based algorithm. As the value of *min_sup* is small, a large number of frequent sequences are generated, leading to a large number of sequential rules, the tree-based algorithm can save a lot of processing time as compared to the list-based algorithm from a few times to a few hundred times.

In short, our work has provided a tree-based sequential rule mining algorithm to discover sequential rules with explicit timestamps in a temporal educational database where the tree-based algorithm is more efficient than the list-based algorithm. This also confirms that with an appropriate prefix tree to manage frequent sequences with explicit timestamps, a sequential rule mining process can be speeded up.

4 Conclusion

In this paper, we have taken into consideration a sequential rule mining process on temporal educational databases to discover sequential rules with explicit timestamps. As compared to traditional sequential rules with no timestamp, our resulting sequential rules with explicit timestamps are more informative and helpful for decision makers in the education system. In order to facilitate the rule mining process, a tree-based algorithm has been defined to generate sequential rules with explicit timestamps automatically while the existing works just focused on prefix trees of frequent sequences with no timestamp. In addition, the experimental results have shown that as the size of the database is larger or the minimum support is smaller, the tree-based algorithm is much more efficient than the brute-force list-based algorithm to tackle a larger number of frequent sequences as well as sequential rules generated.

In the future, we plan to make the most of the resulting sequential rules in an educational decision support system for its effectiveness in educational decision making support. Also, more experiments on larger datasets in various application domains will be conducted to examine how scalable the proposed algorithm is. Besides, more interestingness measures will be considered to filter the resulting rules.

References

1. Academic Affairs Office, Ho Chi Minh City University of Technology, Vietnam, http://www.aao.hcmut.edu.vn/dhcq.html (2014)
2. Campagni, R., Merlini, D., Sprugnoli, R.: Sequential patterns analysis in a student database. In: ECML-PKDD workshop, Mining and exploiting interpretable local patterns, I-Pat 2012, Bristol (2012)
3. Chen, Y.C., Jiang, J.C., Peng, W.C., Lee, S.Y.: An efficient algorithm for mining time interval-based patterns in large database. In: Proceedings of the 19th international conference on information and knowledge management, pp. 49–58. ACM, New York (2010)
4. Djenouri, Y., Drias, H., Habbas, Z.: Hybrid intelligent method for association rules mining using multiple strategies. Int. J. Appl. Metaheuristic Comput. (IJAMC) **5**(1), 46–64 (2014)
5. Fournier-Viger, P., Faghihi, U., Nkambou, R., Nguifo, E.M.: CMRules: Mining sequential rules common to several sequences. Knowl. Based Syst. **25**(1), 63–76 (2012)
6. García, E., Romero, C., Ventura, S., Castro, D.C.C., Calders, T.G.K.: Association rule mining in learning management systems. In: Romero, C., Ventura, S., Pechenizkiy, M., Baker,R. (eds.) Handbook of Educational Data Mining, pp. 93–106. CRC Press (2011)
7. Gharib, T.F., Nassar, H., Taha, M., Abraham, A.: An efficient algorithm for incremental mining of temporal association rules. Data Knowl. Eng. **69**(8), pp. 800–815. Elsevier Press (2010)

8. Hirate, Y., Yamana, H.: Generalized sequential pattern mining with item intervals. J. comput. **1**(3), pp. 51–60. Academy Publisher (2006)

9. Hoang, T.H.V., Vo, T.N.C., Nguyen, H.P.: Frequent temporal pattern mining incrementally from educational databases in an academic credit system. In: Advanced technologies for communications (ATC) international conference, pp. 315–320 (2014)

10. Martín, D., Rosete, A., Alcalá-Fdez, J., Herrera, F.: QAR-CIP-NSGA-II: A new multi-evolutionary algorithm to mine quantitative association rules. Inf. Sci. **258**, 1–28 (2014)

11. Mashat, A.F., Fouad, M.M., Philip, S.Y., Gharib, T.F.: Discovery of association rules from university admission system data. Int. J. Mod. Educ. Comput. Sci. (IJMECS) **5**(4), 1 (2013)

12. Pham, T.T., Luo, J., Hong, T.P., Vo, B.: An efficient method for mining non-redundant sequential rules using attributed prefix-trees. Eng. Appl. Artif. Intell. **32**, 88–99 (2014)

13. Spiliopoulou, M.: Managing interesting rules in sequence mining. Principles of Data Mining and Knowledge Discovery, pp. 554–560. Springer, Berlin (1999)

14. Van, T.T., Vo, B., Le, B.: Mining sequential rules based on prefix-tree. New Challenges for Intelligent Information and Database Systems, pp. 147–156. Springer, Berlin (2011)

15. Verma, K., Vyas, O.P.: Efficient calendar based temporal association rule. ACM SIGMOD Rec. **34**(3), 63–70 (2005)

16. Winarko, E., Roddick, J.F.: ARMADA–An algorithm for discovering richer relative temporal association rules from interval-based data. Data Knowl. Eng. **63**(1), 76–90 (2007)

17. Yin, K.C., Hsieh, Y.L., Yang, D.L., Hung, M.C.: Association rule mining considering local frequent patterns with temporal intervals. Appl. Math. **8**(4), 1879–1890 (2014)

18. Yuan, D., Lee, K., Cheng, H., Krishna, G., Li, Z., Ma, X., Zhou, Y., Han, J.: CISpan: Comprehensive incremental mining algorithms of closed sequential patterns for multi-versional software mining. SDM **8**, 84–95 (2008)

Design and Implementation of Mobile Travel Assistant

Tomas Pochobradsky, Tomas Kozel and Ondrej Krejcar

Abstract Paper deal with a development of application designed for users of the Android operating system. Its main objective is to alert the user to the arrival of the train in a station while the transfers are also taken into account. We use position sensor to determine the exact GPS location. Based on the travelled distance exact approach of speed train to a station is calculated as well as subsequent wake up of user/traveller. Thanks to this fact, application is able to respond to any train delays. Delayed departure of the train from the initial station is also taken into account. The user can choose how many minutes or how many kilometres before the station wants to wake up.

Keywords Android · Public transport · Wakeup notice · LBS · GPS

1 Introduction

This project deals with the development of applications for the Android operating system. Its name is a mobile travel assistant. It is intended to facilitate travel on public transport. The application use a parsing data from websites using modules XmlPullParser and JSoup parser. They are used to parse connected devices to the Internet. The application also implement a feature that detects the position of device using GPS. Therefore, application can be called as location-dependent (LBS) [1–9].

T. Pochobradsky · T. Kozel · O. Krejcar (✉)
Faculty of Informatics and Management, Center for Basic and Applied Research,
University of Hradec Kralove, Rokitanskeho 62, 500 03 Hradec Kralove, Czech Republic
e-mail: ondrej@krejcar.org

T. Pochobradsky
e-mail: tomas.pochobradsky@uhk.cz

T. Kozel
e-mail: tomas.kozel@uhk.cz

© Springer International Publishing Switzerland 2016
D. Król et al. (eds.), *Recent Developments in Intelligent Information
and Database Systems*, Studies in Computational Intelligence 642,
DOI 10.1007/978-3-319-31277-4_37

These types of applications are currently within the trend. Actions such as obtaining information from GPS, are often time consuming. For this reason we implement asynchronous thread, which is secured. Such architecture allow the smooth running of the final application.

2 Problem Definition

Application of Mobile Travel Assistant should use timetables obtained from the Internet to be able to locate the desired train or bus. For communication with the server it should be available the following information:

- Name of stations, which are traversed
- The time when the selected connection is present in those stations.

With this information, you can alert the user to the many events that may occur within this issue. The latter can be, for example, departures or arrivals of chosen transportation method. In the real world, however, times obtained from the timetables can be very different with the real ones. It often happens that a public transport delays. With the help of a GPS system that each device with the afore-mentioned operating system contains, should the resulting alert the user to adapt to this type of event. In order to implement this adjustment, you need to know three basic information:

- Actual location of device
- Location of traversing stations
- Actual time at device.

Current position will be gathered using the built-in GPS [6, 7]. Obtaining the position of stations can be done in two ways:

1. Firstly we can use pre-implemented android.location.Geocoder class that can determine its GPS coordinates based on the name of the city.
2. Second way consist of connecting directly to the application interface of Google Maps (http://maps.google.com/maps/api/geocode/) [10] and communicate with it using JSON [11].

Thanks to this fact, application becomes a location-dependent. Based on these obtained information we can evaluate in which location the user should be notified. The application also includes its own App Widget [12], which alerts the user to the home screen, where the train or bus is currently located and when it should arrive [13, 14].

3   New Solution

The overall implementation of the application is dependent and thus limited by the platform for which it is developed. In this case, the application is adapted for the use on mobile devices running Android. The application is developed in Java object language, which complements the Android SDK which provides access to the aforementioned source.

Because it is an application developed for mobile devices, developer is limited by the power, limited memory capacity as well as the battery life. Although performance in portable devices continues to grow, it is important that application use only a few computing resources and thus conserve battery. Running applications must be as smooth as possible, hence the use of asynchronous fiber. For this reason the main thread is not so busy and does not freeze or in the worst case of a crash. If the main thread does not correspond to a longer time (it is about 5 s), the user is prompted by the operating system, whether the task should not be terminated.

The application will be developed as a location-dependent, hence there is a need to detect the position of the device [7]. As mentioned in Sect. 2, there are two kinds of location detection. Using GSM and GPS system [15]. Since our research showed obtaining location using network as inaccurate and unsuitable for this application [13, 14], we implement a location system using GPS. Although this is a fairly accurate method it must be taken into account the constraints in terms of signal loss in buildings [16].

4   Implementation

Implementation of all mentioned problems is quite extensive so we cover only basic problems that are closely associated with the implementation.

4.1   Services and GPS

Services can be implemented in several ways. The first way is to use the IntentService class that inherits from android.app.Service. This is a simple service in which the operation is performed one after the other. After completing their service ends. For purposes of this application, however, it was necessary to use multiple-fiber processing. This can be achieved by creating services that inherits directly from class Service. There is a need to implement a new thread that is used to run the application. This fiber is frequently type of HandlerThread. This means that the resulting fiber is controlled via messages sent by handlers.

Developed application implemented also one service. This service facilitates communication with GPS. It detects the position of the stations, which passes

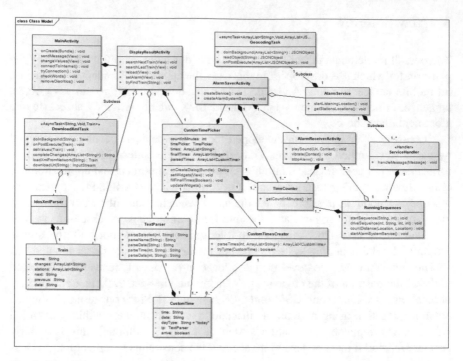

Fig. 1 Class diagram of developed application

through the connection. In next step it update the times to initialize notice based on collected information. For this reason, it is one of the most important parts of the application. Because the service must perform many operations, it is composed of a total of three fibers. As for all parts the main fiber is presented also. This arises immediately after initializing the service while the methods can be found here similar to those located in the activities. One example would be the method onCreate(), which is creating a new type of fiber HandlerThread (hereinafter referred to as fiber A), which cares for the entire life cycle of the main services. Class diagram solution is found in Fig. 1.

4.2 Initialisation and Data Retrieval

Another important method is the main thread method called onStartCommand(). By this method it is passed to the service intent that carries data essential for the running of the whole service. Directly it can then perform initialization of needed parts and subsequently by sending messages via Handler it can run previously mentioned fiber A. Consequently, it also start third thread as asynchronous. This thread established GPS coordinates of stations, which passes through the selected transport connection. This is done through communication with the web application interface of Google Maps. There are obtained JSON objects. After performing this operation, the resulting data are sent to the fiber A. This method of obtaining the position was used mainly for that reason, because the class Geocode, mediated through the Android SDK, had problems obtaining the position through an emulator. This would result in a worsening of conditions for testing the application. Furthermore, in the method onStartComand() it is registered LocationListener which is acquiring information at regular intervals about the current position of the device. These data are similarly sent for processing in thread A, as in the asynchronous thread case [12].

4.3 Sequentions

Fiber A is divided into a several sequences, divided into two types.

The first type is the starting sequence. It includes a scenario of operations to be performed before the train departures from the station. This sequence is performed only if the alarm is activated before departure from the starting station. Diagram of threads and sequences are showed in Fig. 2.

Starting sequence contain these steps:

- Waiting until a certain time—3 min before departure of train (This is a reserve time needed to identify the location)
- Waiting for the current location, if not already known
- Waiting until the train is not going to move from the starting station—compared to the change in position
- Depending on the departure time of train a calculation of the default delay is made

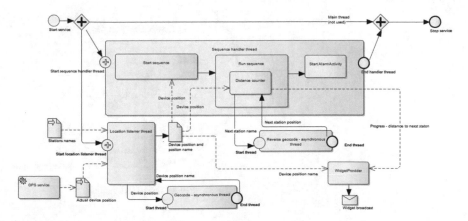

Fig. 2 Threads and sequences diagram

- The second type is a sequence which runs while driving. This sequence can be started several times depending on the number of intermediate stations.
- Sequence of trip contain these steps:
- If the arrival time of the next station is longer than the time that the user has chosen to notice—algorithm pending until this time.
- Three minutes before the scheduled user-chosen time (reflected delay) algorithm starts to compare the current position of the train and the position of the next station.
- When the connection enters the border zone the notice is initialized.

4.4 Computation of Border Zone

The border has a circular shape, as shown in Fig. 6. It is known position of the center of the circle, which represents the position of the station or the city [3]. The radius of this circle is calculated based on the speed of the train and minutes [4–9]. How much sooner the user should be notified. Furthermore, the location of the train is known. The distance of the train from the center of the circle is calculated using the Pythagorean Theorem. This distance is sort of a hypotenuse of a triangle, with the remaining two sides are differences between the latitudes and longitudes of the two known positions. This abstraction of the real situation was used for saving computational time. Calculation ranges is mostly in units of kilometers, what mean that it would be unnecessary to use Vincent's formula [17]. Representation of used calculation is shown in Fig. 3.

Fig. 3 Schema of border zone computation

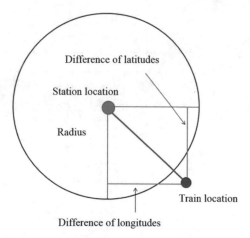

5 Testing of Developed Application

This solution is relatively unique. It was not found any other similar solution on same idea. Therefore, there will not be presented a comparison with other author's solutions.

Due to application functionality testing was carried out mainly through the emulator in the development environment. This was carried out as follows:

- The application was started.
- The developer chose the required inputs
- Coordinates of the mobile device was emulated manually.
- It was examined whether the application responds correctly to the change of coordinates (Figs. 4, 5 and 6).

After debugging of any problems occurred the second phase of testing, which was attended by a total sum of five people. On their mobile phones application was installed, while they travel with application for a week by train. It was found that users most frequently used a mode of application without GPS (notice according to timetables, i.e. by regular train timetable).

They used this mode mainly to conserve battery life. In this mode, no problems were detected. When using the GPS mode the application also run without any major problems. Only error which was reported was an intermittent signal loss what is not problem of application design. After consultation regarding the use of knowledge in the application were tuned last bits.

Fig. 4 User interface for entering the entry and destination stations

Fig. 5 User interface
(DisplayResultActivity)

Fig. 6 Set of wake up 5 min
before train arrive to
destination

6 Conclusion

Developed application deal with the alarms on the arrivals and departures of trains. During testing period it proved to be very useful application. According to actual state the similar application as developed one is still missing in the Google Play store.

The application can is a very perspective in this was. However, the application was developed purely for academic purposes. That fact is due to the account of the license terms that apply to your use the data relating to timetables. Data timetables of Czech Republic is owned by company CHAPS spol. s.r.o.. For deployment into production it would require data from them to buy or rent. If it happen, it would simplify the way of data acquisition. Already there would be no parsing site, but would apply some of their API.

Acknowledgment This work and the contribution were supported by project "SP/2016—Smart Solutions for Ubiquitous Computing Environments" from University of Hradec Kralove, Faculty of Informatics and Management.

References

1. Steiniger, S., et al.: Foundations of LBS. CartouCHe. [Online] 26 Jan 2012. (Cited 11 Feb 2015). http://www.e-cartouche.ch/content_reg/cartouche/LBSbasics/en/html/index.html
2. Chen, Z., Xia, J.C., Irawan, B., Caulfied, C.: Development of location-based services for recommending departure stations to park and ride users. doi:10.1016/j.trc.2014.08.019
3. Benikovsky, J., Brida, P., Machaj, J.: Proposal of user adaptive modular localization system for ubiquitous positioning. LNCS **7197**, 391–400 (2012)
4. Brida, P., Machaj, J.: A novel enhanced positioning trilateration algorithm implemented for medical implant in-body localization. Int. J. Antennas Propag. **2013**(819695), 10 (2013)
5. Machaj, J., Brida, P.: Optimization of rank based fingerprinting localization algorithm. In: 2012 International Conference on Indoor Positioning and Indoor Navigation (IPIN) 2012, pp. 1–7, Sydney, Australia, 13–15 Nov 2012
6. Brida, P., Machaj, J., Gaborik, F., Majer, N.: Performance analysis of positioning in wireless sensor networks. Przegląd Elektrotechniczny (Electrical Review), **87**(5), 257–260 (2011). ISSN 0033-2097
7. Machaj, J., Brida, P.: Performance comparison of similarity measurements for database correlation localization method. In: Nguyen, N., Kim, C.-G., Janiak, A. (eds.) 3rd Asian Conference on Intelligent Information and Database Systems, ACIIDS 2011, vol. 6592, ch. 46, pp. 452–461, Daegu, South Korea, ser. Lecture Notes in Computer Science, Springer, Berlin, Apr 2011. ISBN 978-3-642-20041-0
8. Brida., P., Benikovsky, J., Machaj, J.: Performance investigation of WifiLOC positioning system. In: 34th International Conference on Telecommunications and Signal Processing, TSP 2011, pp. 203–207, Budapest, Aug 2011. ISBN 978-1-4577-1409-2
9. Machaj, J., Brida, P., Majer, N.: Novel criterion to evaluate QoS of localization based services. In: 4th Asian Conference on Intelligent Information and Database Systems, ACIIDS 2012, pp. 381–390, Kaohsiung, Taiwan (2012)

10. Lee, S., Tewolde, G., Kwon, J.: Design and implementation of vehicle tracking system using GPS/GSM/GPRS technology and smartphone application. doi:10.1109/WF-IoT.2014. 6803187

11. Lee, S., Tewolde, G., Kwon, J.: JSON on mobile: is there an efficient parser? doi:10.1109/WF-IoT.2014.6803187

12. Open Handset Alliance.: Android NDK. Android Developers. [Online] 2013. (Cited 11 Feb 2015). http://developer.android.com/tools/sdk/ndk/index.html

13. Kim, I.H., Shin, S.Y., Song, Y.H., Chung, K.D., Jeong, G.M.: A capacity modeling of bluetooth access points for location based service with mobile phone and bluetooth. J. Electr. Eng. Technol. **8**(1), 183–189 (2013). doi:10.5370/JEET.2013.8.1.183

14. Krejcar, O.: Problem Solving of low data throughput on mobile devices by artefacts prebuffering. EURASIP J. Wireless Commun. Network. **2009**(802523), 8 (2009). doi:10.1155/2009/802523

15. Elumalai, G., Bhaskar, G.B., Subash, R.: Controlling of faulty vehicles using GSM and GPS technology. doi:10.4028/www.scientific.net/AMM.591.180, ISBN 978-303835178-8

16. Poslad, Stefan: Ubiquitous Computing: Smart Device, Enviroment, and Interactions. Wiley, Chichester (2009). ISBN 978-0-470-03560-3

17. Chris Veness, Vincenty solutions of geodesics on the ellipsoid in JavaScript | Movable Type Scripts [Online] 2015. http://www.movable-type.co.uk/scripts/latlong-vincenty.html

Design and Implementation of an Algorithm for System of Exposure Limit in Radiology

David Oczka, Marek Penhaker, Lukas Knybel, Jan Kubicek
and Ali Selamat

Abstract This thesis describes the implementation of application for effective planning process of CyberKnife system by processing CT images. CyberKnife system, which focuses therapeutic target using two X-ray tubes and takes images of the patient at an angle of 45°, allows localization of lung irradiated tumor bearing based on the difference in density between the bearing and the surrounding tissue. Some tumor bearings are difficult to localize due to summation and overlay of monitoring area with other high densities structures. The application is designed as a client created in C#, connected to the COM server of computing system MATLAB, which provides most of the calculations. The application was tested on a single core chip system and speed of acquisition and processing is in average around 1 s. The application was developed, implemented and now successfully tested on the Oncology Clinic at FN Ostrava.

Keywords Radiotherapy · Radiodiagnostics · Cyberknife · CT image processing · MATLAB

D. Oczka · M. Penhaker (✉) · L. Knybel · J. Kubicek · A. Selamat
Faculty of Electrical Engineering and Computer Science, VSB—Technical University
of Ostrava, Ostrava, Czech Republic
e-mail: marek.penhaker@vsb.cz

D. Oczka
e-mail: david.oczka@vsb.cz

J. Kubicek
e-mail: jan.kubicek@vsb.cz

L. Knybel
Faculty Hospital, Oncology Clinic, Ostrava, Czech Republic
e-mail: lukas.knybel@vsb.cz

A. Selamat
UTM-IRDA Digital Media Center of Excellence, UTM & Faculty of Computing,
Universiti, Teknologi, 81310 Johor Bahru, Johor, Malaysia
e-mail: aselamat@utm.my

© Springer International Publishing Switzerland 2016   433
D. Król et al. (eds.), *Recent Developments in Intelligent Information
and Database Systems*, Studies in Computational Intelligence 642,
DOI 10.1007/978-3-319-31277-4_38

1 Introduction

Radiation therapy is a treatment method that uses to treat tumor and non-tumor tissue effects of ionizing radiation. This radiation passes through tissue emits its energy and thereby triggers a series of processes leading to the degeneration of the tissue. Ionizing radiation can be emitted radioactive material or device for that purpose.

The aim of radiotherapy is to deliver as much energy ionizing radiation to the target area of the tumor, while the surrounding tissue is affected as little as possible.

After surgery radiation therapy is the most effective method of oncology treatment and is used today only half the cases. This method is often combined with chemotherapy or immunotherapy [1–3].

1.1 CyberKnife Robotic System

CyberKnife robotic system is irradiator, stereotactic method which uses radiation to destroy tumor masses. It is able to irradiate tumors almost throughout the body. Management system is based on technology management picture. Image data obtained by means of two X-ray tubes the system located on the right and left side of the patient. These X-ray of the patient scanned at an angle of 45° [4].

Due to the continuous target tracking, CyberKnife system is able to irradiate and moving targets, such as the lungs or liver. Whet the tumor is not clearly visible, in soft tissues it may be implanted into a target area of gold grains that are system CyberKnife related points. The irradiation accuracy for the moving targets is 2 mm. The entire process of irradiation is planned in advance by the planning CT or MRI.

Collimators are screens provide output radiation from the radiation in the correct direction and shape. There are primary and secondary collimators. The primary collimators are permanently placed inside the feed and ensures that the radiation did not get out somewhere else than where it is required [4–6].

1.2 Imaging System

It is a stereotactic X-ray imaging system that consists of two orthogonally positioned relative to each X-ray tubes and flat-panel imaging. X-ray tubes are suspended from the ceiling over the sides of the beds at an angle of 45°. Flat-panels can then be on stands or embedded in the floor.

The central beam of X-ray tubes going through each item that has a value of zero for all axes. This point is called imaging centers and irradiated bearing should always be at this point or in its immediate vicinity [7].

The system produces images at the beginning of the procedure for determining the correct position of the patient and then during the procedure, which is controlled by the position of the patient. The correct position is checked by comparing the digitally reconstructed images (DRR) generated from a planning CT images from X-ray tubes [8–10].

2 Problem Definition

The goal of this work is to solve the possibility of summation and thus overlap irradiation targets tissues with a higher density. In this case the tumor bearing system CyberKnife visible on only one X-ray tube or completely invisible. This problem can be solved wedges, which are subject to planning CT subsoil, and it will be rotated about the angle required to reveal targets for sighting shots.

Wedge that will be used, is currently collecting estimate and it is possible that the target will not be visible even after the rotation, leading to a further repeat CT scan, which is not only time consuming but also increases the radiation dose to the patient. Therefore it is necessary to ensure the proper selection of the angle by which the patient will be rotated in order to ensure the best visibility for both targets X-ray tubes.

Currently there CyberKnife system software extension that is called one view. When using this extension it is possible to control information obtained from only one X-ray tube, therefore bypass invisibility targets on one of the X-ray tubes, but at the cost of efficiency in the form of enlargement of the protective rim and thus a larger volume of irradiation of healthy tissue. In addition, this expansion is an expensive affair, resulting in an oncology clinic only disadvantages and no further plus.

3 Design and Implementation

The idea was to create application solutions in MATLAB computing system that retrieves, processes the CT data, and based on them creates images at an angle of 45°, which will also have a wedge angle rotated. This angle could then engineer change and instantly see the change sighting shots [11, 12].

Part of the application should also be the basic processing and image editing such as changing the contrast, brightness, image inversion and the possibility of dandruff, therefore enlarge the selected image area. An important part should be also possible to display the contours targets for easier orientation technology to lighting. The entire application should then have a graphical interface developed in MATLAB system.

3.1   The Proposal Solution

The application is designed as a graphical client implemented in C#, connected to a COM server MATLAB, which means triggering functions in the MATLAB language provides retrieval and processing CT data but also calculate images at an angle of 45°. In addition, the system provides a MATLAB calculation also rotated contours goals, which in turn allows the client to render generated images.

Additional image processing functions, which are contrast and brightness adjustment, image inversion, the possibility of dandruff and gamma adjustment, provided directly by the client, so they are implemented in C#.

Any time data, if not stated otherwise, in the calculation to one of the core chips, under the most favorable conditions, and are each the average of several measurements of 389 frames [13, 14].

3.2   Data Handling

The basis of the whole solution was to ensure proper CT load files that are in DICOM format. Thanks to a package of functions for image processing (Image Processing Toolbox), MATLAB supports retrieval of files in DICOM format using "Dicomread".

Another important feature of the above package is "Dicominfo" feature that allows you to retrieve all the information hidden in the structure of DICOM file format. Between this information can be found, for example, personal data of the patient data on the hospital and the device that created the images, technical information about the image, such as size, bit depth, deviation or multiple data.

The aim was to shorten the existing algorithm for a long time calculating the preview display from different angles. Since version with loading and holding in memory of two huge three-dimensional matrix of 512×512 pixels large images with a number of cuts greater than 100 images occupy a large portion of physical memory, it was necessary to find a way not to use when calculating three-dimensional matrix, thereby reducing the burden on the physical computer memory thereby speeding up the calculation.

During the experiments it was found that after the data is loaded into MATLAB for data type uint16 speed and load increases if the variable is preformed, namely nut, to be subsequently retrieved data is written. This matrix is created by the function zeros, which is available as an input parameter to specify the desired data type of the output variable. For this procedure, however, it is necessary to know in advance the dimensions of the image that are not in our case known, and are always 512×512 pixels.

The user enters via a graphical wizard path to the folder containing the DICOM files. The folder selection function is mediated through uigetdir. The algorithm for reading and checking CT images (CT_load), then based on the path checks the files inside the folder and create lists of paths to files, images and contours that marked as valid. List of routes (The way, is usually connected only one set of contours) to the contours of the moment is not used and is therefore discarded.

A list of image files in entering into the very algorithm for generating images from an angle of 45°. At the beginning of a blank frame is created, which will be subsequently filled with the resultant values. Through the loop is loaded into memory only one file and it's taken most of every diagonal. This procedure is performed the same way as in the previous version of the algorithm, using the functions of diagnostic and max. From each set of values in the vector, which is stored into the previously prepared image that was created at the beginning, as they go files consecutively. After performing this procedure for all files in the resulting image can be loaded and rendered.

For the second X-ray image of the process is almost identical. The only difference is a horizontal flip each loaded frame which provides flip function.

Upload time frame is about 5 s (speed algorithm ct load) and create two images takes about 8 s.

3.3   Calculation of the Graphics Card

Due to the relatively long processing times have been considered in a calculation option frame on the graphics card. This option is implemented in MATLAB using "GpuArray" that converts into a variable data type of the same name (GpuArray) as these variables are automatically at runtime working on the graphics card. Transfer to the original format provides functions gather.

Unfortunately, this algorithm slowed down the computation and it was not used the graphic card for computation in future (Fig. 1).

Since computation on graphics card algorithm accelerate, nuclear development, another attempt to accelerate the algorithm was support for parallel computing. Parallel computation works that share tasks among multiple processors, thus speeding up the calculation. In MATLAB is to run parallel support functions mediated through Parpool and subsequent block of code to be distributed and the calculation is determined by the cycle Parfor. To accelerate the algorithm adds support parallel computing to accelerate computation nearly doubled with each added core. Resulting speeds for different numbers of nuclei are shown below (Table 1) [15].

For simplicity, the first graphical interface MGUI_v1, was created a new graphical interface, which is more sophisticated and user-friendly. This interface was a modular processing thus consists of individual modules. Among other things, it has this interface supporting English and Czech.

The modules are written in the language of MATLAB and the fact it is a common feature that, based on input data, creates a new window and it displays the result of the operation (e.g. Picture Viewer, 3D Viewer) and returns the result of the

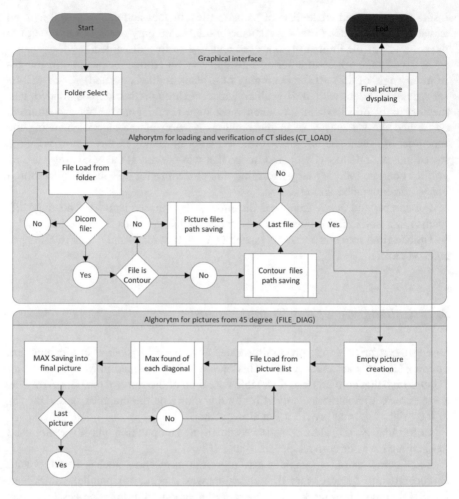

Fig. 1 Flowchart application for creating images from an angle of 45° composed of algorithms Ct_load, file_diag a graphical environment

Table 1 Computing time for the quad-core chip Intel i7-2600 k

Number of kernels	Time for each picture	
	Picture A (non flipping) (s)	Picture B (with flipping)
1	3.770568	3.84427
2	2.29597	2.04572
4	1.262843	1.120922

operation of a new window, back to the main window graphical user interface (e.g. graphical selection guide folder).

- custom modules
- graphical guide to choosing ingredients
- Picture viewer with a simple post processing
- Browser-dimensional model of the patient
- Logging panel with support for exporting to file

4 The Algorithm for Finding the Approximate Direction of the Irradiation

As an additional function by the algorithm for finding approximate direction of irradiation, thus speeding up the calculation algorithm, which is now the CyberKnife system is implemented, and thus more effective therapy.

The user enters via a graphical wizard path to the folder containing the DICOM files. Selecting a folder is mediated through module graphical guide to choosing ingredients MGUI_v2 graphical interface that is written in MATLAB using Java. Difference from uigetdir function is to display files inside a folder and choice ingredients using either right clicking on the folder or file within that folder. The algorithm for reading and checking CT images (ct_load), then based on the path checks the files inside the folder and create lists of paths to files, images and contours that marked as valid [16].

List of routes (i.e. The way, is usually connected only one set of contours) to the contours is given as input to the algorithm functions dicomrt2matlab and waiting to create a file with the results, which will be located in the active component of the algorithm.

After you create the contours are loaded and sorted on the bearing contour and contour bodies. Contours of the body are joined in a contour to them to be able to work more easily. Based on the bearing contour is found, the approximate center of the bearing in all axes and is also calculated compensation in all axes so as to deposit it in all axes in the middle of the 3D matrix.

A list of image files in entering into the very algorithm for finding approximate direction of irradiation. It created an empty three-dimensional matrix data type uint16. Data individual images are loaded, compensated so that the bearing was at the center of the image are added artificial density based on the contours that symbolize the critical organs, and the image is added to the matrix. When the matrix is filled, it is still offset in number of frames, that is so that the center bearing the X-axis (axis of the spine of the patient) in the middle. Subsequently, the nut enters into a cycle, where it is rotated, using imrotate, in these ranges of angles in the X and Z. This rotation simulates the motion of the irradiation head around the patient, but in our case, the patient rotates and the head is static. In each step the algorithm

Fig. 2 Passage beam (*red*) through the bearing (*orange*), including the display contours bodies (*white*)

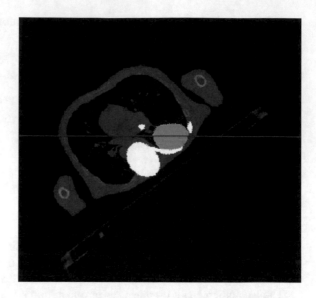

goes through the cycle line, the index value is also the center axis Z. This is simulated by passing the beam through the center of the bearing. Before we start the algorithm go through the line, the control-numbered using the sum, which immediately find out if the value is the sum of the high, the beam path is a high density structure (bone) or body whose density was based on the contours changed to artificial and extremely high value. If the value is below a specified threshold sum, the algorithm scans the value and evaluate their suitability passage of the tabulated data on densities of individual organs or contours. The results of evaluation of individual rotations are then stored in a structure containing information about the rotation axis X, Z and suitability to the exposure determined by the logical values 0 and 1. For an idea of the passage of the beam is shown in Fig. 2. Time calculating beam passage for a given rotation is about 0.8 s.

4.1 The Algorithm for Processing Contours

One of the last steps were to be displayed in the generated rotated images also rotated contour. That's why I wrote the algorithm in conjunction with the dicom-rt2matlab and 3D_index algorithm retrieves, processes and draws outlines rotated contours.

Contours are loaded and processed using dicomrt2matlab and output data of functions are stored on the hard drive. Data file is read as a three-dimensional matrix of the same dimensions as the three-dimensional CT data matrix containing a logical one in each element, which was drawn contour.

Nut subsequently enters 3D_index algorithm which is rotated by the required angle, and exactly the same manner are taken maxima (i.e. logic one) in each row. Give us a view of the rotated contour.

In the final image the algorithm 3D_index are using the edge found the edges of the plotted contour and the resulting data is then processed using the find, which searches the matrix, according to the specified conditions, indexes satisfying the conditions.

The result of the algorithm is a vector of points which determine the indexed position of the edge contours in the resulting image. Time for calculating rotated two contours is the same as the velocity algorithm 3D_index, or about 0.5 s.

After completing computational algorithm was needed to render the final images of sufficient size and resolution. For this purpose it was necessary to create a more user-friendly graphical interface, which in MATLAB could not be reached because it is a computing system, and it created a graphical environment is better suited to work with graphs than image processing. It was therefore, a new graphical interface, implemented in C# that acts as a client connected to a COM server system MATLAB.

Creating rotated images and contours provides MATLAB by function calls within the embedded client code. These images are loaded into the client in two ways.

The first method is easy to store on disk from the system MATLAB and retrieve client. The second method is direct transfer of data from the server to the client, when the recovered data as a two-dimensional matrix with values for individual pixels (voxels). Such data must be converted to a stream of pixels and then assemble the frame. The outcome is both ways almost as quickly and at the moment it is used way to read through a direct transfer (Fig. 3).

After completing a graphical client computing time frame has been rotated patient and calculating contours rotated about 2 s (1 s on two shots and one second on the contour). This speed was acceptable, but still it was from my perspective a little.

In further attempts to speed up the effort once again to try to support the implementation of the calculation of the graphics card, but this time on the graphics

Fig. 3 Views contours of the graphical interface implemented in C#

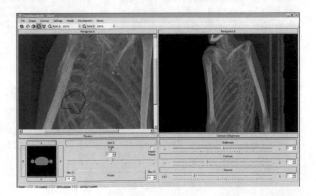

card enumerated only a certain part of the code. This process accelerated entire algorithm twice, and the resulting calculation time at the moment is 0.25 s per frame and also to calculate one rotated contours. The speed algorithm, the calculation of two frames and two contours is around 1 s.

5   Conclusion

The aim of this study was to design and implement an algorithm that could, on the basis of image data from CT scans create images 45° angle of the patient, which is rotated by a certain angle. The algorithm would then in practice, facilitated addressing overlap irradiated targets in the lung tissue with high density, thereby increasing the efficiency of the planning process.

The algorithm has been designed and developed in MATLAB computing system. During the development of the algorithm has gone through several versions, when the gradually decreasing time calculation of the resulting images. Calculation time was two shots in the first version of the algorithm to 3.5 min and in the final version of a mere 0.5 s. This final rate of two images is the result of a long path of development, which has been tested a number of different approaches to solutions.

The result of this work was discovered that the graphical interface developed in MATLAB as inappropriate and was replaced by a graphical interface in C# because of the speed. With this solution, however, a problem with communication and transfer of data between the MATLAB and C# language. These problems, however, were soon resolved and the resulting graphical interface works as a client implemented in C# is connected to the COM server system MATLAB. The client sends to the server requirements for the calculation and the results returned to him. In this case, images obtained from the server at an angle of 45°. Additional image processing is then provided to the client.

Acknowledgment This article has been supported by financial support of TA ČR PRE SEED: TG01010137 GAMA PP1. The work and the contributions were supported by the project SP2015/179 'Biomedicínské inženýrské systémy XI' and this work is partially supported by the Science and Research Fund 2014 of the Moravia-Silesian Region, Czech Republic and this paper has been elaborated in the framework of the project "Support research and development in the Moravian-Silesian Region 2014 DT 1—Research Teams" (RRC/07/2014). Financed from the budget of the Moravian-Silesian Region.

References

1. Nuyttens, J.J. et al.: Tumor tracking system of the cyberknife for early stage of non-small-cell lung cancer. Robotic Radiosurgery: Treating Tumors that Move with Respiration. pp. 81–87. ISBN 978-3-540-69885-2; ISSN 0942-5373
2. Seppenwoolde, Y., Heijmen, B.: Accuracy of tumor motion compensation algorithm from a robotic respiratory tracking system: a simulation study. Med. Phys. **34**(7), 2774–2785 (2007)

3. Giraud, P., et al.: Conformal radiotherapy (CRT) planning for lung cancer: analysis of intrathoracic organ motion during extreme phases of breathing. Int. J. Radiat. Oncol. Biol. Phys. **51**(4), 1081–1092 (2001)
4. Accuray inc. Příručka pro ozařování—Treatment Delivery Manual. Sunnyvale: Accuray Incorporated, CA 94089 USA, 640 p. (2010)
5. International Specialty Products Inc., Gafchromic EBT2 [online]. c2011 [cit. 2012-01-15]. Available from: http://www.gafchromic.com
6. Fox, J.: Applied Regression Analysis and Generalized Linear Models. Sage Publications. (2008). 665 pp. ISBN 0761930426
7. Heijmen, B., et al.: Clinical Accuracy of the Respiratory Tumor Tracking System of the CyberKnife: Assessment by Analysis of Log Files. Int. J. Radiat. Oncol. Biol. Phys. **74**(1), 297–303 (2009)
8. Ozhasoglu, C., et al.: Synchrony—cyberknife respiratory compensation technology. Med. Dosim. **33**(2), 117–123 (2008)
9. Seppenwoolde, Y., Jansen, D., Marijnissen, H.: Accuracy of predicting respiratory tumor motion with the synchrony. CyberKnife tumor tracking system. Radiother. Oncol. **76**(2), S91 (2005)
10. Pustkova, R., Kutalek, F., Penhaker, M., Novak, V.: Measurement and Calculation of Cerebrospinal Fluid in Proportion to the Skull (2010)
11. Michalski, D., et al.: Four-dimensional Computed Tomography-based Interfractional Reproducibility Study of Lung Tumor Intrafractional Motion. Int. J. Radiat. Oncol. Biol. Phys. **71**(3), 714–724 (2008)
12. Kasik, V., Penhaker, M., Novak, V., Bridzik, R., Krawiec, J.: User interactive biomedical data web services application. In: Yonazi, J.J., Sedoyeka, E., Ariwa, E., ElQawasmeh, E. (eds.) E-Technologies and Networks for Development, vol. 171, pp. 223–237 (2011)
13. Kasik, V., Penhaker, M., Novak, V., Pustkova, R., Kutalek, F.: Bio-inspired Genetic Algorithms on FPGA Evolvable Hardware. Intelligent Information and Database Systems (Aciids 2012), Pt Ii, 7197, 439–447LeBLANC, D. C. Statistics: Concepts and Applications for Science. Jones & Bartlett Learning, USA. 2004. 382 pp. ISBN 0-7637-2220-0 (2012)
14. Penhaker, M., Krawiec, J., Krejcar, O., Novak, V., Bridzik, R., Society, I.C.: Web system for electrophysiological data management. In: 2010 second international conference on computer engineering and applications: ICCEA 2010, Proceedings, vol. 1, 404–407 (2010). doi:10.1109/iccea.2010.85
15. Weiss, E. Et al. Tumor and normal tissue motion in the thorax during respiration: analysis of volumetric and positional variations using 4D CT. Int. J. Radiat. Oncol. Biol. Phys. **67**(1), 296–307 (2007)
16. Kubicek, J., Penhaker, M., Feltl, D., Cvek, J., IEEE.: Guidelines for modelling BED in simultaneous radiotherapy of two volumes: tpv(1) and tpv(2) (2013)

Application of the Characteristic Objects Method in Supply Chain Management and Logistics

Wojciech Sałabun and Paweł Ziemba

Abstract This paper presents a new multi-criteria decision-making method: the Characteristic Objects method. This approach is an alternative for AHP, TOPSIS, ELECTRE or PROMETHEE methods. The paper presents the possibility of using the Characteristic Objects Method (COMET method) in supply chain management (SCM) and Logistics. For this purpose, a brief review of the literature is shown. Then the COMET method is presented in detail. At the end of the paper, a simple problem is solved by using COMET method.

Keywords Fuzzy set theory · Characteristic objects method · AHP · ELECTRE · Supply chain management · Logistics · TOPSIS · MCDA

1 Introduction

Methods of multi-criteria decision support are widely used in supply chain management and solving complex problems associated with the wider logistics [1–7, 8, 12, 13, 17, 22, 23, 27]. This happens because of the complexity of the same processes as well as the entire transport systems and logistics. In consideration issues there is a very large number of parameters and measurement evaluations, where we can specify, among other things measures: economic, technical, environmental and social [9–11, 24–26, 28, 29]. These measures typically include opposing, and often conflicting interests. A good example would be to assess the economic and environmental assessment, which usually have opposite goals.

W. Sałabun (✉)
West Pomeranian University of Technology, al. Piastów 17, 70-310 Szczecin, Poland
e-mail: wojciech.salabun@zut.edu.pl

P. Ziemba
The Jacob of Paradyż University of Applied Sciences in Gorzów Wielkopolski,
ul. Chopina 52, 66-400 Gorzów Wielkopolski, Poland
e-mail: pziemba@pwsz.pl

© Springer International Publishing Switzerland 2016
D. Król et al. (eds.), *Recent Developments in Intelligent Information and Database Systems*, Studies in Computational Intelligence 642,
DOI 10.1007/978-3-319-31277-4_39

445

The rest of the paper is organized as follows. In Sect. 2 is presented a brief review of the literature in terms of the used methods and the research issues in the field of supply chain management and logistics. Section 3 presents a new approach in assisting decision, which is the Characteristic Objects method (COMET method). This method was described previously in [18–21]. This is an approach which haven't got many of the disadvantages identified in the multi-criteria decision analysis methods and, above all, resistant to the phenomenon of rank reversal. Rank reversal is a reversal of the rankings when adding or removing an alternative or alternatives from the collection of considered objects [21]. Section 4 shows a simple example intended to better illustrate the operation of the COMET method. Finally, Sect. 5 presents concluding remarks.

2   Literature Review

This section presents selected research papers related to multi-criteria decision-making decision support in supply chain management and logistics. These works are presented in terms of the method and subject of the research problem. Papers were presented in the outline in Table 1. This shows a great variety of problems that are solved by using methods of multi-criteria decision analysis, and at the same time presents the problem of selecting the appropriate method.

Table 1 Illustrative applications and methods of multi-criteria decision support in supply chain management and logistics

Application area	Method	Reference
The choice of location of burdensome objects	ANP	[23]
The choice of location for the construction of a new facility for the company	fAHP	[22]
The choice of location of the new facility for textile companions	AHP fTOPSIS	[6]
Selecting the location of the bus station	fPROMETH	[12]
The choice of location for the construction of the factory	fTOPSIS	[5]
Selecting the location of military stores	TOPSIS	[7]
The choice of location for urban distribution centers	fTOPSIS	[1]
The choice of a global vendor for a company	fAHP	[4]
The choice of a new supplier of products for the telecommunications company	fANP fTOPSIS	[13]
Customers' rating by the supplier	fTOPSIS	[3]
The choice of scenario for changes to fuel used for transportation	AHP	[17]
The choice of transport infrastructure development scenario	ANP	[27]
Rating sustainable performance of suppliers (Green SCM)	fTOPSIS	[8]
Performance Evaluation of national transport systems in terms of impact on the economy, environment and society	ELECTRE	[2]

A multitude of methods and their modification leads to the next problem, which method to use. This article presents the new COMET fuzzy method, which eliminates above all the problem of rank reversal, and in addition to the existing research shows greater accuracy than other methods of multi-criteria decision support [18–21].

In the literature, the most frequently considered problems are the tasks related mainly to: a choice location [1, 5–7, 12, 22, 23], choice of supplier [4, 13], selection strategies (scenario) [17, 27] or performance rating (utility) [2, 3, 8]. In order to solve them, authors use to classical methods or their fuzzy extension. These are mainly methods such as: Analytic Hierarchy Process (AHP) [6, 17], fuzzy AHP (fAHP) [4, 22], Analytic Network Process (ANP) [23, 27], fuzzy ANP (fANP) [13] Technique for Order of Preference by Similarity is Ideal Solution (TOPSIS) [7], fuzzy TOPSIS (fTOPSIS) [3, 6, 8, 13], Elimination and Choice Expressing Reality (ELECTRE) [2] and Fuzzy Preference Ranking Organization Method for Enrichment Evaluations (fPROMETHEE) [12].

3 The Characteristic Objects Method

The COMET is a very easy and useful approach, but to be able to understand this, the basic knowledge on the Fuzzy Sets is necessary [14–16]. The formal concept and notation of the COMET method can be presented by using five steps. The COMET method was described previously in [15, 16, 18–21] by the following steps, where the fuzzy model is created in the result [14]:

Step 1: Define the space of the problem as follows:
An expert determines the dimensionality of the problem by selecting number r of criteria, $C_1, C_2, \ldots, C_r$. Subsequently, the set of triangular fuzzy numbers for each criterion C_i is selected, i.e., $\tilde{C}_{i1}, \tilde{C}_{i2}, \ldots \tilde{C}_{ic_i}$. In this way, the following result is obtained (1):

$$
\begin{aligned}
C_1 &= \{\tilde{C}_{11}, \tilde{C}_{12}, \ldots, \tilde{C}_{1c_1}\} \\
C_2 &= \{\tilde{C}_{21}, \tilde{C}_{22}, \ldots, \tilde{C}_{2c_2}\} \\
&\ldots\ldots\ldots\ldots\ldots\ldots\ldots\ldots\ldots \\
C_r &= \{\tilde{C}_{r1}, \tilde{C}_{r2}, \ldots, \tilde{C}_{rc_r}\}
\end{aligned}
\tag{1}
$$

where $c_1, c_2, \ldots, c_r$ are numbers of the fuzzy numbers for all criteria.

Step 2: Generate the characteristic objects
The characteristic objects (CO) are obtained by using the Cartesian Product of triangular fuzzy numbers cores for each criterion as follows (2):

$$
CO = C(C_1) \times C(C_2) \times \ldots \times C(C_r)
\tag{2}
$$

As the result of this, the ordered set of all CO is obtained (3):

$$CO_1 = \{C(\tilde{C}_{11}), C(\tilde{C}_{21}), \ldots, C(\tilde{C}_{r1})\}$$
$$CO_2 = \{C(\tilde{C}_{11}), C(\tilde{C}_{21}), \ldots, C(\tilde{C}_{r2})\}$$
$$\ldots\ldots\ldots\ldots\ldots\ldots\ldots\ldots\ldots\ldots\ldots\ldots\ldots$$
$$CO_t = \{C(\tilde{C}_{1c_1}), C(\tilde{C}_{2c_2}), \ldots, C(\tilde{C}_{rc_r})\}$$

(3)

where t is the number of CO (4):

$$t = \prod_{i=1}^{r} c_i$$

(4)

Step 3: Rank and evaluate the characteristic objects.
Determine the Matrix of Expert Judgment (MEJ). This is a result of the comparison of the characteristic objects by the knowledge of the expert. The MEJ structure is as follows (5):

$$MEJ = \begin{pmatrix} \alpha_{11} & \alpha_{12} & \ldots & \alpha_{1t} \\ \alpha_{21} & \alpha_{22} & \ldots & \alpha_{2t} \\ \ldots & \ldots & \ldots & \ldots \\ \alpha_{t1} & \alpha_{t2} & \ldots & \alpha_{tt} \end{pmatrix} \begin{matrix} CO_1 \\ CO_2 \\ \ldots \\ CO_t \end{matrix}$$
$$\quad\quad CO_1 \quad CO_2 \quad \ldots \quad CO_t$$

(5)

where α_{ij} is a result of comparing CO_i and CO_j by the expert. The more preferred characteristic object gets one point and the second object get a null point. If the preferences are balanced, the both objects get a half point. It depends solely on the knowledge and opinion of the expert and can be presented as (6):

$$\alpha_{ij} = f(CO_i, CO_j) = \begin{cases} 0.0, f_{exp}(CO_i) < f_{exp}(CO_j) \\ 0.5, f_{exp}(CO_i) = f_{exp}(CO_j) \\ 1.0, f_{exp}(CO_i) > f_{exp}(CO_j) \end{cases}$$

(6)

where f_{exp} is an expert judgment function. The most important properties are described by the formulas (7) and (8):

$$\alpha_{ii} = f(CO_i, CO_i) = 0.5$$

(7)

$$\alpha_{ji} = 1 - \alpha_{ij}$$

(8)

On the basis of formulas (7) and (8), the number of comparisons is reduced from t^2 cases to p cases (9):

$$p = \binom{t}{2} = \frac{t(t-1)}{2}$$

(9)

Afterwards, we obtain a vertical vector of the summed Judgments (SJ) as follows (10):

$$SJ_i = \sum_{j=1}^{t} \alpha_{ij} \tag{10}$$

The last step assigns to each characteristic object the approximate value of the preference. In the result, we obtain a vertical vector P, where i-th row contains the approximate value of preference for CO_i. This algorithm is presented as a fragment of pseudo code:

```
1: k = length(unique(SJ));
2: P = zeros(t,1);
3: for i = 1:k
4:    ind = find(SJ == max(SJ));
5:    P(ind) = (k - i) / (k - 1);
6:    SJ(ind) = 0;
7: end
```

In line 1, we obtain number k as a number of unique value of the vector SJ. In line 2, we create vertical vector P of zeros (with identical size as vector SJ). In line 4, we obtain index with the maximum value of vector SJ. This index is used to assign the value of the preference to an adequate position in vector P (based on the principle of indifference of Laplace'a). In line 6, the maximum value of the vector SJ is reset.

Step 4: The rule base.
Each one characteristic object and value of preference is converted to a fuzzy rule as follows, general form (11) and detailed form (12):

$$IF \quad CO_i \ THEN \ P_i \tag{11}$$

$$IF \ C(\tilde{C}_{1i}) \ AND \ C(\tilde{C}_{2i}) \ AND \ldots THEN \ P_i \tag{12}$$

In this way, the complete fuzzy rule base is obtained, which can be presented as (13):

$$
\begin{array}{lll}
IF & CO_1 & THEN \quad P_1 \\
IF & CO_2 & THEN \quad P_2 \\
\ldots\ldots\ldots\ldots\ldots\ldots\ldots \\
IF & CO_t & THEN \quad P_t
\end{array}
\tag{13}
$$

Step 5: Inference in a fuzzy model and final ranking.
The each one alternative is a set of crisp number, which corresponding with criteria
$C_1, C_2, \ldots, C_r$. It can be presented as follows (14):

$$A_i = \{a_{1i}, a_{2i}, \ldots, a_{ri}\} \tag{14}$$

where condition (15) must be satisfied.

$$
\begin{aligned}
a_{1i} &\in [C(\tilde{C}_{11}), C(\tilde{C}_{1c_1})] \\
a_{1i} &\in [C(\tilde{C}_{21}), C(\tilde{C}_{2c_2})] \\
&\ldots\ldots\ldots\ldots\ldots\ldots\ldots \\
a_{ri} &\in [C(\tilde{C}_{r1}), C(\tilde{C}_{rc_r})]
\end{aligned}
\tag{15}
$$

Each one alternative activates the specified number of fuzzy rules, where for
each one is determined the fulfillment degree of the conjunctive complex premise.
Fulfillment degrees of all activated rules sum to one. The preference of alternative is
computed as the sum of the product of all activated rules, as their fulfillment
degrees, and their values of the preference. The final ranking of alternatives is
obtained by sorting the preference of alternatives.

4   Experimental Study

The purpose of this section is to determine the fuzzy model for a simple example to
choose the optimal supplier. For this aim, the presented problem occurs only two
criteria ($r = 2$). These criteria will be the cost (C_1) and quality of service (C_2).
Values for both criteria will be normalized in the range from 0 to 1. Figure 1
presents three linguistic values as triangular fuzzy numbers. In this way, nine
characteristic objects are obtained, which are presented in Table 2. Then, the expert
make pairwise comparisons in accordance with the formula (6) thus forming a
matrix *MEJ*. Subsequently, this knowledge is aggregated to the vector *SJ*. For
considered problem, *SJ* vector and matrix *MEJ* is presented as (16). Creating a *MEJ*
matrix requires only 36 pairwise comparisons, which allows to identify the decision
maker's preferences and build a fuzzy model. Each set of alternatives can be
evaluated automatically because the fuzzy rule base (17) is a perfect substitute for
our expert. In reality, this decisional model is calculated exactly in the same way as
the Mamdani fuzzy model.

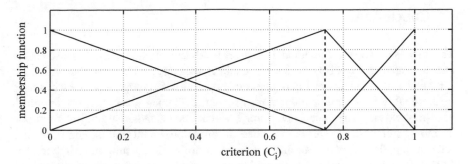

Fig. 1 The membership functions for the low (about 0), the average (about 0.75) and high (about 1) for the i-th criterion

Table 2 List of 9 characteristic objects CO_i

CO_i	CO_1	CO_2	CO_3	CO_4	CO_5	CO_6	CO_7	CO_8	CO_9
C_1	0.00	0.00	0.00	0.75	0.75	0.75	1.00	1.00	1.00
C_2	0.00	0.75	1.00	0.00	0.75	1.00	0.00	0.75	1.00

$$
MEJ = \begin{pmatrix}
0.5 & 0 & 0 & 1 & 0 & 0 & 1 & 0 & 0 \\
1 & 0.5 & 0 & 1 & 1 & 0 & 1 & 1 & 1 \\
1 & 1 & 0.5 & 1 & 1 & 1 & 1 & 1 & 1 \\
0 & 0 & 0 & 0.5 & 0 & 0 & 1 & 0.5 & 0 \\
1 & 0 & 0 & 1 & 0.5 & 0 & 1 & 1 & 1 \\
1 & 1 & 0 & 1 & 1 & 0.5 & 1 & 1 & 1 \\
0 & 0 & 0 & 0 & 0 & 0 & 0.5 & 0 & 0 \\
1 & 0 & 0 & 0.5 & 0 & 0 & 1 & 0.5 & 0 \\
1 & 0 & 0 & 1 & 0 & 0 & 1 & 1 & 0.5
\end{pmatrix} ; \quad SJ = \begin{pmatrix}
2.5 \\ 6.5 \\ 8.5 \\ 2.0 \\ 5.5 \\ 7.5 \\ 0.5 \\ 3.0 \\ 4.5
\end{pmatrix}
$$

$$(16)$$

$$R_1 : IF\ C_1 \sim 0.00\ AND\ C_2 \sim 0.00\ THEN\ 0.250$$
$$R_2 : IF\ C_1 \sim 0.00\ AND\ C_2 \sim 0.75\ THEN\ 0.750$$
$$R_3 : IF\ C_1 \sim 0.00\ AND\ C_2 \sim 1.00\ THEN\ 1.000$$
$$R_4 : IF\ C_1 \sim 0.75\ AND\ C_2 \sim 0.00\ THEN\ 0.125$$
$$R_5 : IF\ C_1 \sim 0.75\ AND\ C_2 \sim 0.75\ THEN\ 0.625$$
$$R_6 : IF\ C_1 \sim 0.75\ AND\ C_2 \sim 1.00\ THEN\ 0.875$$
$$R_7 : IF\ C_1 \sim 1.00\ AND\ C_2 \sim 0.00\ THEN\ 0.000$$
$$R_8 : IF\ C_1 \sim 1.00\ AND\ C_2 \sim 0.75\ THEN\ 0.375$$
$$R_9 : IF\ C_1 \sim 1.00\ AND\ C_2 \sim 1.00\ THEN\ 0.500$$

$$(17)$$

5   Conclusions

The presented COMET method has several advantages that allow its use in the issues related to logistics and supply chain management. The identified model is fully resistant to the reversal rank phenomenon. Pairwise comparison, which is used in the COMET, is significantly more resistant to human error than the approach used in the AHP. The expert determines, which one characteristic object from pair is more preferred. There is no need information about the strength of this relationship. This further simplifies the process of decision support. Applying the COMET method provides reproducible results and their persistence in respect to a particular expert in the space of the problem. This method, in many cases, allows a more accurate assessment of alternatives than other methods of multi-criteria decision support. In future work, more complex problems should be solved in the area of SCM and logistics. Afterwards, received results should be used to analysis of local weights of significance.

References

1. Awasthi, A., Chauhan, S.S., Goyal, S.K.: A multi-criteria decision making approach for location planning for urban distribution centers under uncertainty. Math. Comput. Model. **53**(1–2), 98–109 (2011)
2. Bojkovic, N., Anic, I., Pejcic-Tarle, S.: One solution for cross-country transport-sustainability evaluation using a modified ELECTRE method. Ecol. Econ. **69**(5), 1176–1186 (2010)
3. Chamodrakas, I., Alexopoulou, N., Martakos, D.: Customer evaluation for order acceptance using a novel class of fuzzy methods based on TOPSIS. Expert Syst. Appl. **36**(4), 7409–7415 (2009)
4. Chan, F.T., Kumar, N.: Global supplier development considering risk factors using fuzzy extended AHP-based approach. Omega **35**(4), 417–431 (2007)
5. Chu, T.-C.: Facility location selection using fuzzy topsis under group decisions. Int. J. Uncertainty Fuzziness Knowl. Based Syst. **10**(6), 687–701
6. Ertgrul, I., Karakasoglu, N.: Comparision of fuzzy AHP and fuzzy TOPSIS mewthods for facility location selection. Int. J. Adv. Manuf. Technol. **39**, 783–795 (2008)
7. Farahani, R.Z., Asgari, N.: Combination of MCDM and covering techniques in a hierarchical model for facility location: a case study. Eur. J. Oper. Res. **176**(3), 1839–1858 (2007)
8. Kannan, G., Khodaverdi, R., Jafarian, A.: A fuzzy multi criteria approach for measuring sustainability performance of a supplier based on triple bottom line approach. J. Clean. Prod. **47**, 345–354 (2013)
9. Kawa, A.: Simulation of dynamic supply chain configuration based on software agents and graph theory. In: Distributed Computing, Artificial Intelligence, Bioinformatics, Soft Computing, and Ambient Assisted Living, pp. 346–349. Springer, Berlin (2009)
10. Kawa, A., Golińska, P.: Supply chain arrangements in recovery network. In: Agent and Multi-Agent Systems: Technologies and Applications, pp. 292–301. Springer, Berlin (2010)
11. Kawa, A., Ratajczak-Mrozek, M.P: Supply chain configuration in high-tech networks. In: Intelligent Information and Database Systems, pp. 459–468. Springer, Berlin (2012)
12. Martín, J.M., Fajardo, W., Blanco, A., Requena, I.: Constructing linguistic versions for the multicriteria decision support systems preference ranking organization method for enrichment evaluation i and ii. Int. J. Intel. Syst. **18**, 711–731 (2003)

13. Onut, S., Kara, S.S., Isik, E.: Long term supplier selection using a combined fuzzy MCDM approach: a case study for a telecommunication company. Expert Syst. Appl. **36**(2), 3887–3895 (2009)
14. Piegat, A., Sałabun, W.: Comparative analysis of MCDM methods for assessing the severity of chronic liver disease. Artif. Intel. Soft Comput. LNAI **9119**, 228–238 (2015)
15. Piegat, A., Sałabun, W.: Identification of a multicriteria decision-making model using the characteristic objects method. Appl. Comput. Intel. Soft Comput. (2014)
16. Piegat, A., Sałabun, W.: Nonlinearity of human multi-criteria in decision-making. J. Theor. Appl. Comput. Sci. **6**(3), 36–49 (2012)
17. Poh, K.L., Ang, B.W.: Transportation fuels and policy for Singapore: an AHP planning approach. Comput. Ind. Eng. **37**(3), 507–525 (1999)
18. Sałabun, W.: Application of the fuzzy multi-criteria decision-making method to identify nonlinear decision models. Int. J. Comput. Appl. **89**(15), 1–6 (2014)
19. Sałabun, W.: The use of fuzzy logic to evaluate the nonlinearity of human multi-criteria used in decision making. Przegląd Elektrotechniczny **88**(10b), 235–238 (2012)
20. Sałabun, W.: Reduction in the number of comparisons required to create matrix of expert judgment in the Comet Method. Manage. Prod. Eng. Rev. **5**(3), 62–69 (2014)
21. Sałabun, W.: The characteristic objects method: a new distance-based approach to multicriteria decision-making problems. J. Multi-Criteria Decis. Anal. **22**(1–2), 37–50 (2015)
22. Tabari, M., Kaboli, A., Aryanezhad, M.B., Shahanaghi, K., Siadat, A.: A new method for location selection: a hybrid analysis. Appl. Math. Comput. **206**(2), 598–606 (2008)
23. Tuzkaya, G., Onut, S., Tuzkaya, U.R., Gulsun, B.: An analytic network process approach for locating undesirable facilities: an example from Istanbul, Turkey. J. Environ. Manage. **88**(4), 970–983 (2008)
24. Wątróbski, J., Jankowski, J.: Guideline for MCDA method selection in production management area. In: Intelligent Systems Reference Library, vol. 98. Springer, Heidelberg, pp. 119–138 (2015)
25. Wątróbski, J., Jankowski, J.: Knowledge management in MCDA domain. In: Annals of Computer Science and Information Systems, vol. 5, pp. 1445–1450, IEEE (2015)
26. Wątróbski, J., Ziemba, P., Wolski, W.: Methodological aspects of decision support system for the location of renewable energy sources. In: Annals of Computer Science and Information Systems, vol. 5, pp. 1451–1459 IEEE (2015)
27. Wey, W.M., Wu, K.Y.: Using ANP priorities with goal programming in resource allocation in transportation. Math. Comput. Model. **46**(7–8), 985–1000 (2007)
28. Ziemba, P., Piwowarski, M., Jankowski, J., Wątróbski, J.: Method of criteria selection and weights calculation in the process of web projects evaluation. LNAI **8733**, 684–693 (2014)
29. Żak, J., Redmer, A., Sawicki, P.: Multiple objective optimization of the fleet sizing problem for road freight transportation. J. Adv. Transp. **45**(4), 321–347 (2011)

Personalization of Learning Content and Its Structure, Based on Limitations of LMSs

Aneta Bartuskova and Ondrej Krejcar

Abstract This paper evaluates learning management systems as the standardized solutions for e-learning. We present an overview of already researched limitations of these systems, accompanied by new findings from an analysis of existing learning object repositories. Several interesting novel approaches in the area of personalization, adaptation, collaboration and evaluation of learning systems are discussed. Detected shortcomings are used as a basis for the new learning system solution, focused on personalization of both learning content and its organization structure. This paper includes a conceptual proposal and an outline of the technical solution of this system.

Keywords Learning environment · Learning management systems · Adaptation · Personalization · Learning content

1 Introduction

Web-based learning became an important way to enhance learning and teaching, offering many learning opportunities [1]. Learning management systems (LMSs) belong to the most common e-learning solutions. LMSs include a wide variety of features that can be utilized to support both distance and traditional teaching [2].

It is apparent that e-learning needs an adequate management of educational resources to promote quality learning [3, 4]. However LMSs have some limitations especially in the area of usability and personalization. Also learning systems often neglect users' knowledge management requirement [5]. We present an overview of

A. Bartuskova (✉) · O. Krejcar
Faculty of Informatics and Management, Center for Basic and Applied Research,
University of Hradec Kralove, Rokitanskeho 62, Hradec Kralove 500 03,
Czech Republic
e-mail: Aneta.Bartuskova@uhk.cz

O. Krejcar
e-mail: ondrej@krejcar.org

© Springer International Publishing Switzerland 2016 455
D. Król et al. (eds.), *Recent Developments in Intelligent Information
and Database Systems*, Studies in Computational Intelligence 642,
DOI 10.1007/978-3-319-31277-4_40

already researched limitations of LMSs with some new ideas. Some of these limitations were already dealt with in novel approaches in the area of personalization, adaptation, collaboration and evaluation of learning systems. These approaches are discussed and evaluated. Persisting limitations of learning systems are used as a basis for the new solution, focused on personalization of both learning content and its organization structure, which can lead to the collaborative knowledge building.

2 Standardized Learning Systems

Standardized re-usable solutions for e-learning make use of learning objects (LOs). With appropriate metadata descriptions, LOs can be modular units that can be assembled together to form lessons and courses [6]. The IEEE Standard for Learning Object Metadata (LOM) is an open standard for LOs [7]. LOM is represented as an hierarchy of elements, which are used for description of LO. Shareable Content Object Reference Model (SCORM) is a collection of standards and specifications for the packaging and sequencing of learning and assessment material in the form of shareable, reusable content objects [8]. Among standardized solutions belong LMSs and LORs (learning object repositories).

LMSs are used with an increasing frequency to support the basic needs of administration and higher-education teaching [9]. Most universities now use LMSs to support and improve learning and teaching processes [2]. These features are typical for LMSs: course content organization, user accounts with different roles, posting announcements, threaded discussion forums, assessment system for exams, grade book or email system [10]. Moodle, distributed as an open source, is the leading learning management system. Another popular learning platforms are Edmodo, Blackboard or SumTotal Systems [11].

LCMSs (LORs) provide the flexibility to have an online content organized in ways other than a traditional online course [12]. This system allows users to search and retrieve LOs from the repository, it typically supports simple and advanced queries, as well as browsing through the material by subject or discipline [13].

3 Limitations of LMSs

3.1 Fixed Structure of the Course

LMSs are course-based systems, which present learning materials and other course-related information to user in a fixed set and visual composition, with no significant means to adapt the information to one's needs. Similarly as with reading from textual material where the order is set, traditional LMSs offer only fixed structure of materials for student to go through. However, people have different levels of prior domain knowledge, they can be learning under time pressure or they can find it difficult to recognize the important concepts and the relationships among them [14]. LMS offer their users "one size fits all" service, that means that all learners taking an LMS-based course, regardless of their knowledge, goals and interests, receive access to the same educational material and set of tools with no personalized support [15].

3.2　Adaptive Learning

Personalization is often connected to adaptive learning, which is defined as a style of learning that uses student successes as the basis for developing future learning directions while a student is participating in the e-learning course [16]. Adaptive techniques are examples of user-centred techniques for approaching a range of serious usability problems found in conventional non-adaptive web-based e-learning systems, usually related to present homogeneous content and navigation scheme for all students, without focusing on a more adequate for each student [17].

However an adaptive course authoring has one major drawback, which negatively affects the spread of adaptive learning systems, and that is the complexity of metadata used for the personalization in both definition of such metadata as well as their further maintenance, which is even more challenging in a collaborative environment [18]. We are therefore more interested in personalization which can be controlled and optimized by learners themselves and not by busy teachers or automated processes.

3.3　Customization and Personalization

Customization refers to the structure or style of the webpage (or system), while user personalization usually refers to the content itself [19]. Customization in LMS systems is possible by limited choices e.g. of colour scheme or composition of widgets, which shows new posts in discussion forum, active assignments, calendar, announcements etc. However learning materials themselves cannot be organized or filtered by students, not even can they add personal comments. As Hwang pointed out, limitation of traditional web-based learning is the restricted ability of students to personalize and annotate the learning materials [1].

LMS should provide students with means for their active contribution to the presented content (in the form of tagging, commenting and other annotating mechanisms), its sharing and organization [18]. Peng et al. [5] suggested that teachers need to collect interactive information while students need to select the useful content from a course, consume it and re-organize it by themselves.

3.4   Content Management and User Interface

LMSs offer seemingly limitless e-space, where teachers can add content. This arrangement often supports unadvisable behaviour of continuously accumulating content and just making it available to learners, who may eventually face similar information overload as with the Google search. Without any support, the student can only with difficulties identify which parts of the course are relevant and which are presenting only additional, not that important information [18].

There is also a frequent issue with keeping old files along with the new ones. More effective approach for learner would be refinement of existing resources and keeping their amount at a reasonable level. However user interface of these systems usually does not encourage this desirable behaviour in teachers. In consequence students are forced to choose the right materials among those offered in learning repository. However with the increasing number of available learning materials, it is becoming crucial to be able to support students in their way through the course, to locate, recognize and understand information, which is the most relevant, considering the given time and progress of the student [18].

Furthermore, LMS systems are loaded with too many functionalities in a complex user interface, which often discourage teachers and students from exploration of both basic and advanced functions.

3.5   Summarization

To conclude, web-based learning systems used by colleges aim to display course resources and often neglect users' knowledge management requirement [4, 5]. Resources in such course-based learning are organized by teachers in fixed course scope, which is not sufficient for higher education today [5]. As the main issues related to this type of learning systems were consequently identified:

- an insufficient personalization support for:

  - organization of courses
  - organization of learning resources
  - annotating mechanisms
  - sharing mechanisms

- an insufficient customization support
- the system's interface does not encourage:

  - regular revising of existing content
  - disposing of outdated content

4   The Novel Systems

Some of the discussed shortcomings were already investigated and new systems were proposed to replace the current systems, enhance them or complement them.

We can see that there is a great effort to make learning systems more adaptive, flexible and personalized. Many of the solutions are focused on automated adaptation and personalization, i.e. the systems algorithms should deliver more relevant content on the basis of metadata and data from users. Some of the novel solutions are based more on human factor instead of automated processes. They emphasize the need to deliver quality content through evaluations, recommendations and collaboration.

The general problem with the novel solutions is that the majority of these projects did not expand beyond the authors´ institution and traditional LMSs are still in common use. This section will present some of the interesting novel systems and their contribution, along with a brief assessment of their long-term sustainability.

4.1   Personalization

Peng et al. [5] presented a knowledge management system which would support web-based learning in higher education. The main idea was to integrate different course resources and let students select the part of resources helpful for them and organize them in their own way. Implemented system KMS-THU manages knowledge organization by different form (tree structure and tags) and different scope (individual, group and public knowledge). This proposed system so far dealt with learning resources as digital files and focused mostly on a technical solution.

However personalization require a solid foundation of structure and organization and is really difficult to do well [20]. Most of all, the complexity of metadata needed for personalization, their definition and further maintenance, are the major drawbacks, which negatively affect the spread of adaptive learning systems [18]. Also because we don't have time to teach our systems, or because we prefer to maintain our privacy, we often don't share enough information to drive effective personalization [20]. Even if we do, our needs and interests change. Past performance is no guarantee of future results [20]. Badly performed personalization can even hide useful information from users because of incomplete or outdated metadata or badly constructed algorithms.

4.2   Adaptation

Gasparini et al. [17] proposed e-learning system AdaptWeb—an adaptive hypermedia system aiming to adapt the content, the presentation and the navigation in web-based courses, according to the student model. The AdaptWeb's educational contents were modelled through a hierarchical structure of concepts stored in XML format. This system's evaluation however identified some weaknesses, system is not being updated lately and also demo site does not exist anymore. Šimko et al. [18] introduced a schema of adaptive web-based learning for future LMSs, with these key principles: (1) domain modelling, (2) extensible personalization and course adaptation and (3) student active participation in a learning process. Adaptive and intelligent web-based educational systems are based upon domain and user models [21]. Domain model of ALEF contains two main parts—metadata (concepts, tags, comments) and educational content (LOs) [18]. The limitations which were discussed for personalization, apply for adaptation as well, as both approaches deal with metadata.

4.3 Collaboration and Evaluation

Rego et al. [3] proposed an evaluation collaborative system in which experts and teachers analyze LOs and give them an individual evaluation. After this individual evaluation, all the persons that evaluated the LO gather in a sort of forum to reach to its final evaluation. Similarly, Redondo et al. [22] designed a learning platform called Educateca, where learners as "prosumers" are supposed to tag and rate learning objects as they use the platform and like teachers they can also create or modify content to contribute to the learning distributed repository.

This concept would be however in practice very time-demanding for all participants, not mentioning problems with insufficient knowledge, experience and most of all motivation towards quality evaluation or organizing learning content.

4.4 Learning as a Service

Several suggestions on novel learning systems were made with the use of cloud computing. Redondo et al. suggested that learning organizations should publish and even share their material in the cloud so that learners can access them directly. The LMS is then becoming another SaaS (Software as a Service) in the cloud so that the student can select a LMS according to the LMS's features and his/her preferences [22]. User preference is of the great importance in a competitive environment of various products, interaction systems and websites [23].

This approach however requires that every learner has knowledge needed for distinguishing the right learning materials for his/her purposes, as well as already explored and clarified preferences for choosing the right LMS. Redondo et al. [22] also mentioned that e-learning in the cloud should give the illusion of infinite resources available on demand. This concept however threaten to overwhelm student with a large amount of resources and at the same time fail in delivering the right materials.

5 Proposal of the New Solution

5.1 Conceptual Foundation

The most frequently mentioned limitation of LMSs was the lack of personalization support, which can be divided into these two main categories:

- organization of learning content (regrouping, adding, deleting)
- modification of learning content (annotation, highlighting, notes)

In order to solve limitations of LORs, we suggest implementing a learning environment within the scope of an individual institution. This would ensure a maintainable number of learning resources and promote collaboration and greater interestedness of participants, both teachers and students, which would lead to better management of learning content. A LMS can be viewed as a suitable learning environment for this purpose, however traditional LMSs face many issues, related mainly to already discussed insufficient personalization. The discussed novel systems focused mainly on automated adaptation techniques and collaborative evaluations of content. These approaches however also have their limitations and consequently we suggest that personalization should be controlled solely by the user. Personalization in content and its organization can further extend towards group personalization and collaborative knowledge building, which again would support engagement of users. The following section brings a closer look on our proposed solution.

5.2 Overview of the Architecture

To overcome the discussed limitations, we propose a new solution for managing learning content. The key principle of our solution is facilitating personalized organization of learning content, which would lead to collaborative knowledge building. In this paper we present only basic architecture based on the previous analysis, focused on the first category of personalization (organization of content). The second category (modification of content) is contained in the "update" function later in the text and will be implemented as the functionality within the system.

The first issue to consider is the extent of possible personalization. In current LMSs, a learner can only "read" the content. As Peng et al. [5] stated, learning systems used by colleges aim to display course resources and often neglect users' knowledge management requirement. Active students are then forced to create their own repository of learning materials, either as a local copy of course materials or they are accumulating their own resources or the combination of both.

We suggest supporting all basic functions of persistent storage—"create", "read", "update" and "delete" (commonly called "CRUD"). This means that a learner could add his own learning content to the course, update the existing resource or delete the extra resources. Of course, these changes would take effect only in the individual user's view, not in the core repository of learning resources. As far as a learner does not personalize the course, he will only "read" the core repository. When the learner starts manipulating learning content, the changes are logged, new content is stored, and his view of the learning course becomes the personalized learning course. The option should be however kept for accessing the original view of the course.

The way we organize, label, and relate information influences the way people comprehend that information [20]. However organizing is usually a subjective process, when done by ambiguous schemes, and language used for labeling is also often ambiguous [20]. Therefore learners should be allowed not only to add and modify files, but also change labels or the position of files and thus adapt the learning environment to their needs. With this possibility, they could e.g. relocate the learning content they use most to the more convenient place in the course.

The possibilities of personalized organization of learning content are depicted in Fig. 1. The learning resources are for simplicity presented as files in folders, labels

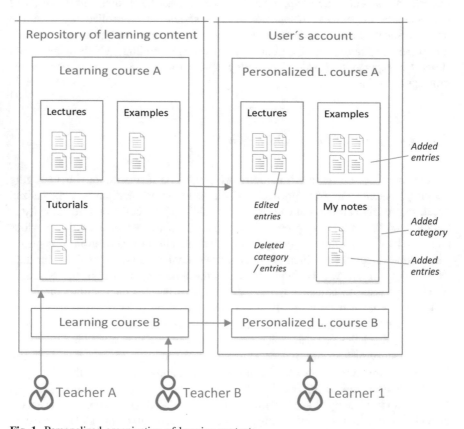

Fig. 1 Personalized organization of learning content

were selected arbitrarily for demonstration. This system supports unlimited number of both teacher and learner roles.

Technical implementation of personalization has to be carefully planned in order to avoid performance issues and duplicities. We propose creating a personalized learning course only when a learner initiates its creation by adding or updating content. Additionally changes in the course will be stored as instructions, by which the personalized course can be reconstructed, not as a hard copy. New and modified content will be then saved into a "group repository", so user accounts will contain only reference to the files and thus they will be as lightweight as possible. This way all references of one piece of learning content will be associated with one file, which can be managed efficiently with version control.

This arrangement facilitates the second phase of our solution—knowledge building. The group repository should reflect all performed personalization, including the new content, modified content, information about deleted content and changed labels and position of the content. Individual single-purposed personalization is this way transformed into a reverse process. Teachers receive feedback from students and as we assume it will inspire them to regular revising of existing content and disposing of outdated content, which was one of the main discussed issues of current LMSs.

This idea is depicted in Fig. 2. Students can access learning courses from the core repository. They can personalize the courses, by which they refine existing

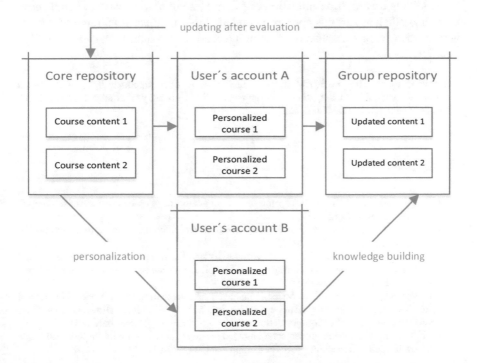

Fig. 2 The process of personalization leading to knowledge building

content and add new content. This process leads to spontaneous knowledge building, which can be used by teachers (original creators of courses) as a form of feedback. Ideally teachers are inspired to refine the learning courses based on student's personalization.

6 Conclusion and Future Work

We have investigated weaknesses of traditional LMSs, which are commonly being used in web-based education. Some novel approaches in the area of personalization, adaptation, collaboration and evaluation were discussed. Finally the need for a novel solution was substantiated and personalization was selected as one of its key principles. The proposal of an architecture focused on personalized organization of learning content, accessed from the core repository. All changes (individual personalizations) would be reflected in the group repository as a spontaneous knowledge building. The updated content is offered back to teachers of the learning courses, who are expected to be motivated to refine the courses based on this feedback from students.

We would like to continue in introducing our proposal in more detailed stages of development, along with the actual implementation and testing. Our proposal is based on overcoming the limitations of current solutions, however overall feasibility and efficiency of this approach compared to traditional solutions needs to be verified, as well as the actual benefits of introduced personalization and knowledge building.

Acknowledgment This work and the contribution were also supported by project "Smart Solutions for Ubiquitous Computing Environments" FIM, University of Hradec Kralove, Czech Republic (under ID: UHK-FIM-SP-2016).

References

1. Hwang, W.-Y., et al.: A study of multimedia annotation of Web-based materials. Comput. Educ. **48**(4), 680–699 (2007)
2. Islam, A.K.M.N.: Investigating e-learning system usage outcomes in the university context. Comput. Educ. **69**, 387–399 (2013)
3. Rego, H., Moreira, T., Garcia, F.J.: A Web-based learning information system resource and knowledge management. Knowl. Manag. Inf. Syst. E-Learn. Sustain. Res. Commun. Comput. Inf. Sci. **111**, 591–599 (2010)
4. Bartuskova, A., Krejcar, O.: Knowledge management and sharing in e-learning: hierarchical system for managing learning resources. In: Proceedings of the 6th international conference on knowledge management and information sharing, pp. 179–185, Rome, Italy (2014)
5. Peng, J., et al.: Design and implement a knowledge management system to support Web-based learning in higher education. Procedia Comput. Sci. **22**, 95–103 (2013)

6. McGreal, R.: Learning objects: a practical definition. Int. J. Instr. Technol. Distance Learn. **1**(9), 21–32 (2004)
7. IEEE Learning Technology Standard Committee (IEEE-LTSC): WG12 Learning Object Metadata. IEEE standard 1484.12.1. Retrieved from: http://ieeeltsc.org/wg12LOM/ (2002)
8. Gonzalez-Barbone, V., Anido-Rifon, L.: Creating the first SCORM object. Comput. Educ. **51**(4), 1634–1647 (2008)
9. Alvarez, A., Martin, M., Fernandez-Castro, I., Urretavizcaya, M.: Blended traditional teaching methods with learning environments: experience, cyclical evaluation process and impact with MAgAdI. Comput. Educ. **68**, 129–140 (2013)
10. iNACOL (International Association of K-12 Online Learning): Learning Management System. Retrieved from: http://www.onlineprogramhowto.org/admin/learning-management-systems/ (2010)
11. Capterra Inc.: Top LMS Software Solutions Infographic: Retrieved from: http://blog.capterra.com/top-lms-software-solutions-infographic/ (2012)
12. iNACOL (International Association of K-12 Online Learning): Learning content management system. Retrieved from: http://www.onlineprogramhowto.org/admin/learning-content-management-system/ (2011)
13. Neven, F., Duval, E.: Reusable learning objects: a survey of LOM-based repositories. In: Proceedings of the Tenth CM International Conference on Multimedia, pp. 291–294. ACM (2002)
14. Lee, J.H., Segev, A.: Knowledge maps for e-learning. Comput. Educ. **59**, 353–364 (2012)
15. Brusilovsky, P.: KnowledgeTree: a distributed architecture for adaptive E-Learning. In: Proceedings of the 13th International World Wide Web Conference, WWW 2004, pp. 104–113. ACM Press (2004)
16. Mason, R.T., Ellis, T.J.: Extending SCORM LOM. Informing Sci. Inf. Technol. **6**, 863–875 (2009)
17. Gasparini, I., Pimenta, M.S., Palazzo, J., Kemczinski, A.: Usability in an adaptive e-learning environment: lessons from AdaptWeb. IEEE Learn. Technol. Newsl. **12**(2), 13–15 (2010)
18. Simko, M., Barla, M., Bielikova, M.: ALEF: A framework for adaptive Web-based learning 2.0. In: Reynolds, N., Turcsányi Szabó, M. (eds.) KCKS 2010, IFIP Advances in Information and Communication Technology, vol. 324. Held as Part of World Computer Congress 2010, pp. 367–378. Springer, Berlin (2010)
19. Bouras, C., Poulopoulos, V.: Enhancing meta-portals using dynamic user context personalization techniques. J. Netw. Comput. Appl. **35**, 1446–1453 (2012)
20. Morville, P., Rosenfeld, L.: Information Architecture for the World Wide Web, 3rd edn. O'Reilly Media, Inc. ISBN-13: 978-0-596-52734-1 (2006)
21. Bielikova, M., Simko, M., Barla, M., Tvarozek, J., Labaj, M., Moro, R., Srba, I., Sevcech, J., ALEF: From application to platform for adaptive collaborative learning. In: Recommender Systems for Technology Enhanced Learning: Research Trends and Applications, pp. 195–225 (2014)
22. Redondo, R.P.D., Vilas, A.F., Arias, J.J.P.: Educateca: a Web 2.0 approach to eLearning with SCORM. In: IFIP Advances in Information and Communication Technology, vol. 341, pp. 118-126 (2010)
23. Bartuskova, A., Krejcar, O.: Evaluation framework for user preference research implemented as Web application. In: Computational Collective Intelligence. Technologies and Applications, pp. 537–548. Springer, Berlin (2013)

Author Index